Optimal Operation and Planning of Smart Power Distribution Networks-Volume I

Optimal Operation and Planning of Smart Power Distribution Networks-Volume I

Collection Editor

Pavlos S. Georgilakis

MDPI

Basel • Beijing • Wuhan • Barcelona • Belgrade • Novi Sad • Cluj • Manchester

Collection Editor
Pavlos S. Georgilakis
School of Electrical and
Computer Engineering
National Technical University
of Athens
Athens
Greece

Editorial Office
MDPI AG
Grosspeteranlage 5
4052 Basel, Switzerland

This is a reprint of the Topical Collection, published open access by the journal *Electricity* (ISSN 2673-4826), freely accessible at: https://www.mdpi.com/journal/electricity/topical_collections/power_networks.

For citation purposes, cite each article independently as indicated on the article page online and as indicated below:

Lastname, A.A.; Lastname, B.B. Article Title. *Journal Name* **Year**, *Volume Number*, Page Range.

ISBN 978-3-7258-3763-2 (Hbk)
ISBN 978-3-7258-3764-9 (PDF)
https://doi.org/10.3390/books978-3-7258-3764-9

Contents

electricity

Review

Comparison of DSOGI-Based PLL for Phase Estimation in Three-Phase Weak Grids

Jorge Pinto [1,2,*], Adriano Carvalho [1], Agostinho Rocha [1] and Armando Araújo [1]

1 Faculty of Engineering, University of Porto, Rua Roberto Frias, 4200-465 Porto, Portugal; asc@fe.up.pt (A.C.); agostinho@fe.up.pt (A.R.); asa@fe.up.pt (A.A.)
2 Vasco da Gama CoLAB, Rua Roberto Frias, 4200-465 Porto, Portugal
* Correspondence: jorgemsp@fe.up.pt

Abstract: The paper presents a summary of different double second-order generalized integrator (DSOGI)-based phase-locked loop (PLL) algorithms for synchronization with three-phase weak grids. The different methods are compared through simulation under a variety of grid conditions, such as unbalanced phase voltages, high low-order harmonics distortion, frequency steps, phase jumps, and voltage sags. Following the simulation results, the three methods that have shown the overall best results are compared through an experimental setup for further results validation under operation with a voltage-source converter. Based on the obtained results, a benchmark table is presented that allows ranking the performance of the tested methods for different expected grid conditions.

Keywords: synchronism; PLL; DSOGI; positive sequence; fundamental frequency; weak grids; frequency; phase

Citation: Pinto, J.; Carvalho, A.; Rocha, A.; Araújo, A. Comparison of DSOGI-Based PLL for Phase Estimation in Three-Phase Weak Grids. *Electricity* **2021**, 2, 244–270. https://doi.org/10.3390/electricity2030015

Academic Editor: Francois Auger

Received: 27 April 2021
Accepted: 28 June 2021
Published: 8 July 2021

1. Introduction

Microgrids are currently an extensive topic of discussion since such a concept presents a viable solution to allow the integration of distributed energy resources near loads without disturbing or contributing to the stability of the main grid. Those microgrids can operate either when connected to the main grid or disconnected (island mode). During island mode operation, the inertia provided from the main grid is no longer available, and the local grid becomes very sensitive to power changes (weak grid). To guarantee the adequate operation of the connected power converters, one of the key aspects is to ensure synchronization under the mentioned conditions.

The power converters control loop is normally fully or partially performed as in the synchronous reference frame (SRF), where voltage estimation becomes even more critical as the phase estimation errors are propagated to the synchronous voltage and current measurements through the Park transform. Additionally, methods such as droop control or the virtual synchronous generator demand stable frequency measurement/estimation to achieve robust and proper power-sharing. The phase-locked loop (PLL) is designed primarily to estimate the phase of the voltage fundamental frequency positive sequence, but it also allows to obtain a stable voltage amplitude and frequency as discussed in this paper. The importance of the phase angle estimation is addressed in [1], where it is studied the dynamic impact of phase jumps in synchronous generators and power converters, which may be caused by the estimation error during the connection with the grid.

There are currently several techniques to achieve grid synchronization, though it is not possible to point to a single synchronization algorithm that can simultaneously be the most accurate, fastest, most robust and lightest for every grid condition. Therefore, it is naturally necessary to establish a compromise between different goal objectives or grid conditions before choosing a particular synchronization algorithm.

In accordance with [2], there are two main methods to achieve grid synchronization: open-loop or closed-loop. In an open-loop approach, the phase angle is directly estimated from a filtered signal. In the closed-loop, there may exist a filtering stage but angle estimation is obtained through a closed-loop structure.

An open-loop and simple technique for grid synchronization is zero cross-detection, which allows synchronizing with the grid at every instant that the grid voltage crosses a reference zero. The synchronization is achieved through the detection of the zero-cross events and the generation of a rectangular waveform, accordingly. The resulting waveform allows to estimate the phase and frequency but only during the zero-cross instants; for that reason, it presents phase tracking slow dynamics and it is susceptible to external interference, leading to phase tracking errors [3,4]. With such a technique, synchronization is achieved for a single-phase system, twice per cycle. Additionally, voltage harmonics or measurement noise may lead to undesired zero-crossing detection if no filters are added. Another open-loop approach is the use of filters in the $\alpha\beta$ frame and prior angle calculation through a trigonometric relationship. Some of those filters include the Butterworth low-pass filter, space vector filter, adaptive notch filter, *etc.* As an alternative to digital filters, there can be found methods that suggest the use of discrete Fourier transform, Kalman filters, least mean square [5], or sliding mode control [6].

Another method, and the most common synchronization technique in grid-tied power converters, is the closed-loop approach known as the synchronous reference frame phase-locked loop (SRF-PLL). The SRF-PLL presents fast dynamics with a very low steady-state error when no low order harmonics exist and balanced conditions are considered [7]. Such assumptions are unrealistic in most real cases, especially when considering weak grids. Hence, most of the advanced PLL algorithms consist of the application of techniques to extract the fundamental frequency positive sequence (FFPS) of the grid voltage and further phase tracking through the SRF-PLL [8–10].

To extract the FFPS, the authors in [11] suggest the application of a double synchronous reference frame: one rotating with the positive frequency and another in the opposite direction (negative sequence). A decoupling network is presented that allows the extraction of the positive sequence from the double SRF. Furthermore, a second-order low pass filter is applied at the output to reduce harmonics interference in the positive sequence synchronous voltage estimation.

Another method is based on the single-phase enhanced PLL (EPLL). In this method, it is applied a quadrature signal generator based on the second-order generalized integrator (SOGI-QSG) to obtain a lagging 90 degrees phase shift of the input signal, offering simultaneously a filtering capability at frequencies beyond the filter central frequency [12]. An extension of this method [13] consists of considering a double SOGI-QSG (DSOGI-QSG), i.e., two SOGI-QSG are used to obtain the direct and quadrature signals of the alpha and beta components, respectively, for further positive sequence calculation. This method allows obtaining the positive sequence with intrinsic filtering capability. Since the DSOGI-QSG central frequency is adjusted accordingly with the PLL frequency, the DSOGI is a frequency-adaptive filter. Additionally, in [10], it is shown that the DSOGI-PLL is equivalent to the DSRF-PLL under certain conditions; therefore, the analysis on this paper disregards the second. The DSOGI-QSG can be applied with phase tracking (PLL) or frequency tracking, becoming a frequency-locked loop (FLL).

In [10,14–18] different PLL methods, including the DSOGI-based PLL, are compared and qualified where the DSOGI is highlighted as one of the most performant. Though it does not discriminate the performance of different DSOGI-based methods and in most of the tests, it lacks numerical performance quantification.

The paper is separated into eight sections. In Section 2 it is introduced the basic principle of work of the closed-loop PLLs, based on the synchronous reference frame (SRF-PLL). The generation of the quadrature signal, its extension to form the double second-order generalized integrator, and the extraction of the positive and negative phases sequence are presented in Section 3. In Section 4, it is presented the complete structure of the PLLs for synchronization with the grid-positive sequence and in Section 5, synchronization through the frequency-locked loop (FLL). Section 6 presents the simulation and experimental results and the respective discussion. Finally, in Section 7, it is discussed the main chapter conclusions and the PLL implementation chosen for the thesis is justified.

2. Synchronous Reference Frame Phase-Locked Loop

The SRF-PLL is based on the *dq0* frame, and it is illustrated in Figure 1. The principle of operation consists of orienting the voltage vector in the *dq0* frame to be aligned with a reference phase of the three-phase voltage system. To achieve this, the quadrature component of the voltage in the *dq0* frame is controlled to zero, typically through a proportional-integral (PI) controller. It is important to notice that in the synchronous frame (*dq0* frame), all vectors rotating at synchronous speed are represented in a steady state by a constant, direct and quadrature component. The output of the PI is further integrated to obtain the phase angle that is used as feedback for the calculation of the *dq0* transformation matrix. In the steady state, the PI output and the forward compensation represent the grid frequency ω'.

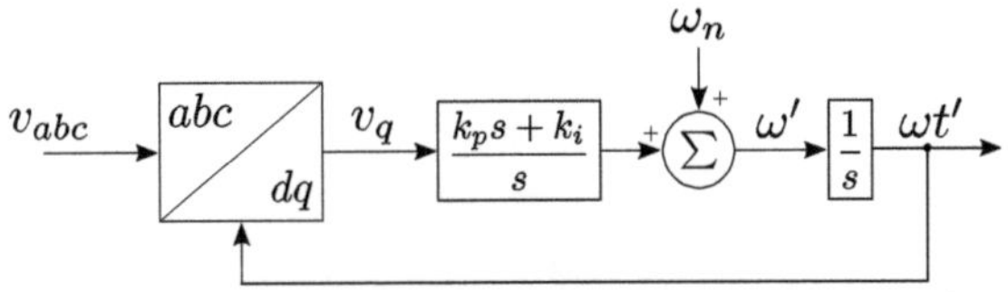

Figure 1. Block diagram of the SRF-PLL.

The Clarke and Park (1) transformation matrix, presented in Figure 1, are in the power invariant form and aligned with the cosine of phase v_a. Notably, the zero component is disregarded since it does not influence the synchronization loop. Additionally, the PI gains k_p and k_i are tuned in accordance with [3] and synthesized in Equation (2), where ω_c represents the filter natural (or controller) frequency, ζ the damping, and E_g the grid peak voltage amplitude.

$$T_{abc\to\alpha\beta} = \sqrt{\frac{2}{3}}\begin{bmatrix} 1 & -\frac{1}{2} & -\frac{1}{2} \\ 0 & \frac{\sqrt{3}}{2} & -\frac{\sqrt{3}}{2} \end{bmatrix}, \quad T_{\alpha\beta\to dq} = \begin{bmatrix} \cos(\omega t) & \sin(\omega t) \\ -\sin(\omega t) & \cos(\omega t) \end{bmatrix} \tag{1}$$

$$\begin{cases} k_p = & \dfrac{2\omega_n \zeta}{E_g} \\ k_i = & \dfrac{E_g}{4\zeta^2} \end{cases} \tag{2}$$

3. Fundamental Frequency Positive Sequence Extraction

This section reviews the principle of operation of the SOGI filter, its extension to the DSOGI, and the decoupling network that allows extracting the positive and negative grid phase sequence.

3.1. *Second-Order Generalized Integrator-Quadrature Signal Generator*

The SOGI-QSG is based on a band-pass filter tuned at a center frequency ω_0 and quality factor k $(= 1/Q)$. There are two main particularities on this filter: one is that it presents at the center frequency two outputs with unitary gain (the input signal with 0° and with 90° phase delay); the second is that allows to adapt the center frequency. Hence, at the center frequency, the SOGI-QSG generates the direct and quadrature signals of the input filtered signal. Such behavior can be further understood by analyzing the filter block diagram of Figure 2, the transfer functions (3) and the respective bode plots as illustrated in Figure 3.

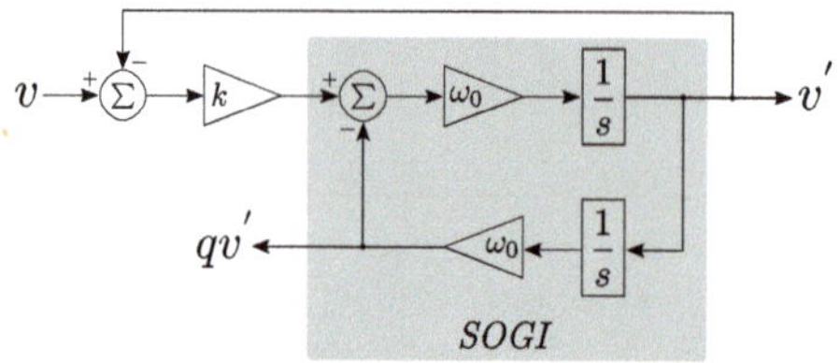

Figure 2. SOGI-QSG block diagram.

$$
\begin{aligned}
D(s) &= \frac{v'}{v}(s) = \frac{k\omega' s}{s^2 + k\omega' s + \omega'^2} \\
Q(s) &= \frac{qv'}{v}(s) = \frac{k\omega'^2}{s^2 + k\omega' s + \omega'^2}
\end{aligned}
\tag{3}
$$

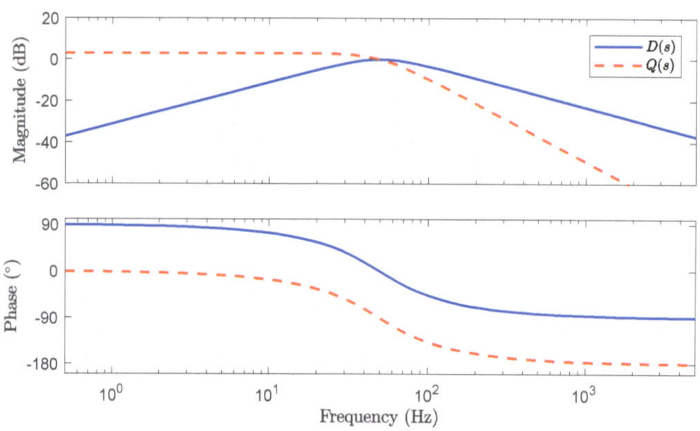

Figure 3. SOGI-QSG bode plot for a centre frequency of 50 Hz and a quality factor $k = \sqrt{2}$.

In Figure 3, it can be noticed that at the center frequency, the gain of both direct $D(s)$ and quadrature $Q(s)$ is unitary, while the phase of $Q(s)$ is lagging 90°. Another interesting feature of this filter, highlighted in [13], is the fact that, independent of the input signal frequency, the output qv' always lags 90° from v'; therefore, when the frequency of the input signal deviates from the center frequency, the filter output signals differ only in magnitude.

The SOGI-QSG, as shown before, allows to filter the input signal with an attenuation of 20 dB/dec and simultaneously generates the direct and quadrature component of the filtered signal. Additionally, it is possible to ensure unitary gain if the filter center frequency is properly adjusted.

3.2. Double SOGI-QSG and Positive Sequence Extraction

The SOGI-QSG can be extended to extract the fundamental frequency of an unbalanced phase system. To achieve it, we consider the symmetrical components method (4) to extract both positive and negative sequences of the unbalance three-phase system. As suggested in [12], the DSOGI consists of two SOGI-QSG (one for each component of the Clarke transform), which allows extracting the positive sequence of the three-phase voltage through the positive sequence calculator (PSC) (5), where qv_α and qv_β are obtained through two different SOGI-QSGs. In Figure 4, it is represented the DSOGI-QSG block diagram.

$$v^+_{abc} = T_p v_{abc}, \quad T_p = \frac{1}{3}\begin{bmatrix} 1 & a & a^2 \\ a^2 & 1 & a \\ a & a^2 & 1 \end{bmatrix}, \quad T_n = \frac{1}{3}\begin{bmatrix} 1 & a^2 & a \\ a & 1 & a^2 \\ a^2 & a & 1 \end{bmatrix}, \quad a = e^{j\frac{2\pi}{3}} \tag{4}$$
$$v^-_{abc} = T_n v_{abc}$$

$$v^+_{\alpha\beta} = [T_{\alpha\beta}]v^+_{abc} = [T_{\alpha\beta}][T_p]v_{abc} = [T_{\alpha\beta}][T_p][T_{\alpha\beta}]^T v_{\alpha\beta} \Leftrightarrow$$
$$\Leftrightarrow v^+_{\alpha\beta} = \frac{1}{2}\begin{bmatrix} 1 & -q \\ q & 1 \end{bmatrix} v_{\alpha\beta}, \quad q = e^{-j\frac{\pi}{2}} \tag{5}$$

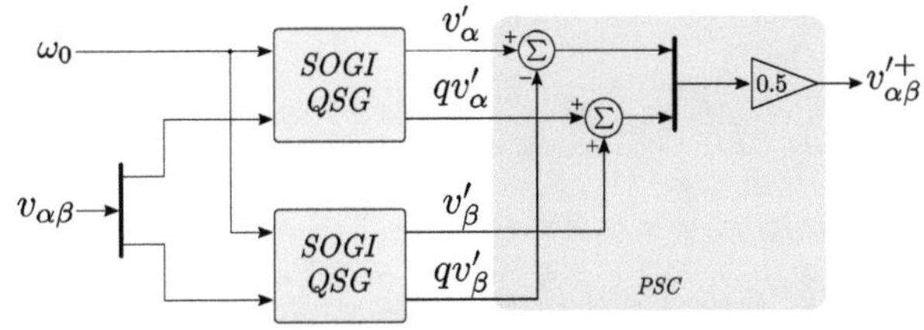

Figure 4. DSOGI-QSG block diagram.

As shown before, it is possible to extract the FFPS of the three-phase voltages represented in the static orthogonal frame (Clarke). Furthermore, the same procedure can be applied to obtain the negative sequence (6), despite its omission in Figure 4. It is worth noting that $v_{\alpha\beta}$ is obtained directly from the three-phase voltage measurement, while ω_0 must be set or estimated in accordance with the grid conditions.

$$v^-_{\alpha\beta} = \frac{1}{2}\begin{bmatrix} 1 & q \\ -q & 1 \end{bmatrix} v_{\alpha\beta} \tag{6}$$

4. DSOGI Based Phase-Locked Loop

In previous sections, it was presented the base of the DSOGI filter and it was shown how it can be used for extraction of the three-phase voltage FFPS.

In the current section, it is discussed two methods for the filter center frequency feedback: the DSOGI-PLL with a low pass filter in the frequency feedback (here simply called DSOGI-PLL) and the frequency-fixed DSOGI-PLL (FFDSOGI-PLL). Both algorithms consist of adding the SRF-PLL to the positive sequence obtained from the DSOGI-QSG, for frequency and phase estimation. The methods differ in how the DSOGI-QSG center frequency is adapted.

4.1. DSOGI-PLL

In the DSOGI-PLL, the estimated frequency of the SRF-PLL is added as a feedback path to the DSOGI filter center frequency. Furthermore, to smooth the frequency feedback from the PLL estimated frequency transients, it is proposed to add a low-pass filter to the frequency feedback path. The estimated frequency is provided by the SRF-PLL as a result of the PI controller keeping the phase angle error to zero. During voltage transients, such as phase jumps, the frequency estimation spikes to bring the phase error back to zero, independent of whether the actual grid frequency changed or not. The added LPF attenuates the interference of such transients as well the normal ripple produced by the PI controller. This scheme is represented in Figure 5, where the SRF-PLL estimated frequency ω'' is filtered by the LPF (ω') before being fed to the DSOGI-QSG. Hence, there are four tunable parameters: DSOGI damping (k), SRF-PLL gains (kp, k_i), and the LPF cut-off frequency (ω_c).

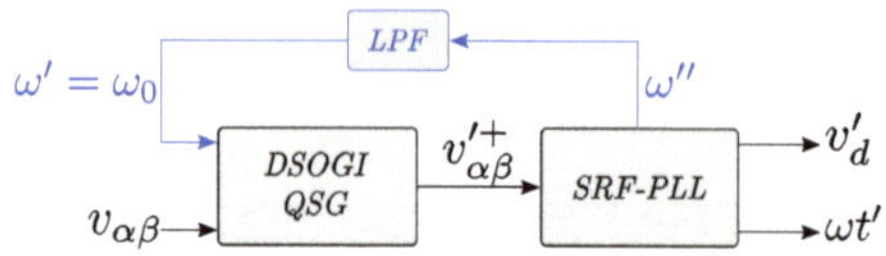

Figure 5. DSOGI-PLL with LPF block diagram.

4.2. FFDSOGI-PLL

A more recent approach, suggested in [19,20], considers a SOGI filter with a fixed center frequency ($\omega_0 = \omega_n$) for single-phase systems (FFSOGI). The method was further extended in [21] to a double SOGI, becoming an FFDSOGI. Contrary to the DSOGI, the FFDSOGI operates with a fixed frequency (there is no frequency feedback path to the SOGIs). Instead, any frequency deviance ($\omega' - \omega_0$) is further compensated in the filter output magnitude and phase. To achieve so, the authors present a linearized small-signal model of the DSOGI filter, considering small deviance $|\omega_0^2 - \omega'^2| << k\omega'\omega_0$. Under these circumstances, deviance can be compensated for by adding a gain correction to the quadrature signal of both $\alpha\beta$ component SOGIs and adding an angle correction to the estimated angle of the SRF-PLL (7). These changes result in the schematic of Figure 6, where grey areas highlight the amplitude and angle compensations. Comparing it with the previous method, the FFDSOGI-PLL has one less tunable parameter (the LPF time constant), as it is suggested to extract the frequency from the PI error integral [19]. Notably, in a steady-state, the error is zero and the frequency is given by the PI error integrative component. Furthermore, without frequency filtering, the estimated frequency oscillations are feed-forward to both frequency and angle compensations. In the paper, it is proposed to filter the estimated frequency as in the case of the DSOGI-PLL before feeding it to the compensation blocks, as shown in Figure 7. The option of adding the LPF, instead of using the PI error integral output, allows decoupling the synchronization dynamics (angle) from the frequency estimation. With this change, the FFSDOGI-PLL and DSOGI-PLL present the same number of tunable parameters, tuned with the same parameters for a fair comparison of both methods.

$$\begin{cases} qv' = \dfrac{\omega'}{\omega_0} qv' \\ \omega t' = \omega t' + \delta \\ \delta = \dfrac{\omega'^2 - \omega_0^2}{k\omega'\omega_0} \end{cases} \tag{7}$$

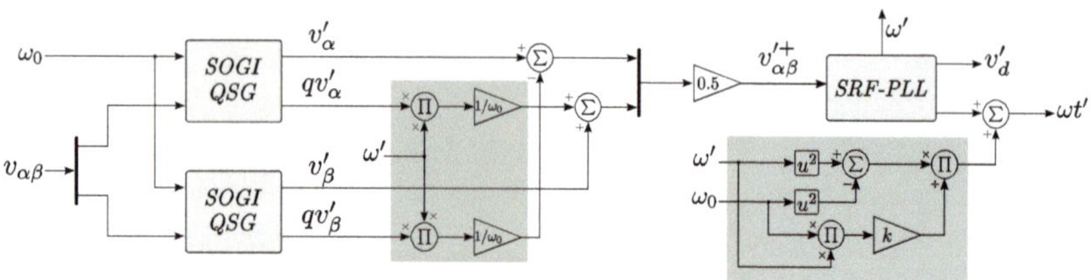

Figure 6. FFDSOGI-PLL block diagram.

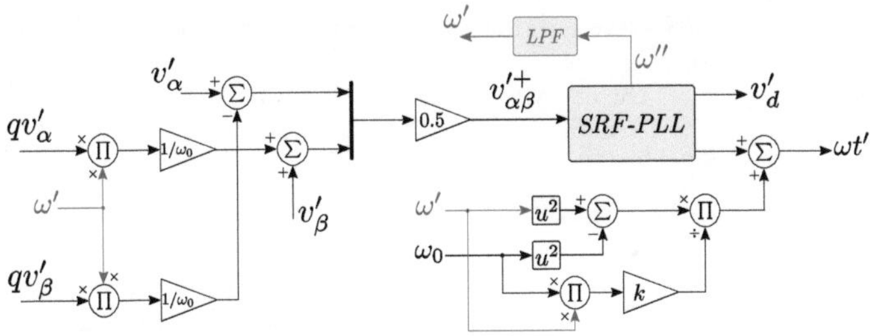

Figure 7. FFDSOGI with frequency LPF.

5. DSOGI Based Frequency-Locked Loop

Another approach to adapt the DSOGI-QSG center frequency ω_0, is proposed and well described in [22]. The author suggests an extension of the EPLL frequency adaptive structure to be applied in the SOGI-QSG. Such modifications result in the block diagram presented in Figure 8, where ω_N is the nominal grid frequency, and the grey area represents the gain normalization GN (9). The input error $\varepsilon_{\alpha\beta}$ is given by (8), considering the SOGI-QSG nomenclature shown in Figure 2). To highlight that the $\sum \varepsilon_{\alpha\beta}$ is negative for $\omega' < \omega$, zero for $\omega' = \omega$ and positive for $\omega' > \omega$. Hence, it is possible to obtain zero steady-state errors ($\omega' = \omega$) with an error integral and gain $-\Gamma$.

$$\begin{cases} \varepsilon_\alpha = (v_\alpha - v'_\alpha)qv_\alpha \\ \varepsilon_\beta = (v_\beta - v'_\beta)qv_\beta \end{cases} \tag{8}$$

$$GN = \frac{k\omega'}{(v'^{+}_{\alpha})^2 + (v'^{+}_{\beta})^2} \tag{9}$$

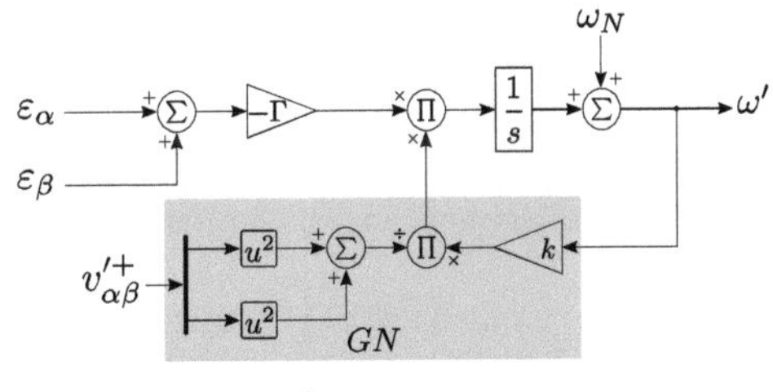

Figure 8. FLL block diagram.

Taking into account the FLL block diagram of Figure 8, it can be noticed that the FLL has only one tuning parameter, Γ, since k is the DSOGI-QSG filter quality factor and ω_N is a known constant. Furthermore, it is shown in [22] that with the gain normalization, the FLL presents a first-order transfer function (10), which allows tuning Γ in accordance with the desired system response settling time t_s.

$$\frac{\omega'}{\omega} = \frac{\Gamma}{s+\Gamma} \Rightarrow \Gamma \approx \frac{5}{t_s} \tag{10}$$

In [23], it is presented an improved frequency locked-loop (IFLL) that differs from the previous FLL, in a slight change on how the gain is normalized. Instead of considering only the positive sequence, the negative sequence is also considered (11). Such modification only improves the method, when the negative sequence becomes significant compared to the positive sequence (very high unbalancing). Another method variation is presented in [24], where a third-order generalized integrator (TOGI) is combined with an FLL for the elimination of the DC component. Such an approach is suited for single-phase systems, where the DC component is not eliminated by the Clarke transform. Additionally, any

SOGI referred on this work can be turned into a TOGI for further implementation in a single-phase system.

The DSOGI-FLL (or the DSOGI-IFLL) has only two tunable parameters: the DSOGI-QSG damping k and the FLL gain Γ.

$$GN = \frac{k\omega'}{(v_{\alpha}'^{+})^2 + (v_{\beta}'^{+})^2 + (v_{\alpha}'^{-})^2 + (v_{\beta}'^{-})^2} \tag{11}$$

To compare the synchronization of both PLL and FLL, it is suggested to add a phase estimation stage to the previously discussed FLL scheme. To achieve so, it is suggested to use the FFPS ($v_{\alpha\beta}'^{+}$) filtered by the DSOGI-FLL to estimate the phase angle. Three different approaches are considered: zero-cross detection, SRF-PLL, and the arc-tangent function. Notice that, contrary to the PLL schemes, in the FLL, both the frequency and phase are tracked independently, despite the phase tracking accuracy being dependent on the FLL performance.

5.1. DSOGI-FLL with Zero-Cross Detection

The zero-cross detection method employed here consists of integrating the frequency estimated by the FLL, but with an angle reset at every zero-cross detected, i.e., when $v_{\alpha}'^{+}$ or $v_{\beta}'^{+}$ crosses zero, the frequency integral is reset to the known angle. This way, it is possible to correct the estimated phase four times per period. Every time a zero-cross occurs, a flag is triggered ($\delta_{tr,\alpha}$ or $\delta_{tr,\beta}$) and feeds into a lookup table (LUT) to obtain the respective reset angle value accordingly with Table 1. Though it only allows synchronizing during four instants per wave period, this method is simple with a low processing demand, though it may lead to significant phase deviations during transients. This method does not add tunable parameters to the DSOGI-FLL.

Table 1. Zero-cross detection: angle reset LUT.

	$v_{\alpha}'^{+} > v_{\beta}'^{+}$	$v_{\alpha}'^{+} < v_{\beta}'^{+}$
$\delta_{tr,\alpha} > 0$	$\frac{3\pi}{2}$	$\frac{\pi}{2}$
$\delta_{tr,\beta} > 0$	0	π

5.2. DSOGI-FLL with SRF-PLL

The SRF-PLL can be added to the DSOGI-FLL to further estimate the phase frequency. Different from the PLL methods, in the FLL, the PI dynamics of the SRF-PLL do not influence the positive sequence extraction since the center frequency feedback of the DSOGI-FLL does not depend on the PI but on the FLL block diagram instead.

This scheme is here called the DSOGI frequency and phase-locked loop (DSOGI-FPLL). The method adds two tunable parameters to the original DSOGI-FLL (kp and ki), resulting in a total of four tunable parameters.

5.3. DSOGI-FLL with Atan2

The last considered method consists of calculating the arc-tangent of $v_{\alpha\beta}'^{+}$ to obtain the respective angle at any time instant. To reduce the processing burden associated with the typical arc-tangent methods, the fast arc-tangent suggested in [25] is employed. The proposed function is implemented through a LUT (or array) with limited angle representation, i.e., $\omega t' = \arctan(u) \wedge u \in [0,1]$ or $\omega t' \in [0, \pi/4]$. To extend its operation for the four quadrants, the functional properties (12) and (13) are considered.

$$\begin{cases} \arctan(-u) = -\arctan(u) \\ \arctan(u) + \arctan\left(\frac{1}{u}\right) = \frac{\pi}{2}, \quad u > 0 \\ \arctan(u) + \arctan\left(\frac{1}{u}\right) = -\frac{\pi}{2}, \quad u < 0 \end{cases} \tag{12}$$

$$\operatorname{atan2}(y,x) = \begin{cases} \arctan(\frac{y}{x}), & x > 0 \\ \pi + \arctan(\frac{y}{x}), & y \geq 0, x < 0 \\ -\pi + \arctan(\frac{y}{x}), & y < 0, x < 0 \\ \frac{\pi}{2}, & y > 0, x = 0 \\ -\frac{\pi}{2}, & y < 0, x = 0 \\ \text{NaN}, & y = 0, x = 0 \end{cases} \tag{13}$$

6. Results and Discussion

To compare the different SOGI-based PLL performance under severe grid conditions of a weak grid, several simulations are performed in Matlab/Simulink. Furthermore, all aforementioned methods were implemented in the discrete-time domain with a sampling rate of 10 kHz to emulate the micro-controller behavior. Discretization was performed considering the forward Euler integration method.

The performed tests are divided into three different subgroups: unbalancing and harmonics (test #1), frequency and phase jumps (test #2), and voltage sags (test #3). However, in the last two, harmonics and unbalancing are also considered but with a smaller impact. In Table 2, it is shown the harmonic content considered in the different tests as a percentage of the FFPS voltage.

The grid voltage is generated directly from pre-defined positive and negative sequences, while harmonics are posteriorly added. Hence, at any instant, it is possible to know the exact frequency, phase, and magnitude of the fundamental positive (and negative) sequence and the harmonics magnitude.

Additionally, it is considered that the PLL is locked when the estimated frequency reaches a steady state. Therefore, if both the FLL and PLL dynamic responses present the same frequency settling time, it is possible to carry a fair comparison between the methods. Hence, PI was tuned (2) considering $\omega_c = 314$ rad/s and $\zeta = 1/\sqrt{2}$, while the FLL gain was tuned (10) with $t_s = 0.1$ s. The SOGI damping, in all different methods, was set as $k = \sqrt{2}$. The tuned gains are summarized in Table 3.

Table 2. Harmonics magnitude ($\%V^+$).

h	Highly Polluted [23] (Test #1)	Limits in EN 50160 [26] (Tests #2 and #3)
5th	20%	6%
7th	15%	5%
11th	10%	3.5%
13th	8%	3%

Table 3. Tuned gains.

k_p	k_i	ω_c	Γ	k
1.37	163	78.5	40	$\sqrt{2}$

6.1. *Test #1—Unbalancing and Harmonics*

This test considers balanced three phases (positive sequence only) at 50 Hz, with all the methods already in a steady state at the simulation start. Unbalancing is achieved by injecting a negative sequence component, followed by harmonics with an amplitude, as presented in Table 2. Moreover, for the phase estimation, the results in this test are first applied to the FLL (DSOGI-IFLL) and further compared with the PLL (DSOGI-PLL and FFDSOGI-PLL).

The analysis of the three-phase voltages are illustrated in Figure 9, where t_1 to t_4 are the time instants that mark the added changes to the subsequent conditions listed in

Table 4. The estimated frequency ω' and the respective errors $(\omega - \omega')$ are illustrated in Figures 10 and 11, respectively. In Figure 10, it can be noticed that all methods remain near the nominal frequency with the exception of the SRF-PLL. Its performance is significantly poor when facing unbalanced conditions and further deteriorates when high low-order harmonics are present in the three-phase voltage. For these reasons, the following analysis disregards the SRF-PLL and justifies the need for methods capable of extracting the FFPS.

Table 4. Introduced voltage unbalancing and harmonics during test #1.

$t = 0$	Positive sequence only with $v^+_{\alpha\beta} = [325\ 0]\ V$ (balanced three phases)
t_1	Negative sequence is added with $v^-_{\alpha\beta} = [25\ 12]\ V$
t_2	Injection of the 5th and 7th order harmonics (highly polluted)
t_3	Injection of the 11th and 13th order harmonics (highly polluted)
t_4	Negative sequence is changed to $v^-_{\alpha\beta} = [100\ 0]\ V$

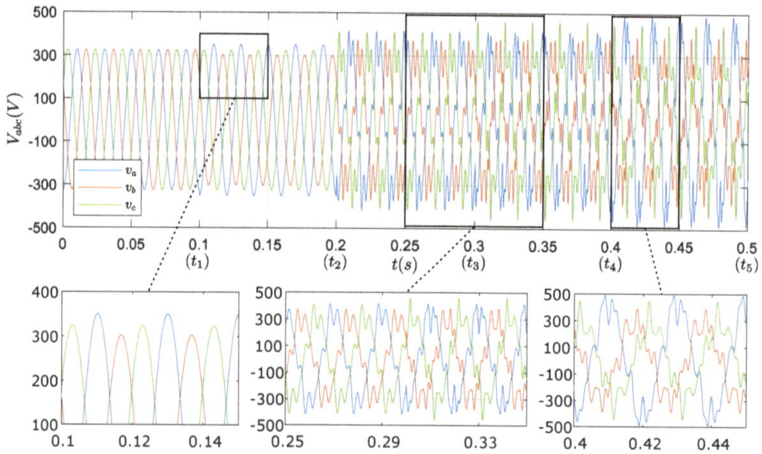

Figure 9. Three-phase voltage and the respective changes introduced during test #1.

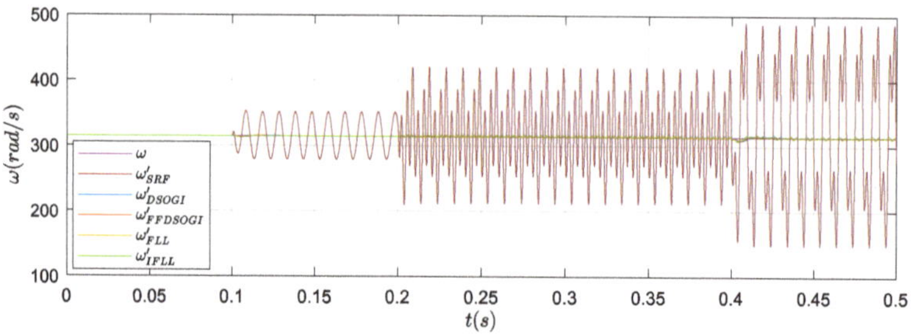

Figure 10. Estimated frequency for the different methods during test #1.

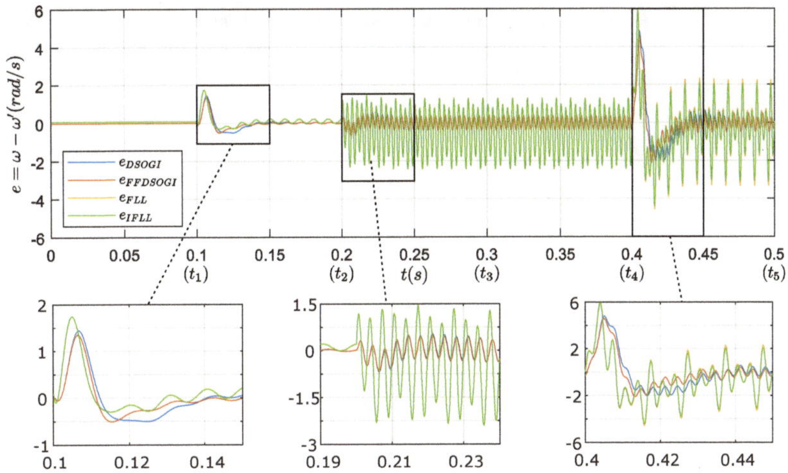

Figure 11. Estimated frequency errors for the different methods during test #1.

In Figure 11 it is noticeable that the DSOGI and FFDSOGI (PLL) presents similar behavior, both with lower estimated frequency steady-state oscillations when compared with the FLL methods. Furthermore, both FLL and IFLL perform very similarly, mostly until t_4, where the negative sequence is reduced. After t_4, unbalancing is more noticeable and the IFLL presents lower oscillations than the FLL, though such a difference may be neglected. To allow a quantification analysis of the results, in Table 5, it is presented the root mean square error (RMSE) as a measure of the error oscillations, and the mean error (ME) as its DC value during steady-state stages. In Table 5, $T_N = 0.02\ s$ and refers to the FFPS voltage waveform time period.

Table 5. Steady-state estimated frequency RMSE and ME obtained during test #1.

	SRF		DSOGI		FFDSOGI		FLL		IFLL	
	RMSE	ME	RMSE	ME	RMSE	ME	RMSE	ME	RMSE	ME
$t_1 - 2T_N \Rightarrow t_1$	8.3×10^{-13}	1.8×10^{-9}	1.1×10^{-10}	1.8×10^{-9}	1.2×10^{-10}	1.9×10^{-9}	4.9×10^{-14}	90×10^{-3}	5.2×10^{-14}	90×10^{-3}
$t_2 - 2T_N \Rightarrow t_2$	26	-54×10^{-3}	27×10^{-3}	6.1×10^{-4}	25×10^{-3}	-2.5×10^{-3}	78×10^{-3}	0.10	78×10^{-3}	0.10
$t_3 - 2T_N \Rightarrow t_3$	56	-2.5×10^{-3}	0.26	-7.2×10^{-3}	0.26	-6.5×10^{-3}	1.1	−0.56	1.1	−0.56
$t_4 - 2T_N \Rightarrow t_4$	56	-2.5×10^{-3}	0.26	-7.1×10^{-3}	0.26	-7.2×10^{-3}	1.1	−0.56	1.1	−0.56
$t_5 - 2T_N \Rightarrow t_5$	105	33×10^{-3}	0.27	-9.0×10^{-3}	0.27	-14×10^{-3}	1.3	−0.40	1.2	−0.40

Analyzing data from Table 5, it can be noticed that under ideal conditions $t = [0;\ t_1]$ the FLL and IFLL present a higher steady-state error (ME) when compared with the DSOGI-based PLL structures, though under such conditions, all methods perform very well. With the introduction of the three-phase unbalancing and harmonics, oscillations are more noticeable in the FLL methods than in the PLL. An interesting behavior of the FLL methods is in the change of the steady-state frequency ME with the changes in grid conditions. Observing the integral error of the FLL, it is concluded that the integral error gain is high enough to guarantee zero steady-state error, which indicates that the small ME is caused by the FLL structure, particularly in the way that the error $\varepsilon_{\alpha\beta}$ is generated for adapting the frequency.

Since both the FLL and IFLL performance are similar, only the IFLL is considered in the following performance evaluation.

In Figure 12, it is presented the errors obtained for the phase estimation $\omega t'$. Note that the different methods presented in Section 5 for phase estimation are applied to the IFLL

(FLL is not shown since it presents similar results). Furthermore, the RMSE and ME were obtained under steady-state conditions for the same intervals of Table 5 and are shown in Table 6.

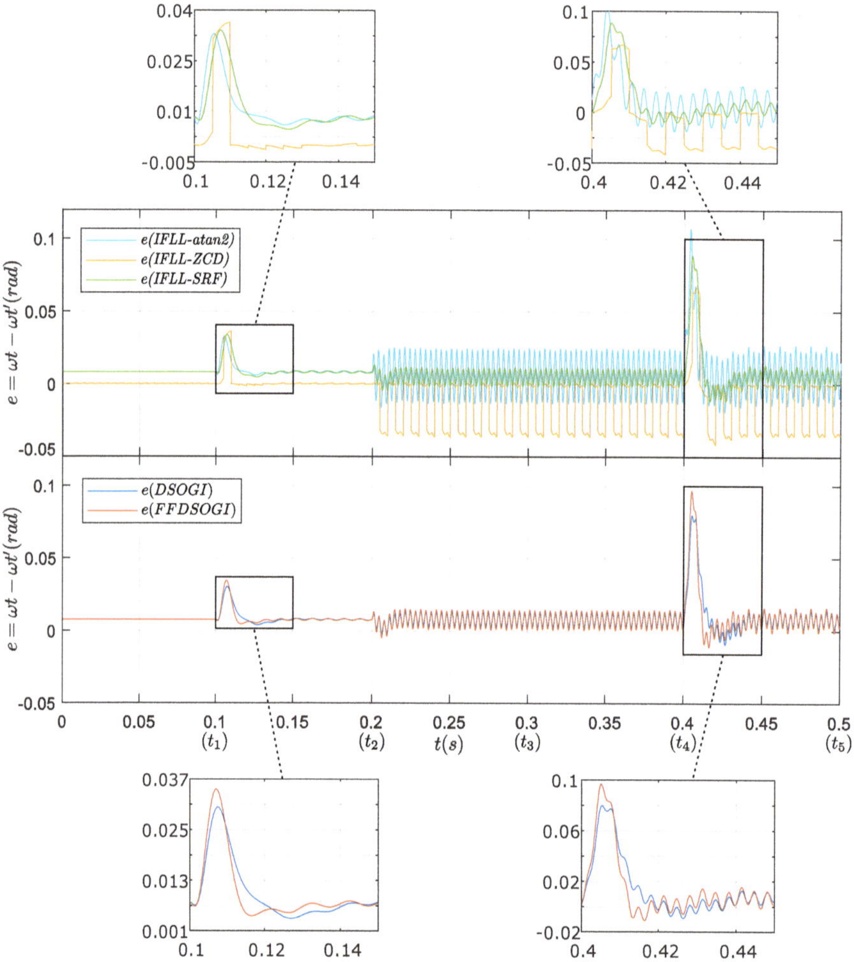

Figure 12. Estimated phase errors for the different methods during test #1. In the top figure, it is shown the different phase estimation methods applied to the IFLL, and in the bottom, the estimated phase by the PLL methods.

Table 6. Steady-state estimated phase angle RMSE and ME obtained during test #1.

	IFLL-Atan2		IFLL-ZCD		IFLL-SRF		DSOGI		FFDSOGI	
	RMSE	ME	RMSE	ME	RMSE	ME	RMSE	ME	RMSE	ME
$t_1 - 2T_N \Rightarrow t_1$	4.4×10^{-6}	8.2×10^{-3}	1.3×10^{-4}	2.3×10^{-4}	8.3×10^{-15}	8.2×10^{-3}	9.0×10^{-13}	7.8×10^{-3}	5.8×10^{-13}	7.8×10^{-3}
$t_2 - 2T_N \Rightarrow t_2$	5.0×10^{-4}	8.2×10^{-3}	2.4×10^{-4}	2.6×10^{-4}	4.6×10^{-4}	8.2×10^{-3}	3.4×10^{-4}	7.8×10^{-3}	4.4×10^{-4}	7.8×10^{-3}
$t_3 - 2T_N \Rightarrow t_3$	13×10^{-3}	5.6×10^{-3}	17×10^{-3}	-18×10^{-3}	4.0×10^{-3}	5.3×10^{-3}	3.3×10^{-3}	7.9×10^{-3}	4.5×10^{-3}	7.8×10^{-3}
$t_4 - 2T_N \Rightarrow t_4$	13×10^{-3}	5.6×10^{-3}	17×10^{-3}	-18×10^{-3}	4.0×10^{-3}	5.3×10^{-3}	3.3×10^{-3}	7.9×10^{-3}	4.5×10^{-3}	7.8×10^{-3}
$t_5 - 2T_N \Rightarrow t_5$	13×10^{-3}	6.1×10^{-3}	17×10^{-3}	-18×10^{-3}	4.2×10^{-3}	5.7×10^{-3}	3.5×10^{-3}	7.9×10^{-3}	4.7×10^{-3}	7.8×10^{-3}

Through analysis of Figure 12, it is clear that the IFLL-SRF results in the most stable angle estimation when compared with the IFLL-atan2 and IFLL-ZCD. Hence, the non-linearity on phase reset at every grid period quarter (IFLL-ZCD) and the high oscillations in the IFLL-atan2 indicate that the $v^{+}_{\alpha\beta}$ components resulting from the DSOGI-FLL are not strictly sinusoidal nor do they present a phase difference of exactly 90°. Such distortions are introduced mostly by the low order harmonics as already concluded before in the frequency analysis. In [7,23], it is suggested to add multiple DSOGIs tuned at different harmonic frequencies for the elimination of the respective low order harmonics; however, such an extension is not explored in this paper.

The PLL methods perform very similarly as expected, though the DSOGI-PLL outperforms all the other methods in terms of overall performance as it can be concluded by analyzing the *RMSE* during steady states, presented in Table 6.

For the test #1 conditions, it can be concluded that the performance of both PLL methods is superior to the FLL. Where the FFDSOGI presents the best results in frequency estimation and the DSOGI in the phase angle, both perform very similarly in both estimations. Despite the differences, as far as harmonics and unbalancing conditions are considered, all methods based in the SOGI-QSG show great performance under unbalancing conditions and good harmonics filtering capability.

To qualify the obtained results for the different methods during test #1, it is considered only the steady-state oscillations (RMSE). During test #1, the ME remains practically constant for both frequency and phase estimations, making it more meaningful to analyze the ME with the results obtained during test #2. Furthermore, the RMSE is applied to the frequency estimation since in the case of the phase angle estimation, the differences can be disregarded. The final classification is shown later in Table 12, where the mark is defined by the averaged ratio of the method RMSE by the minimum RMSE obtained for the different grid distortion intervals.

6.2. Test #2—Frequency Steps and Phase Jump

Test #2 consists of applying frequency steps followed by a phase jump. The test, as already stated before, also considers unbalancing and low-order harmonics in accordance with Table 2. The frequency and phase profile is equally applied to the three phases and is summarized in Table 7.

Table 7. Frequency steps and phase jump applied during test #2.

$t = 0$	Nominal frequency of 50 Hz
t_1	Frequency step change to 55 Hz
t_2	Frequency step change to 45 Hz
t_3	Frequency step change back to 50 Hz
t_4	Phase jump of $\pi/4$ rad

The results obtained for the estimated frequency and respective frequency errors during test #2 are shown in Figure 13.

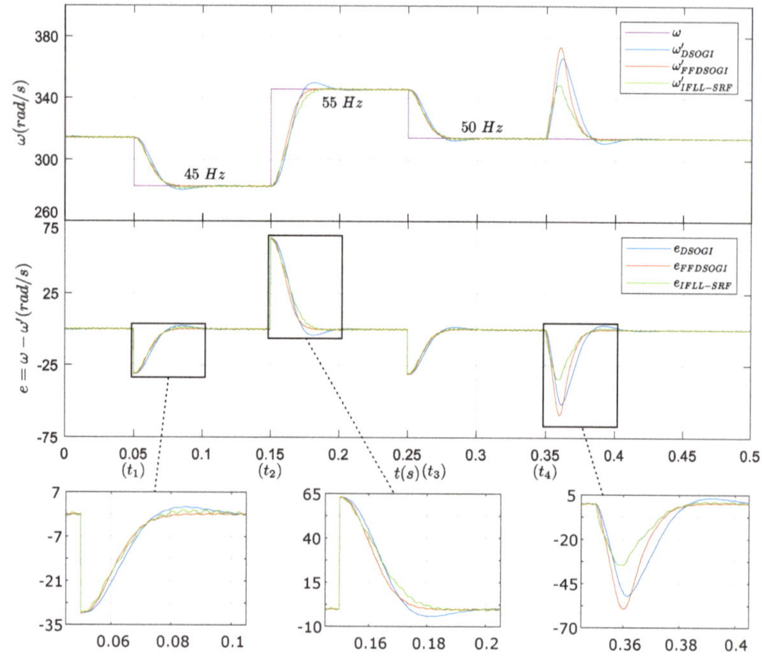

Figure 13. Estimated frequency errors for the different methods during test #2.

In Figure 13, the steady state of the different methods is coherent with the results obtained during test #1, and, therefore, it is not analyzed again. The frequency steps at (t_1) and (t_3) are the same ($\Delta\omega$ = 5 Hz), and it can be seen that the transient response of the FFDSOGI and IFLL are very similar, both in terms of overshoot and settling time. During this step, the DSOGI takes almost 20 ms more to reach the steady state. At (t_2) occurs the biggest frequency step ($\Delta\omega$ = 10 Hz) and during this transient, the FFDSOGI presents a slightly faster response. Finally, at (t_4) occurs the 45° phase shift, and the results, in terms of the response time, follow the same behavior as during (t_1). Nevertheless, the IFLL presents significantly less overshoot than other methods, which indicates that the FLL scheme is less sensitive to phase jumps when compared with the PLL.

To evaluate the performance in terms of phase estimation, the results in Figure 14 are presented. In terms of phase estimation, during frequency step changes, the differences are more obvious, where the FFDSOGI performance is better than the others, both in terms of overshoot and settling time. The FFDSOGI reaches its steady state in $\approx$ 0.03 s, with the following taking 0.02 s more. Still, during the phase jump, the FLL presents less overshoot and takes the same time to reach the steady state, which corroborates the estimated frequency results.

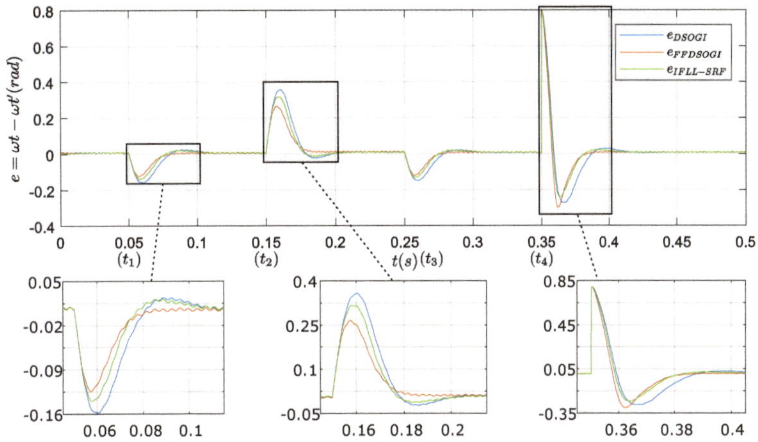

Figure 14. Estimated phase errors for the different methods during test #2.

In Tables 8 and 9, the results are summarized for the methods that presented the best performance during test #2. In Table 8, *OS* is the overshoot and t_s the settling time. In Table 9, it is shown the ME for estimation of the steady-state frequency and phase at different grid frequencies.

Table 8. Results summary obtained for test #2—overshoot (*OS*) and settling time (t_s).

		DSOGI			FFDSOGI			IFLL		
		$\Delta\omega =$ 5 Hz	$\Delta\omega =$ 10 Hz	$\Delta\omega t =$ $\pi/4$	$\Delta\omega =$ 5 Hz	$\Delta\omega =$ 10 Hz	$\Delta\omega t =$ $\pi/4$	$\Delta\omega =$ 5 Hz	$\Delta\omega =$ 10 Hz	$\Delta\omega t =$ $\pi/4$
ω'	t_s (ms)	45.0	45.0	45.0	30.0	30.0	30.0	30.0	33.0	33.0
	OS (rad/s)	31.4	62.8	52.0	31.4	62.8	59.0	31.4	62.8	34.0
$\omega t'$	t_s (ms)	60.0	60.0	65.0	38.0	38.0	40.0	60.0	60.0	65.0
	OS (rad)	0.16	0.36	0.79	0.13	0.27	0.79	0.14	0.32	0.79

Table 9. Results summary obtained for test #2—steady state ME.

		DSOGI			FFDSOGI			IFLL		
		$\omega =$ 45 Hz	$\omega =$ 55 Hz	$\omega =$ 50 Hz	$\omega =$ 45 Hz	$\omega =$ 55 Hz	$\omega =$ 50 Hz	$\omega =$ 45 Hz	$\omega =$ 55 Hz	$\omega =$ 50 Hz
ω'	*ME* (rad/s)	4.8 m	25 m	0.3 m	14 m	6.4 m	0.9 m	22 m	46 m	46 m
$\omega t'$	*ME* (rad/s)	6.8 m	8.4 m	7.8 m	5.0 m	11 m	7.8 m	6.9 m	8.8 m	8.0 m

Based on the results summarized in Tables 8 and 9, the final classification is presented later in Table 12. The classification is based in the average ratio of each evaluated value with the minimum value obtained during that test for the different conditions. Hence, the best result corresponds to the smaller classification value 1.0.

6.3. Test #3—Voltage Sags

During test #3 the grid frequency is kept constant (nominal value), while harmonics and unbalancing are the same as those applied in test #2 (Tables 2 and 4). Hence, the changes applied during the present test focus on the voltage amplitude, i.e., the sags are applied to the symmetrical voltage components and also the harmonics.

The voltage sags periods and amplitudes are summarized in Table 10 and the resulting grid voltage waveforms are shown in Figure 15.

Table 10. Voltage sags applied during test #3.

$t = 0$	No voltage sag
t_1	Voltage sag of 30% with 75 *ms* duration
t_2	Voltage sag of 60% with 150 *ms* duration
t_3	Voltage sag of 90% (permanent)

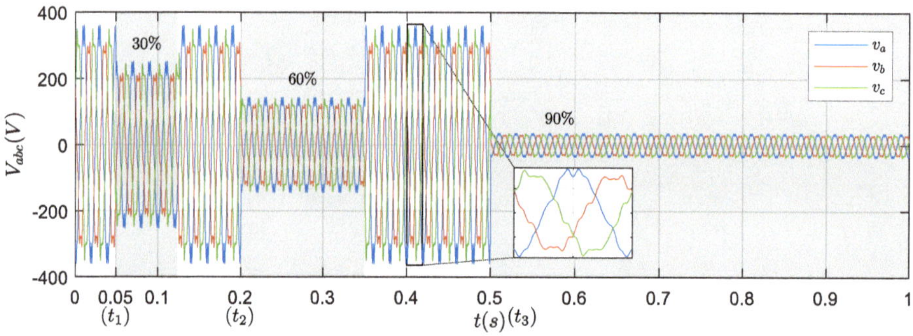

Figure 15. Voltage sags applied during test #3.

The errors associated with frequency and phase estimations, obtained during test #3, are shown in Figure 16, where the top plots refer to the frequency and the bottom ones to the phase.

Analyzing the frequency dynamic response of Figure 16 allows concluding that all methods respond very similarly to the 30% voltage sag (t_1). During the 60% voltage sag (at t_2), both FFDSOGI and IFLL-SRF also perform with the same settling time, despite the first response with larger oscillations. In the same interval, it is noticeable the weaker performance of the DSOGI, which takes $\approx$20 ms more to reach a steady state. The 90% voltage sag is the perturbation with a higher impact of all the analyzed tests, where it can be seen the long settling time of both FFDSOGI and DSOGI, with both taking, respectively, 200 ms and 350 ms to reach a steady state. On the other hand, the IFLL-SRF presents a significantly faster response time, taking 100 ms to reach the steady state.

In terms of phase dynamic response, the results are coherent with the aforementioned comments for the estimated frequency. The only exception is during (t_3), where the IFLL-SRF presents a phase response, similar to the FFDSOGI, of $\approx$200 ms.

The summary of the voltage sag results is presented in Table 11, where the steady-state ME is not shown since it remains approximately constant for all methods.

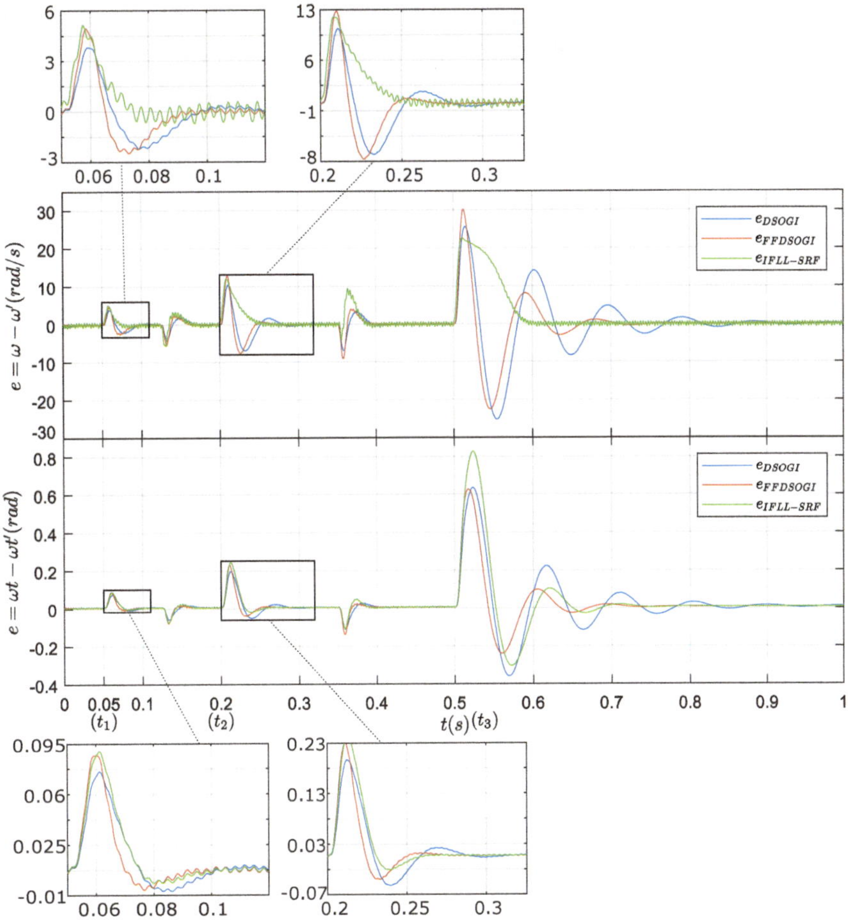

Figure 16. Estimated frequency and phase errors for the different methods during test #3.

Table 11. Results summary obtained for test #3- voltage sags.

		DSOGI			FFDSOGI			IFLL		
		$VS =$ 30%	$VS =$ 60%	$VS =$ 90%	$VS =$ 30%	$VS =$ 60%	$VS =$ 90%	$VS =$ 30%	$VS =$ 60%	$VS =$ 90%
ω'	t_s (ms)	42.0	70.0	360	38.0	60.0	160	38.0	60.0	90.0
	OS (rad/s)	3.81	10.4	25.6	4.94	12.8	30.1	5.15	12.0	22.6
$\omega t'$	t_s (ms)	51.0	113	500	41.0	75.0	280	48.0	75.0	280
	OS (rad)	0.061	0.200	0.635	0.087	0.230	0.627	0.090	0.240	0.827

6.4. Tests Benchmark

Taking into account Tables 5, 8, 9 and 11, classification Table 12 is presented. For classification purposes, the marks were calculated by considering the ratio of the averaged registered values (per method of each test transition) by the minimum average obtained from all tests for each test transition. Therefore, the best result (minimum value) for each test/transition is 1.0 and the choice of the suitable method should be based on the minimum classification value obtained. The table classifications are divided into two categories (frequency and phase), where for each category, the methods are classified in terms of the steady-state performance SS (RMSE or ME), dynamic response to frequency step changes $\Delta\omega$, phase jumps ($p.j.$), and voltage sags VS (evaluating both the overshoot and settling time). It can be noticed that the steady state is based only on the average of the RMSE for the frequency and the average of the ME for the phase angle estimations (both from test #1). The choice of the steady-state error indicators was based on its significance, i.e., the RMSE in the case of the frequency, and ME in the phase estimation.

Table 12. Classification table of the methods' performance during different tests.

			DSOGI	*FFDSOGI*	*IFLL*	*Description*
ω	*RMSE*		1.0	1.0	4.3	Steady-state frequency oscillations—test #1
	$\Delta\omega$	*OS*	1.0	1.0	1.0	Overshoot during frequency step—test #2
		t_s	1.9	1.0	1.6	Settling time during frequency step—test #2
	p.j.	*OS*	1.5	1.7	1.0	Overshoot during phase jump—test #2
		t_s	2.0	1.0	1.7	Settling time during phase jump—test #2
	VS	*OS*	1.0	1.3	1.2	Overshoot during voltage sag—test #3
		t_s	2.1	1.3	1.0	Settling time during voltage sag—test #3
ωt	*ME*		1.0	3.9 *	1.2	Steady-state phase deviance—test #2
	$\Delta\omega$	*OS*	1.3	1.0	1.2	Overshoot during frequency step—test #2
		t_s	1.6	1.0	1.6	Settling time during frequency step—test #2
	p.j.	*OS*	1.0	1.0	1.0	Overshoot during phase jump—test #2
		t_s	1.6	1.0	1.6	Settling time during phase jump—test #2
	VS	*OS*	1.0	1.2	1.3	Overshoot during voltage sag—test #3
		t_s	1.5	1.0	1.1	Settling time during voltage sag—test #3
MCU execution time			1.0	1.0	1.3	MCU processing time— experimental result

* for operation frequency of $\omega = \omega_n \pm 5\%$, consider 1.3 for the FFDSOGI.

For the grid-following operation with the main grid (strong grid) sudden changes are not expected, such as voltage sags or frequency steps, since the grid voltage transients are slow and smooth when compared with the converter and PLL dynamics. Hence, in this case, it is recommended to consider the steady-state phase and frequency estimation (RMSE and ME) and the MCU burden as the selection criteria. In such conditions, the DSOGI-PLL is the best choice by a marginal difference when compared with the FFDSOGI (selection based on the method simplicity and MCU execution time of Figure 23). The IFLL is only recommended if fast-tracking of frequency is important during voltage sags or phase jumps, where mostly because of its gain normalization, a smoother and faster response is achieved when compared with the other methods but at the cost of a higher MCU burden. Such a requirement can be considered in power converters operating with reactive power compensation in inductive weak grids. The FFDSOGI outperforms the other methods in the overall results of the test, making it the most suitable choice when uncertain conditions are expected. The only constraint with high impact in its selection is related to the grid frequency variation, which must be small ($\omega_n \pm 5\%$).

Through the presented classification table, it is possible to apply weights to the different marks and select the best method based on the minimum value obtained.

As an example, in this paper, the choice of a PLL to be applied on a power converter control loop is considered, operating in a low-voltage microgrid as described in [27]. In such conditions, the grid can be either weak (island) or strong (grid connected). In the grid connected (strong grid) condition, the most important requirements are as discussed in the previous paragraph. When operating in island mode (weak grid), the lack of inertia may result in the occurrence of any of the earlier mentioned perturbations. Since the phase angle estimation is the crucial variable of interest, in the operation of the power converter, the following weights (applied to each component in parenthesis) are suggested:

- Phase angle steady-state estimation is of high importance since controllers are designed to operate in the SRF (25%);
- Frequency should be according to EN 50438, i.e., deviance inside the limit $\omega_n \pm 3\%$;
- Phase jumps may occur during re-connection with the main grid (25%);
- Voltage sags may occur during overload conditions or heavy-load connection (25%);
- Low MCU processor burden (25%).

Following the marks and listed criteria in Table 12, the best choice is FFDSOGI-PLL (1.03 pts), followed by DSOGI (1.25 pts), and IFLL (1.30 pts).

6.5. Experimental Results

To evaluate and confirm the synchronistic performance of the methods discussed in the previous section, a voltage source converter (VSC) is configured to generate a local grid voltage (instead of a DAC board) so that the switching EMI can be present in the measurements. The VSC operates in an open-loop as a programmable voltage source with a filter inductance supplying a resistive load that is y-connected. Hence, it is possible to generate any three-phase voltage through VSC modulation signal manipulation.

Additionally, voltage measurements are taken at filter output; therefore, the modulation phase angle leads the measured phase voltage, due to the load angle. As a consequence, the estimated phase angle presents a DC error resulting from the load angle when the modulation phase angle is taken as reference. The resulting setup is presented in Figure 17. The harmonics injected in the modulation are also adjusted so that the voltage, measured at inductance output, contains the harmonics as indicated in Table 2.

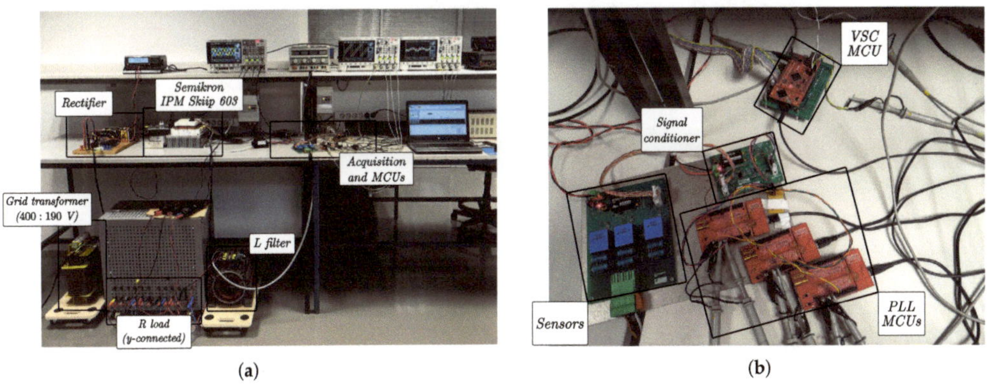

Figure 17. Laboratory setup for three-phase voltage generation and evaluation of the PLL's performance. (**a**) shows the full setup and (**b**) the acquisition board, conditioning board, and MCUs.

For evaluation of the methods, three identical MCUs (Infineon XMC 4500) were programmed, each with a different method. The sensed three-phase voltage was fed into each of the MCU boards and the two DACs available per board were tuned for further evaluation of the results. The PLL MCU DACs generated signals according to the estimated frequency and phase angle of each method and were compared with the VSC MCU modulation frequency and phase. The generated voltages were limited below 70 V_{rms}, though MCU readings were amplified by a factor of 3 so that the measured voltage approached near nominal grid conditions. Notably, the presented setup does not allow to apply steps as applied during the simulation, due to the RL time constant.

In Figure 18 are shown the results obtained for test #1 considering conditions similar to t_3 of Table 4. In the figure, it is shown the voltage waveforms at filter inductance output with unbalanced phase voltages and harmonic content, measured through the oscilloscope FFT function.

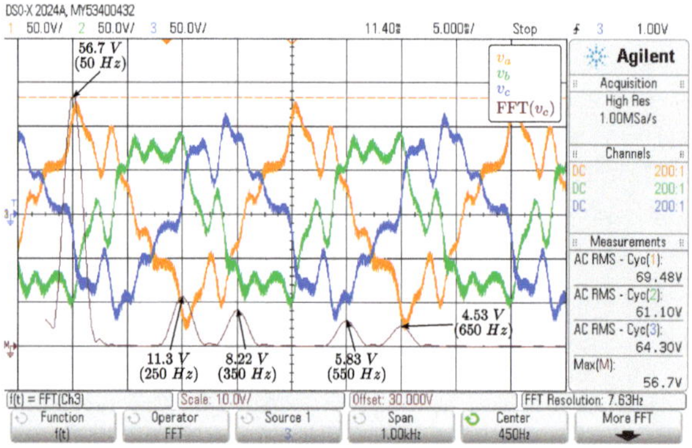

(**a**)

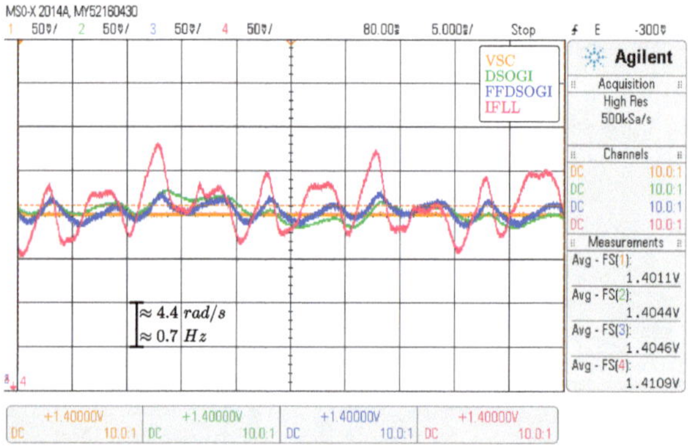

(**b**)

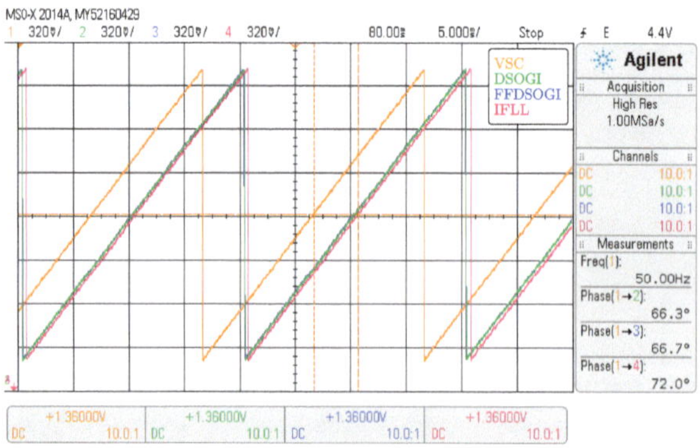

(**c**)

Figure 18. Experimental results for test #1. (**a**) Measured voltages with the FFT of phase v_c, (**b**) estimated frequency and (**c**) estimated phase angle.

Analysis of the results obtained for test #1 allows concluding that, as discussed in Section 6.1, the DSOGI and FFDSOGI methods perform very similarly in terms of frequency estimation. The IFLL presents higher oscillations than the other two, though such differences are not as noticeable in practice as the ones obtained during simulation. The phase estimation of the IFLL presents a higher ME (DC error) than the other methods, as also discussed in Section 6.1. Despite not being noticeable in Figure 18, the steady-state phase estimation oscillations are identical in all three methods. Such a fact was concluded by monitoring the PI outputs of all methods.

For test #2, it was applied a frequency step ($45 \rightarrow 55\ Hz$), as shown in Figure 19, and a phase jump of 45° at nominal frequency (Figure 20). Additionally it is shown in Figure 21 the estimated angle in a steady state for both limit frequencies.

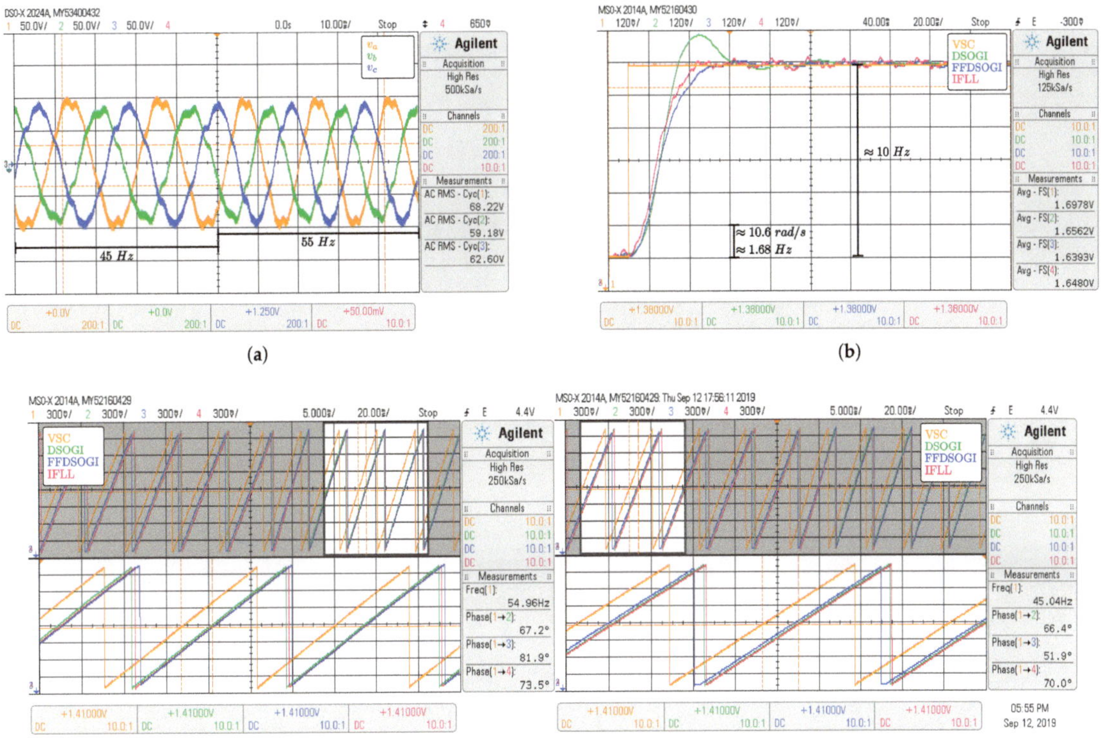

(c)

Figure 19. Experimental results for test #2 frequency step. (**a**) Measured voltages, (**b**) estimated frequency, and (**c**) estimated phase angle.

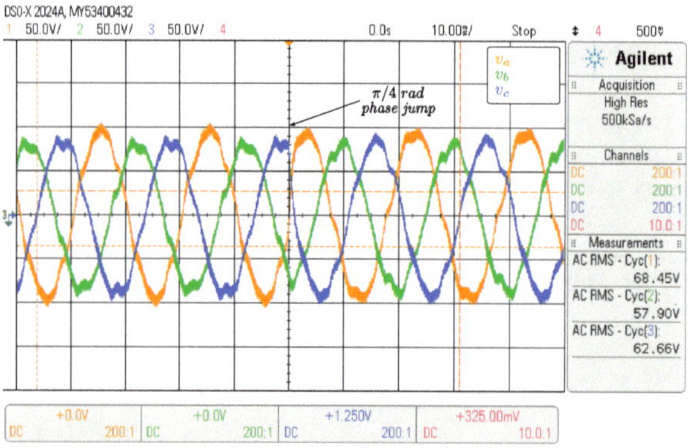

(**a**)

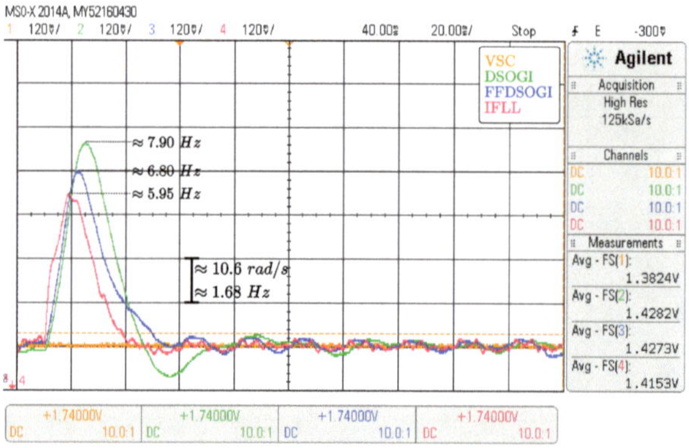

(**b**)

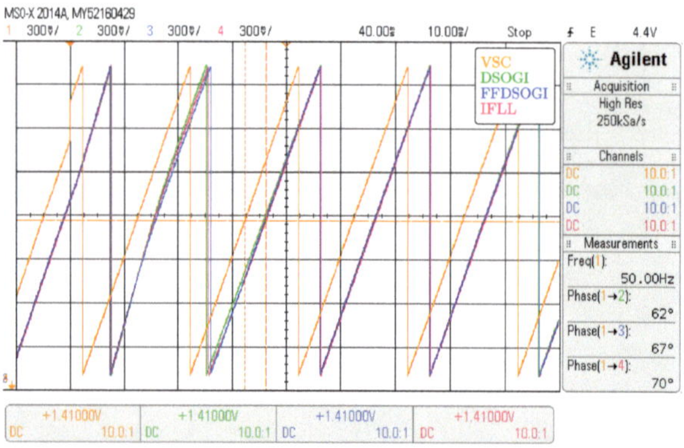

(**c**)

Figure 20. Experimental results for test #2 phase jump. (**a**) Measured voltages, (**b**) estimated frequency, and (**c**) estimated phase angle.

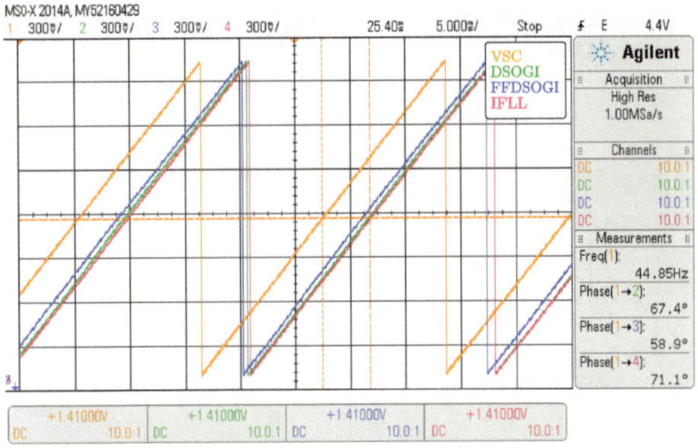

(**a**)

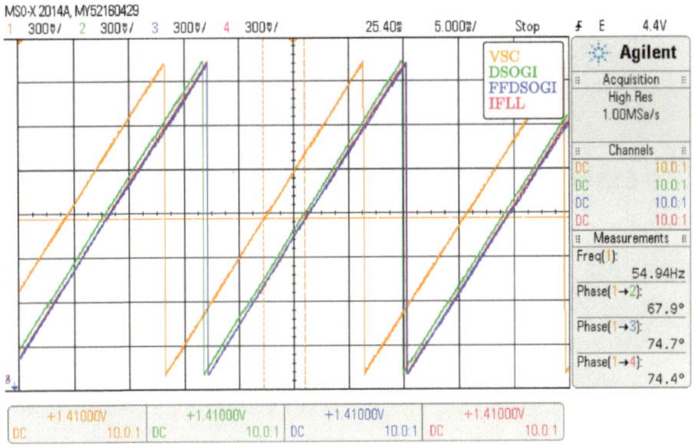

(**b**)

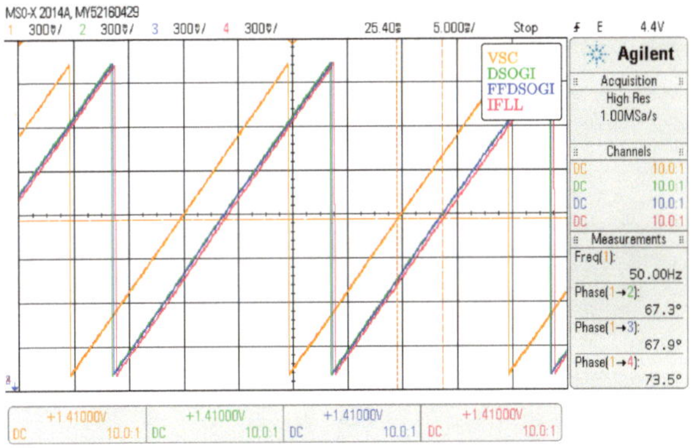

(**c**)

Figure 21. Steady state angle estimation for different frequencies. (**a**) 45 Hz, (**b**) 50 Hz and (**c**) 55 Hz.

The obtained results agree with the discussion in Section 6.2. The estimated frequency of both tests (Figures 19 and 20) follows the results already discussed. The only exception is for the phase estimation by the FFDSOGI during limit frequencies, i.e., there is a DC error of the phase estimation that changes depending on the frequency. Such behavior is justified by the method linearization for the phase estimation and can be visualized in

Figure 21 by checking the change in the phase difference between the modulation phase angle (VSC) and the FFDSOGI.

For the voltage sag of test #3, a 90% sag is applied since it corresponds to the simulated worst case, and the respective obtained results are presented in Figure 22.

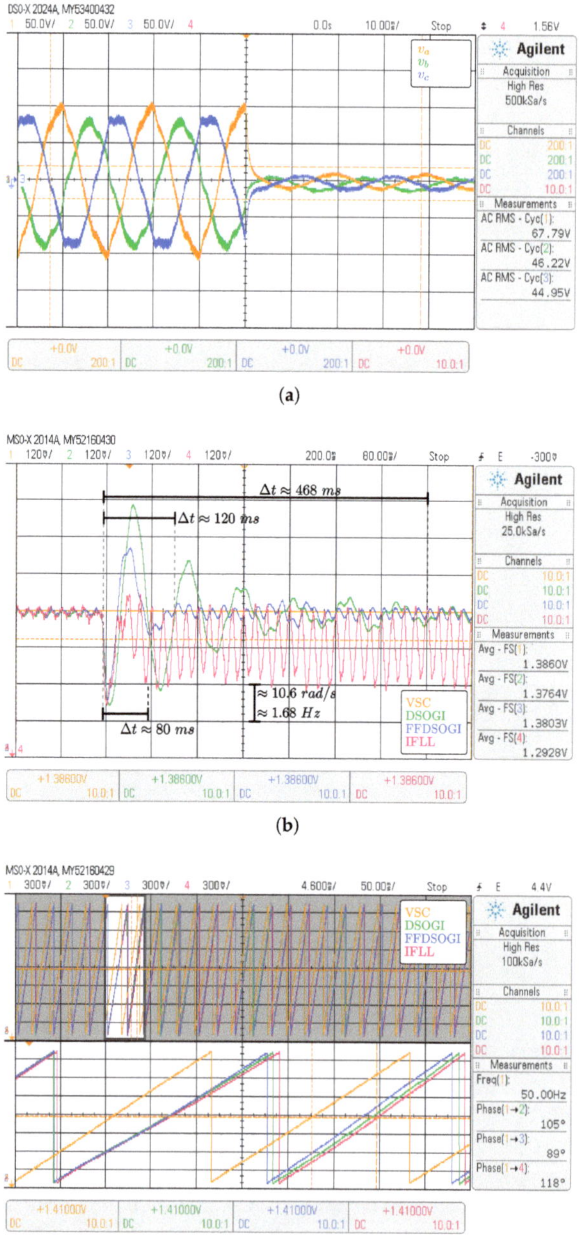

(a)

(b)

(c)

Figure 22. Experimental results for test #3. (**a**) Measured voltages, (**b**) estimated frequency and (**c**) estimated phase angle.

From Figure 22, it can be noticed that the IFLL is the method that responds faster, although it presents significantly higher steady-state oscillations and a change in the average frequency estimation value. The FFDSOGI takes 1.5× more than the IFLL to reach a steady state, though it presents significantly fewer oscillations and the steady-state average error can be neglected. The DSOGI behavior follows the FFDSOGI one with the difference of taking longer, by 4×, to reach a steady state.

The last test is related to the MCU processing times and it is shown in Figure 23. In the figure, it is shown the execution times of each method, i.e., the pulse width of each square wave represents the execution time of the algorithm, including the ADCs acquisition time.

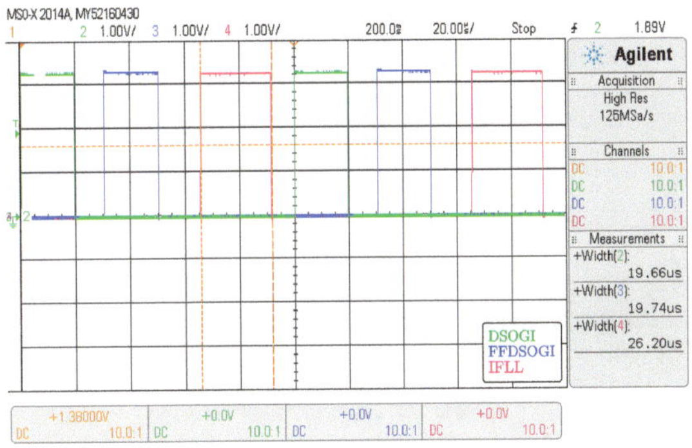

Figure 23. Algorithms process time, including ADCs acquisition time.

7. Conclusions

In the paper, different DSOGI PLL implementations for grid synchronization in weak grids were analyzed. Based on the obtained simulation and experimental results it is possible to conclude that all the DSOGI-based PLL allow to track the three-phase voltage amplitude, frequency, and phase, but show different dynamic and steady-state responses. Based on such results, a benchmark table is presented.

The benchmark table aims to ease the DSOGI-based PLL selection, considering the performance of each implementation for different grid conditions, or events such as low-order harmonics, unbalanced phases, voltage sags, and frequency or phase-step changes. Through the selection of expected individual expected disturbances, it was possible to identify particular scenarios that would result in the choice of each method. The DSOGI is suitable for connection with strong grids, the IFLL in applications where it is required a fast-tracking of the grid frequency during voltage sags or phase jumps, and the FFDSOGI is the most suitable when the phase and/or frequency estimation is required and any of the disturbances can be expected, as far as the grid frequency variations are within the interval of $\omega_n \pm 5\%$.

Criteria based on expected weak-grid disturbances were applied, resulting in the choice of a suitable DSOGI implementation. Therefore, the FFDSOGI-PLL is adopted as the most performant, according to the requirements within the scope of a weak grid.

Author Contributions: Conceptualization, J.P.; methodology, J.P., A.A. and A.R.; software, J.P., A.R.; validation, J.P.; formal analysis, J.P.; investigation, J.P.; resources, A.C. and A.A.; data curation, J.P.; writing—original draft preparation, J.P.; writing—review and editing, J.P., A.C., A.A. and A.R.; visualization, J.P.; supervision, A.C. and A.A.; funding acquisition, A.C. All authors have read and agreed to the published version of the manuscript.

Funding: This research received no external funding.

Institutional Review Board Statement: Not applicable.

Informed Consent Statement: Not applicable.

Data Availability Statement: Not applicable.

Conflicts of Interest: The authors declare no conflict of interest.

Abbreviations

The following abbreviations are used in this manuscript:

ADC	Analog to digital converter
DSOGI	Double second-order generalized integrator
FFPS	Fundamental frequency positive sequence
FFDSOGI	Frequency-fixed double second-order generalized integrator
FLL	Frequency-locked loop
MCU	Micro-controller unit
ME	Mean error
PLL	Phase-locked loop
RMSE	Root-mean-square error
SOGI	Second-order generalized integrator
TOGI	Third-order generalized integrator

References

1. Jarzyna, W. A survey of the synchronization process of synchronous generators and power electronic converters. *Bull. Pol. Acad. Sci. Tech. Sci.* **2019**, *67*. [CrossRef]
2. Karimi-Ghartemani, M.; Reza, I.M. A method for synchronization of power electronic converters in polluted and variable-frequency environments. *IEEE Trans. Power Syst.* **2004**, *19*, 1263–1270. [CrossRef]
3. Chung, S.K. A phase tracking system for three phase utility interface inverters. *IEEE Trans. Power Electron.* **2000**, *15*, 431–438. [CrossRef]
4. Yulin, Y.; Guihua, L.; Hongpeng, L.; Wei, W. A single-phase grid synchronization method based on frequency locked loop for PV grid-connected inverter under weak grid. In Proceeding of the 11th World Congress on Intelligent Control and Automation, Shenyang, China, 29 June–4 July 2014; pp. 5453–5456. [CrossRef]
5. Ramos, C.J.; Martins, A.P.; Carvalho, A.S. Frequency and phase-angle estimation using ordinary least squares. *IEEE Trans. Ind. Electron.* **2015**, *62*, 5677–5688. [CrossRef]
6. Doria-Cerezo, A.; Repecho, V.; Biel, D. Three-Phase Phase-Locked Loop Algorithms Based on Sliding Modes. *IEEE Trans. Power Electron.* **2015**, *62*, 5677–5688. [CrossRef]
7. Rodriguez, P.; Luna, A.; Etxeberría, I.; Hermoso, J.R.; Teodorescu, R. Multiple second order generalized integrators for harmonic synchronization of power converters. In Proceedings of 2009 IEEE Energy Conversion Congress and Exposition, San Jose, CA, USA, 20–24 September 2009; pp. 2239–2246. [CrossRef]
8. Nicastri, A.; Nagliero, A. Comparison and evaluation of the PLL techniques for the design of the grid-connected inverter systems. *IEEE Int. Symp. Ind. Electron.* **2010**, 3865–3870. [CrossRef]
9. Liu, B.; Zhuo, F.; Zhu, Y.; Yi, H.; Wang, F. A Three-Phase PLL Algorithm Based on Signal Reforming Under Distorted Grid Conditions. *IEEE Trans. Power Electron.* **2015**, *30*, 5272–5283. [CrossRef]
10. Golestan, S.; Guerrero, J.M.; Vasquez, J.C.; Abusorrah, A.M.; Al-Turki, Y. Three-Phase PLLs: A Review of Recent Advances. *IEEE Trans. Power Electron.* **2017**, *32*, 1894–1907. [CrossRef]
11. Rodríguez, P.; Pou, J.; Bergas, J.; Candela, J.I.; Burgos, R.P.; Boroyevich, D. Decoupled double synchronous reference frame PLL for power converters control. *IEEE Trans. Power Electron.* **2007**, *22*, 584–592. [CrossRef]
12. Rodríguez, P.; Luna, A.; Ciobotaru, M.; Teodorescu, R.; Blaabjerg, F. Advanced grid synchronization system for power converters under unbalanced and distorted operating conditions. *IECON Proc.* **2006**, *2*, 5173–5178. [CrossRef]
13. Rodríguez, P.; Teodorescu, R.; Candela, I.; Timbus, A.V.; Liserre, M.; Blaabjerg, F. New positive-sequence voltage detector for grid synchronization of power converters under faulty grid conditions. In Proceedings of the PESC Record—IEEE Annual Power Electronics Specialists Conference, Jeju, Korea, 18–22 June 2006. [CrossRef]
14. Ali, Z.; Christofides, N.; Hadjidemetriou, L.; Kyriakides, E.; Yang, Y.; Blaabjerg, F. Three-phase phase-locked loop synchronization algorithms for grid-connected renewable energy systems: A review. *Renew. Sustain. Energy Rev.* **2018**, *90*, 434–452. [CrossRef]
15. Jiang, Y.; Li, Y.; Tian, Y.; Wang, L. Phase-Locked Loop Research of Grid-Connected Inverter Based on Impedance Analysis. *Energies* **2018**, *11*, 3077. [CrossRef]
16. Khatana, V.; Bhimasingu, R. Review on Three-Phase PLLs for Grid Integration of Renewable Energy Sources. In Proceedings of the 2017 14th IEEE India Council International Conference (INDICON), Roorkee, India, 15–17 December 2017. [CrossRef]
17. Jarzyna, W.; Zieliński, D.; Gopakumar, K. An Evaluation of The Accuracy of Inverter Sync Angle during the Grid's Disturbances. *Metrol. Meas. Syst.* **2020**, *27*. [CrossRef]
18. Xu, J.; Qian, H.; Hu, Y.; Bian, S.; Xie, S. Overview of SOGI-Based Single-Phase Phase-Locked Loops for Grid Synchronization Under Complex Grid Conditions. *IEEE Access* **2021**, *9*. [CrossRef]

19. Xiao, F.; Dong, L.; Li, L.; Liao, X.; Al-Turki, Y. A Frequency-Fixed SOGI-Based PLL for Single-Phase Grid-Connected Converters. *IEEE Trans. Power Electron.* **2017**, *32*, 1713–1719. [CrossRef]
20. Golestan, S.; Mousazadeh, S.Y.; Guerrero, J.M.; Vasquez, J.C. A Critical Examination of Frequency-Fixed Second-Order Generalized Integrator-Based Phase-Locked Loops. *IEEE Trans. Power Electron.* **2017**, *32*, 6666–6672. [CrossRef]
21. Hoepfner, B.; Vick, R. Symmetrical Components Detection With FFDSOGI-PLL Under Distorted Grid Conditions. In Proceedings of the 2019 International Conference on Smart Energy Systems and Technologies (SEST), Porto, Portugal, 9–11 September 2019. [CrossRef]
22. Rodríguez, P.; Luna, A.; Candela, I.; Mujal, R.; Teodorescu, R.; Blaabjerg, F. Multiresonant frequency-locked loop for grid synchronization of power converters under distorted grid conditions. *IEEE Trans. Ind. Electron.* **2011**, *58*, 127–138. [CrossRef]
23. Ma, W.; Ouyang, S.; Xu, W. Improved Frequency Locked Loop Based Synchronization Method for Three-Phase Grid-Connected Inverter under Unbalanced and Distorted Grid Conditions. *Energies* **2019**, *12*, 1023. [CrossRef]
24. Du, H.; Sun, Q.; Cheng, Q.; Ma, D.; Wang, Xu An adaptive frequency phase-locked loop based on a third order generalized integrator. *Energies* **2019**, *12*, 309. [CrossRef]
25. Ukil, A.; Shah, V.H.; Deck, B. Fast computation of arctangent functions for embedded applications: A comparative analysis. In Proceedings of the ISIE 2011: 2011 IEEE International Symposium on Industrial Electronics, Gdansk, Poland, 27–30 June 2011; Volume 2. [CrossRef]
26. Nikolic, A.; Naumovic-Vukovic, D.; Skundric, S.; Kovacevic, D.; Milenkovic, V. Methods for power quality analysis according to EN 50160. In Proceedings of the 2007 9th International Conference on Electrical Power Quality and Utilisation, Barcelona, Spain, 9–11 October 2007; pp. 1–6. [CrossRef]
27. Pinto, J.; Carvalho, A.; Morais, V. Power Sharing in Island Microgrids. *Front. Energy Res.* **2021**, *8*. [CrossRef]

electricity

MDPI

Article

Assessment of the Visual Impact of Existing High-Voltage Lines in Urban Areas

Andreas Sumper *, Oriol Boix-Aragones, Joan Rull-Duran, Joan Amat-Algaba and Joachim Wagner

Centre d'Innovació Tecnològica en Convertidors Estàtics i Accionaments (CITCEA-UPC), Departament d'Enginyeria Elèctrica, ETS d'Enginyeria Industrial de Barcelona, Universitat Politècnica de Catalunya, 08028 Barcelona, Spain; oriol.boix@upc.edu (O.B.-A.); joan.rull@upc.edu (J.R.-D.); xuan.amat@gmail.com (J.A.-A.); joachim.wagner@halifax.rwth-aachen.de (J.W.)

* Correspondence: andreas.sumper@upc.edu

Abstract: This article proposes a novel methodology to evaluate the visual impact of high-voltage lines in urban areas based on photographic images. The use of photographs allows for calculating the overall aesthetic impact while eliminating the subjective factors of the observer. To apply the proposed methodology based on photographs, the impact of the position and angle where the photograph was taken was analyzed, and a sensibility analysis was carried out. Moreover, it was applied to an application case, and a comparison with results from a previous study of a visual impact was performed. The methodology shows good performance and a better resolution of the indicator.

Keywords: high voltage lines; environmental impact; urban landscape; social impact

Citation: Sumper, A.; Boix-Aragones, O.; Rull-Duran, J.; Amat-Algaba, J.; Wagner, J. Assessment of the Visual Impact of Existing High-Voltage Lines in Urban Areas. *Electricity* **2021**, 2, 285–299. https://doi.org/10.3390/electricity2030017

Academic Editor: Pavlos S. Georgilakis

Received: 11 June 2021
Accepted: 27 July 2021
Published: 29 July 2021

1. Introduction

The work presented here proposes a methodology to assess the visual impact of existing overhead transmission lines on the environment. Its main objective is a reduction of subjective influences in the assessment procedure. The proposed approach is combinable with the general impact assessment methodology by Sumper et al. [1] and aims to enhance the aesthetic impact assessment of this methodology.

Electricity is a prerequisite for social, industrial, and commercial development [2], and the transmission of electricity is of extreme importance to a proper electricity supply. However, the lack of social and public acceptance of the transmission line infrastructure [3] manifests the need for a deep analysis of the impact of transmission lines. Public opinion is dominated by social syndromes such as 'Not In My Backyard' (NIMBY) [4], 'Build Absolutely Nothing Anywhere Near Anything' (BANANA) [5], and 'Not In My Term of Office' [6], which makes necessary more active participation of stakeholders in the decision processes to prevent and solve conflicts. Mediation techniques are powerful tools to involve the conflicting parties and to find a common, self-provided 'win–win' solution [7]. They often require us to include in the process expert opinion or an external evaluation of the situation by an objective impact analysis or indicators [8].

The measurement of the widespread and diverse impacts presents analytical challenges in developed countries [9]. Transmission lines have several types of impacts on the environment as stated by Bickel and Friedrich [10] and Doukas et al. [11]. The most common impacts are influences on the health of residents, urban development, flora and fauna, and the aesthetics of the landscape. The latter are often controversially discussed because of their highly subjective nature. An assessment of all these impacts is not trivial. Therefore, pairing systematically analytical and experiential methodologies is fundamental to ecological design to include intentional changes of landscapes in cities, their megaregions, and resource hinterlands as shown by Nassauer [12]. The literature commonly contains studies that focus only on a single aspect, such as Torres-Sibille et al. in [13,14]. Hadrian et al. [15] present an automated mapping method enabling us to choose corridors to minimize visual impacts depending on the surrounding landscape. Luken et al. [16] analyze the visual perception on forest edges of power line corridors by surveys and make recommendations to reduce the visual impact by camouflaging corridors in forests. Tempesta et al. [17] show the willingness-to-pay of the Italian population to eliminate the landscape impact of high-voltage overhead transmission lines in rural areas, and it was estimated for the entire national territory for four different landscape contexts. A similar result was shown by Frontuto et al. [18] for agricultural sheds, where residents are willing to pay for mitigation solutions.

A methodology for an environmental impact assessment of existing overhead lines was developed by Sumper et al. [1]. It includes all the above-mentioned impact types and is the first attempt to also account for the aesthetic impact of overhead lines. The approach supports the development of mitigation strategies by identifying the most critical sectors of the transmission line. The aesthetic impact is evaluated by experts. A lack of information can lead to conflicts with residents and lobby groups. Therefore, this work develops a transparent and comprehensive methodology for the aesthetic impact assessment of overhead lines. It reduces the subjective influences of the examiners during the evaluation procedure and aims to improve the acceptance of the assessment.

The contribution of this paper is a methodology for the determination of the aesthetic impact of overhead lines by using images of such lines. For the use of an image processing methodology, the range of observer distances and angles was determined to guarantee comparable results. The proposed methodology improves the overall impact assessment of existing overhead lines presented by Sumper et al. [1] by eliminating the subjective criteria of the observer for the aesthetic impact, which is part of the overall assessment.

2. Environmental Impact of Overhead Power Lines

Sumper et al. [1] propose an overall assessment methodology for the environmental impact of overhead transmission lines in urban areas. The aim of this methodology is to support a decision process for mitigation actions that include not only public authorities but also experts and citizens. The first group is small and usually legitimated by elections. The second group often represents interest groups such as companies or organizations. They are often supposed to act as impartial advisors that provide all necessary information required for the decision process. However, often they are linked to one group with individual interests. The latter group includes residents and other involved citizens. They are considered to be active or common. 'Active' means that these citizens contribute actively to the discussion process. This group is usually very critical and a relatively small group among the population. The 'common' citizens are usually a very heterogeneous group that behaves rather passively. It can include people from all social backgrounds. Therefore, this group often has very diverse interests and opinions. Furthermore, these interests are commonly difficult to gather because of the small willingness to contribute to this group. This is possibly linked to the heterogeneous character of this group and the small perception of common interests.

The methodology presented by Sumper et al. [1] includes eight steps as shown in Figure 1. The first step is the delimitation of the evolution area. In other words, the area that is influenced by the overhead line is defined, which consequently requires the acceptance of all involved parties, such as public authorities, companies, and citizens. In the second step, the area to be studied is defined. This is done by experts from the public side and the utility side. Subsequently, all involved groups determine the impact parameters of the study. This includes three steps. First, the impact types are defined, and then the scale and the weight of each impact parameter are determined. These three steps are very important for the evaluation and, consequently, for the acceptance of the final decision. Once the impact parameters are defined, experts can elaborate on the evaluation. This includes the determination of the individual impacts as well as an analysis of the results. Finally, the results are presented to the other involved parties, and possible mitigation strategies for identified problems can be discussed. Therefore, it is necessary to also agree on mitigation measures to ensure an impartial decision process. A detailed description of the methodology can be found in [1].

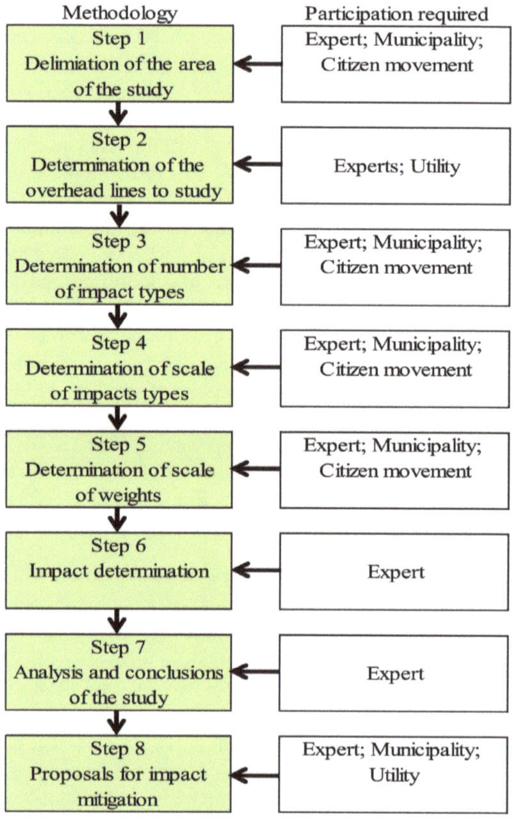

Figure 1. Methodology to evaluate the overall impact of overhead lines in urban areas.

The above-described methodology for impact assessment of transmission lines accounts for several impacts that transmission lines can have on their environment. One important impact type is the visual impact assessment. Three indicators determine this visual impact, as shown in

$$I_{visual} = \sqrt[3]{V \cdot F \cdot IQ} \tag{1}$$

where V is the visibility of the line, F is the fragility of the landscape, and IQ is the intrinsic quality of the area. The visibility accounts for the dimensions of the line. Overhead transmission lines with higher and wider pylons have a larger impact than small distribution lines. The fragility describes the effect the overhead line has on the area. For example, trees of a forest may partially hide the pylon, which consequently reduces the impact the structure has on this area. The intrinsic quality accounts for the characteristics or usage of the area. Obviously, an overhead line that passes through an industrial area is less critically observed than a similar line that passes through a residential area or a singular landscape.

A drawback of the indicators presented above is their subjective nature. This is because of the fact that the presented factors V, F, and IQ are determined by experts, and the values depend on the experience of those experts. The contribution of this paper is to implement a methodology to ensure an objective evaluation of the visual impact by using photographs. Torres-Sibille et al. used the objective aesthetic impact of a wind farm [13] and solar plants [14] located in a defined landscape by a combination of visibility, color, fractality, and continuity, which can be obtained from photographs. Torres-Sibille et al. [14] demonstrated that the calculated indicator correctly represents the order of preference resulting from the perception of impact determined by experts. The present paper adopts the methodology for the objective visual impact assessment of overhead lines and determines the conditions under which the photographs have to be taken.

3. Methodology to Evaluate the Visual Impact of Overhead Power Lines

To create a methodology for the assessment of visual impacts of a structure on the environment it is necessary to identify parameters that influence the aesthetic impression. In order to develop an advanced approach for the visual impact assessment of overhead lines, methodologies from other fields were examined. The new methodology for visual impact assessment was derived from the methodology by Torres-Sibille et al. [13,14]. It aims to extend to the methodology by Sumper et al. presented in [1]. The approach uses four indicators: visibility, color, factuality, and quality of the landscape. As the original approach was used to determine the visual impact of wind turbines and solar power plants, these indicators had to be modified in order to apply this methodology to the evaluation of overhead power lines. A major difference compared with solar power plants and wind turbines, as considered by Torres-Sibille et al., is the influence of color on visibility. The visibility is significantly reduced when the contrast of the overhead line and the environment is small. This is due to the lattice structure of transmission line pylons and the small cross-section of conductors. On the other hand, a solar power plant is not easily overseen, even when its color matches the color of the environment. Consequently, visibility and color are not as independent as in the case of overhead lines. Furthermore, the indicators concurrence and continuity were not included in the presented methodology. Variance in the concurrence and continuity of overhead lines is negligible because the homogeneity and extension of the structure do not vary. Hence, there is no reason to include them in the assessment. Instead of these indicators, the quality of the landscape is accounted for. The new methodology and indicators presented here were developed in [19] and are explained in the following sections.

3.1. The Overall Aesthetic Impact Value

The overall aesthetic impact value (*OAIV*) is the parameter that quantifies the visual impact of an overhead line on a certain area. In order to assess the transmission line, the study zone is divided into sectors. The enhancement reduces subjective effects on the evaluation process. For each sector, an individual *OAIV* is computed. Hence, it is possible to compare different sectors and identify critical sections. Sectors do not overlap and are evaluated based on photos. The overall aesthetic impact value includes the quality of the landscape, the visibility impact, and the fragility impact. The first accounts for the relevance of the environment where the overhead line is built. The visibility impact includes all effects originating from the structure itself, and the fragility impact allows for an evaluation of the vulnerability of the landscape to a disturbance by a transmission line. In order to allow for an approach consistent with the methodology in [1], these impacts are determined by indicators. Thus, the overall aesthetic impact value is defined by

$$OAIV = I_{quali} \cdot \left[\alpha \cdot \left(\beta \cdot \left(I_{vis-pylon} + I_{colour} \cdot I_{vis-wire} \right) \right) + \gamma \cdot I_{fra} \right] \quad (2)$$

where *OAIV* is between 0 and 1, I_{quali} is the quality indicator, β is the climatology coefficient, $I_{vis\text{-}pylon}$ is the visibility indicator for the pylon, $I_{vis\text{-}wire}$ is the visibility indicator for the wires, I_{color} is the color indicator, I_{fra} is the fragility indicator, and α and γ the weighting factors for the fragility and visibility indicators, respectively. All indicators are between 0 and 1, and the sum of the weighting factors α and γ is 1. The *OAIV* does not exceed a value of 1 due to the small contrast between the grey color of the overhead line and the sky. In some cases, conductors are marked with signal orbs; for example, in the vicinity of airports. Then, the color indicator is given a value of 1, which could lead to an *OAIV* larger than 1. In order to make this methodology compatible with the assessment methodology presented in [1], the equation is limited to a maximum value of 1.

3.2. *Quality of the Landscape*

The quality of the landscape evaluates the relevance of the environment around the overhead line, in particular when the landscape is changing by crossing the overhead line, for example, in urban areas. The perception that people have about the landscape changes according to the way the landscape is used and shaped [20]. Obviously, a residential area is more vulnerable than an industrial area, and this has to be accounted for in the assessment procedure. This issue is addressed by the quality of the landscape indicator. However, this assessment is difficult to realize.

For example, consider an overhead line that passes over a historical building. The quality of the landscape should allow for the number of spectators and the time period of their presence. Above that, their personal perception may vary significantly. To account for all this is obviously not trivial. Another approach could use the monetary value of the area around the historical building before and after the construction of the line. Unfortunately, other events could also affect the monetary value, which needs to be considered. Furthermore, such a case is probably not trivial to generalize. Moreover, there are probably other aspects that have yet to be identified.

Due to the complexity of this issue, the methodology presented here uses a robust and rather simple approach. The quality of the landscape is based on photos of the sectors evaluated by a group of experts. In order to guarantee the independence of the indicator value, these people have to be neutral regarding the possible impacts of the line, e.g., neither residents of the zone nor workers of the utility should participate. Each of them classifies the sector into strong, medium, light, and no impact, as depicted in Table 1, following a linear approach. In this study, the approach proposed by Torres-Sibille in [13] was followed, where four different values for the impact measurement are introduced. A higher number of values would increase the granularity of the impact.

Table 1. Strength of the landscape approach.

Strength of the Impact	Indicator Value
Strong impact	1
Medium impact	0.75
Light impact	0.25
No impact	0

Finally, the average of all examiners gives the quality of the landscape of a certain sector. The evaluation of each sector includes the relevance of the area, such as an industrial or a residential area. Furthermore, the density of the population in the area is accounted for. The limitation of this method is that it includes a subjective factor of the criteria of the group of experts. The previous definition of landscape types and their associated values limits this subjective factor, as similar landscape types will be treated the same. Different approaches to assess the quality of the landscape are discussed in [21,22], where a systemized approach to landscape evaluation is discussed for Poland and Alpine regions. Moreover, the findings in [23] argue for the necessity of distinguishing between different ratings and landscape types. For the sake of simplicity, this paper uses the expert approach, as an evaluation of the quality of a landscape is not the primary aim of this paper.

3.3. *Visual Impact*

The visibility of an object depends on two aspects: the area it occupies in the field of view of a spectator and the contrast between the object and the environment. As mentioned above, they are not independent, especially in the case of slender structures such as overhead lines. Therefore, the indicators visibility and color are presented here together. The visibility indicator considers the occupied area in the field of view, and the indicator color considers the contrast between the line and the environment.

Visibility Indicator

The visibility indicator is proportional to the ratio of the area occupied by the overhead line to the whole area of the field of view. Even though this definition is rather elementary, it is difficult to give an exact value for this indicator. In order to do so, the occupied area and the field of view are examined using photos. The overhead line is divided into sectors, and photos are taken. Software packages such as Datinf, Coreldraw, and Photoshop can be used to compute the area of the overhead line in the pictures. This procedure is straightforward in the case of wind turbines or solar power plants. For example, a wind turbine occupies the area of the pylon and the circular area where the turbine blades move. In the presented method, this concept is transferred to transmission line pylons, which usually have a lattice structure. To do so, the occupied area is defined as a polygon of the most extreme points of the structure following the approach of [13] to wind turbines. The resulting surface appears larger than the actual steel surface of the beams and bracings as shown in Figure 2.

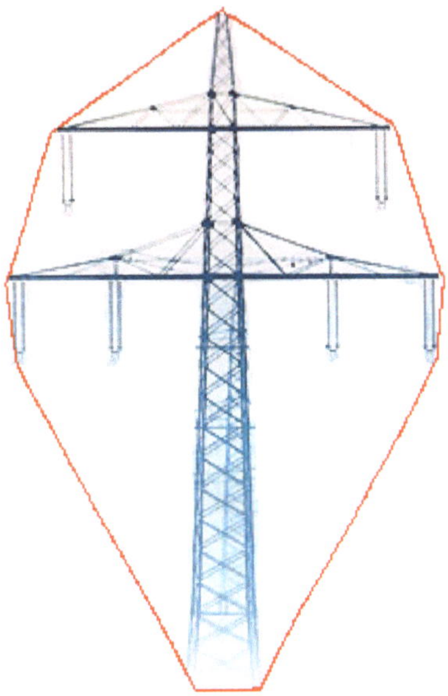

Figure 2. The occupied area of a pylon.

In our case, overhead lines and wind turbines are considered comparable structures. Therefore, the function defining the visibility indicator for wind turbines is considered to also be valid for overhead lines. Obviously, a validation similar to the one performed by Torres-Sibille et al. is desirable; however, at this point, it is a reasonable assumption. The indicator is separately computed for the pylons and the conductors by the following equation

$$I_{vis}(x) = \begin{cases} 0.184 \cdot x & 0 < x \leq 0.7, \\ -0.003 \cdot x^2 + 0.114 \cdot x + 0.051 & 0.7 < x \leq 12.3, \\ 1 & 12.3 < x \leq 20, \end{cases} \tag{3}$$

where

$$x = 100 \cdot \frac{S_{fa}}{S_{ba}}, \tag{4}$$

S_{fa} is the occupied area, and S_{ba} is the whole area of the field of view. Figure 3 shows a graphical representation of this function.

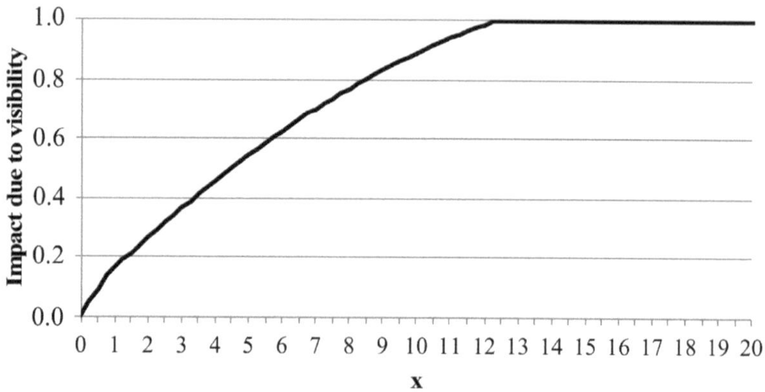

Figure 3. The value function I_{vis}, adopted from [13].

3.4. Color Indicator

The color and the area of the structure have an influence on the visual impact, as already mentioned. In the case of wind turbines and solar power plants with large continuous surfaces, it is reasonable to use separate indicators for color and occupied area. The occupied area of a structure affects only the visual impact if it is noticeable. A grey transmission line pylon in front of a grey sky is still visible, but the conductor's small cross-section is likely to be overseen. Therefore, the color indicator is applied as a factor for the visibility indicator of the conductors. The equation defining the color indicator follows the methodology of Torres-Sibille et al. [14].

$$I_{colour}(\mu) = \begin{cases} 0 & 0 < \mu \leq 5, \\ -356 \cdot 10^{-9} \cdot \mu^2 + 12 \cdot 10^{-4} \cdot \mu - 56 \cdot 10^{-4} & 5 < \mu \leq 1563, \\ 1 & 1563 < \mu \leq 1700, \end{cases} \tag{5}$$

where μ is the difference in the color of the structure and the environment. The parameter μ is defined by

$$\mu = \Delta E_{fa/ba} = \sqrt{\left(L_{fa} - L_{ba}\right)^2 + \left(a_{fa} - a_{ba}\right)^2 + \left(b_{fa} - b_{ba}\right)^2} \tag{6}$$

where $\Delta E_{fa/ba}$ is the color difference, L_{fa} is the hue of the structure, L_{ba} is the hue of the background, a_{fa} is the saturation of the structure, a_{ba} is the saturation of the background, b_{fa} is the brightness of the structure, and b_{ba} is the brightness of the background [14].

3.5. Climatology Coefficient

The visibility of an overhead line also depends on climatology conditions. A cloudy day can reduce the difference in the color of the structure and the sky, for example. Consequently, the visibility of a line will be significantly reduced. This influence is accounted for by an atmospheric coefficient. It reflects the average atmospheric condition in the region where the overhead line is built. Climatology conditions are well documented by weather stations and usually easy to examine. Following the methodology in [14], the weather conditions are classified into four groups: clear, precipitation, fog, and cloudy days. These conditions are a measure of the average atmospheric conditions of a region. Calculation of the climatology coefficient requires the use of a group of experts [14]. Each group is considered to have a certain impact on the visibility. The climatology coefficient determines the average of these weather impacts by

$$\beta = \sum_{i=1}^{n} P_i(m_i) \cdot m_i \tag{7}$$

where P_i is the relative frequency of the weather condition, and m_i is the impact value of the weather condition. Table 2 states the values for m_i.

Table 2. Climatology coefficient values [4].

Climatology	Climatology Value m_i
Clear day	1
Cloudy day	0.75
Precipitation	0.5
Fog	0.25

The methodology of Torres-Sibille et al. [13] weights the indicators of the visual impact and the fragility impact because studies showed that they do not have the same influence on the result. In the case of the visual impact of (1), the weighting factor α is 0.83.

3.6. Fragility Indicator

The fragility aims to allow for an evaluation of the degree of disturbance of the landscape due to the structure. It is difficult to define and to give an exact value for this effect. Therefore, four levels of disturbance are determined: strong, medium, light, and no impact. Their values vary from 0 to 1 with a linear relationship, as shown in Table 3.

Table 3. Fragility impact values.

Strength of the Impact	Indicator Value
Strong impact	1
Medium impact	0.75
Light impact	0.25
No impact	0

As already mentioned, the indicators of the visual impact and the fragility impact are weighted because studies showed that they do not have the same influence on the result [13]. In the case of the fragility impact of (1), the weighting factor γ is 0.17.

4. Impact of the Observation Angle on the Visibility Index

One of the most important drawbacks of the methodology presented in [19] is the fact that the shape of an overhead line has a strong relationship with the angle of vision and the distance of the observation point. With a given distance to the pylon, d, the distance of the camera to the earth, a, and the angle between the horizontal plane and the highest point of the pylon, α, the height, h, of the pylon can be determined by

$$h = (d \cdot tg(\alpha)) + a \tag{8}$$

Once the height of the pylon is known, the observation distance interval can be defined by

$$h \leq d \leq 2.5\, h \tag{9}$$

For each distance, the photographs can be made in relation to the distance and the angle in the intervals shown in Figure 4.

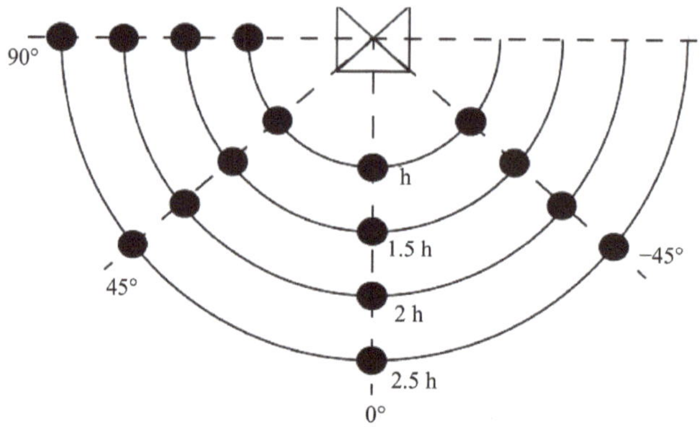

Figure 4. Observation points in the relationship between the height and observation angle intervals.

In order to realize a sensibility analysis to determine how the angle and the observation distance impact upon the visual impact calculation, a 110 kV line pylon of the Castell d'Aro to Vall Llobregat line (Figure 5) was analyzed in Girona Province, Spain [24]. As shown in Figure 4, 16 photographs need to be taken; however, for each photograph, there is the option to take it horizontally or vertically. So, finally, 32 photographs of the pylon were taken. The post analysis was performed by the software Adobe Photoshop CS5® to determine the number of pixels in the photograph and the evolution of the surface of the pylon and the cables, as shown as an example in Table 4. Table 5 shows the resulting percentage of occupation of the pylon and the cables of the photographs in relation to the angle and the height of the pylon for the horizontal and vertical cases.

Figure 5. Pilon T-91 of the 110 kV Castell d'Aro overhead line. The photograph was taken from the observation point at 1.5 h, −45°, horizontal.

Table 4. Example of the values of pixels obtained by the analysis of the photographs for 0° and −45° for the horizontal and vertical cases.

Num. of Pixels		−45° Photograph	−45° Pylon	−45° Cables	0° Photograph	0° Pylon	0° Cables
HORIZONTAL	1 h	13,996,800	718,094	3,701,337	13,996,800	907,989	1,648,112
	1.5 h	13,996,800	340,411	2,520,078	13,996,800	451,266	1,822,646
	2 h	13,996,800	196,106	1,781,716	13,996,800	260,529	1,863,753
	2.5 h	13,996,800	133,522	1,229,322	13,996,800	168,982	1,881,785
Num. of Pixels		**−45° Photograph**	**−45° Pylon**	**−45° Cables**	**0° Photograph**	**0° Pylon**	**0° Cables**
VERTICAL	1 h	13,996,800	684,329	2,838,625	13,996,800	928,726	3,168,071
	1.5 h	13,996,800	337,139	2,006,036	13,996,800	488,690	3,649,559
	2 h	13,996,800	201,857	1,446,967	13,996,800	278,209	3,264,157
	2.5 h	13,996,800	140,260	1,045,124	13,996,800	185,107	3,002,302

Table 5. Percentage of occupation of the pylon and the cables of the photographs in relation to the angle and the height of the pylon for the horizontal and vertical cases.

%		−45° Pylon	−45° Cables	0° Pylon	0° Cables	45° Pylon	45° Cables	90° Pylon	90° Cables
HORIZONTAL	1 h	5.130	26.444	6.487	11.775	5.759	21.194	2.796	28.922
	1.5 h	2.432	18.005	3.224	13.022	2.860	18.780	1.296	20.636
	2 h	1.401	12.729	1.861	13.316	1.718	15.475	0.762	16.130
	2.5 h	0.954	8.783	1.207	13.444	1.145	11.981	0.528	13.023
%		−45° Pylon	−45° Cables	0° Pylon	0° Cables	45° Pylon	45° Cables	90° Pylon	90° Cables
VERTICAL	1 h	4.889	20.281	6.635	22.634	5.932	19.387	2.800	21.557
	1.5 h	2.409	14.332	3.491	26.074	2.800	13.991	1.186	15.414
	2 h	1.442	10.338	1.988	23.321	1.744	10.523	0.756	12.000
	2.5 h	1.002	7.467	1.322	21.450	1.147	8.290	0.533	9.857

By using Table 4, the visibility index I_{vis} was calculated by using (3). In order to perform a sensibility analysis between the same observation point distance and the different angles and possible orientations, the following equation was applied

$$\Delta\,(\%) = \frac{Ivis_{(x)} - Ivis_{ref}}{Ivis_{ref}} \cdot 100 \tag{10}$$

where $Ivis_{(x)}$ is the visibility index, and $Ivis_{ref}$ represents the reference visibility index.

Table 6 shows the differences between the visibility index photographs taken horizontally and vertically using (9). Different angles and pylon and cable visibilities were also analyzed. On the one hand, the pylon visibility suffers from differences of less than ±6.6%, which shows that its influence is low. On the other hand, the cable visibility varies significantly with the distance and angle, with a maximum of 11.7%. Table 7 shows the sensibility analysis of the influence of the angle with respect to the 0° angle. We can observe the high impact of the angle of the photograph taken on the visibility index. We can observe differences of up to 56% in the case of photographs taken vertically at 90°. Tables 8 and 9 show the influence of the angle in relation to the −45° angle and the 45° angle, respectively. On the one hand, we can observe that the comparison between −45° and 45° shows similar values in both tables; on the other hand, some values vary by over 10% depending on the distance and the way the photograph was taken.

Table 6. Sensibility analysis for photographs taken horizontally (H) or vertically (V).

		−45°		0°		45°		90°	
		Pylon	Cables	Pylon	Cables	Pylon	Cables	Pylon	Cables
H vs. V	1 h	3.64	−3.33	−1.66	−2.31	−2.24	0.00	−0.11	0.00
	1.5 h	0.75	0.00	−6.48	0.00	1.64	0.00	6.05	0.00
	2 h	−2.12	9.11	−5.12	0.00	−1.16	8.16	0.46	1.30
	2.5 h	−3.32	10.46	−6.65	0.00	−0.11	19.91	−0.51	11.68

Table 7. Sensibility analysis of the influence of the angle to the 0° position.

		Differences to 0° in %					
		0° and −45°		0° and 45°		0° and 90°	
		Pylon	Cables	Pylon	Cables	Pylon	Cables
HORIZONTAL	1 h	16.16	0.99	8.47	−2.31	47.87	−2.31
	1.5 h	19.84	0.00	9.01	0.00	49.99	0.00
	2 h	18.97	−1.60	5.87	0.00	46.15	0.00
	2.5 h	14.78	17.92	3.62	1.38	40.12	0.00
		Differences to 0° in %					
		0° and −45°		0° and 45°		0° and 90°	
		Pylon	Cables	Pylon	Cables	Pylon	Cables
VERTICAL	1 h	20.54	0.00	7.95	0.00	48.67	0.00
	1.5 h	25.28	0.00	15.94	0.00	55.87	0.00
	2 h	21.29	9.11	9.42	8.16	49.01	1.30
	2.5 h	17.45	26.50	9.54	21.01	43.57	11.68

Table 8. Sensibility analysis of the influence of the angle to the −45° position.

	Differences to −45° in %			
	45° and −45°		90° and −45°	
	Pylon	Cables	Pylon	Cables
HORIZONTAL	−9.18	−3.33	37.82	−3.33
	−13.51	0.00	37.61	11.58
	−16.18	0.00	33.54	1.58
	−13.10	−20.15	29.73	−21.83
	Differences to −45° in %			
	45° and −45°		45° and −45°	
	Pylon	Cables	Pylon	Cables
VERTICAL	−15.84	0.00	35.41	0.00
	−12.50	0.00	40.94	0.00
	−15.08	−1.04	35.22	−8.59
	−9.59	−7.47	31.64	−20.17

Table 9. Sensibility analysis of the influence of the angle to the 45° position.

	Differences to 45° in %			
	45° and −45°		45° and 90°	
	Pylon	Cables	Pylon	Cables
HORIZONTAL	8.41	3.22	43.05	0.00
	11.90	0.00	45.04	0.00
	13.92	0.00	42.80	0.00
	11.58	16.77	37.87	−1.40

Table 9. *Cont.*

	Differences to 45° in %			
	45° and −45°		45° and 90°	
	Pylon	**Cables**	**Pylon**	**Cables**
VERTICAL	13.67	0.00	44.24	0.00
	11.11	0.00	47.50	0.00
	13.11	1.03	43.71	−7.47
	8.75	6.95	37.62	−11.81

Analyzing the results of the sensibility analysis, we can highlight the most stable orientation of 0° and 45° of the horizontal case. This case has an average error of 7.17% for the pylons, while the average is zero for the cables. Choosing 0° and 45° as the preferred angles for taking photographs, the worst case was the distance of 1 h. For this reason, we propose to eliminate this distance.

In conclusion, the recommendation after performing the sensibility analysis is that photographs should be taken horizontally at a distance of 1.5 h, 2 h, and 2.5 h with an angle of 0° or 45°. Choosing these six positions in order to take photographs, the position will not influence significantly the result of the visibility index. Figure 6 shows the selected positions.

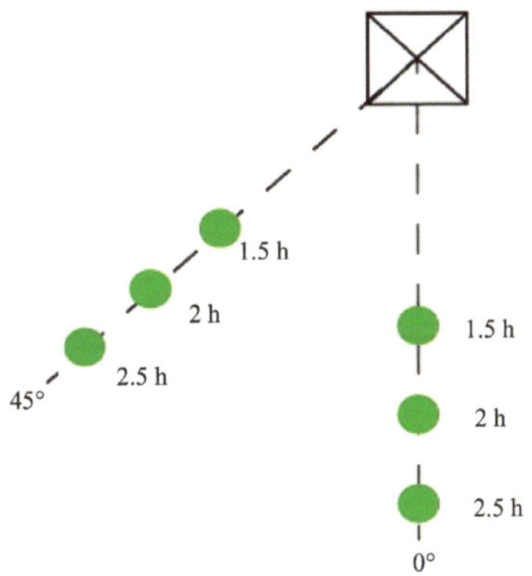

Figure 6. Positions selected for taking photographs.

5. Application Case and Comparison with Previous Results

In this section, the objective is to apply the improvement proposed in this paper of finding the visual impact to tree transmission lines based in Spain. After that, the results are compared with the results of the previous studies [1,25]. The references present a methodology for the assessment of the impact of existing high-voltage lines in urban areas. The impact of transmission lines in urban areas was evaluated by a weighted combination of different factors. One of these factors is the visual impact, which was calculated by following (1). The drawback of this methodology is the subjective nature of the value obtained, as the different input factors of this formula are obtained by the estimations of experts. In this section, we analyze the following transmission lines located in Rubí (Barcelona, Spain) using the proposed methodology:

- the 110 kV transmission line Can Jardí-Collblanc (R1), located at the Avinguda Pep Ventura;
- the 220 kV transmission line Can Jardí-Sant Andreu-Canyet (R2), located at the turnaround Avinguda Electricitat and Passeig de les Torres; and
- the 220 kV transmission line Foix-Mas Figueres (R3), located in the zone Can Fatjó.

Figure 7 shows the transmission line located in the Avinguda Electricitat as an example of the photographs taken. In the above-described locations, the visibility index was determined using (2), and the obtained results are depicted in Tables 10–12.

Figure 7. Photograph taken of the 220 kV transmission line Can Jardí-Sant Andreu-Canyet (R2), located at the turnaround Avinguda Electricitat and Passeig de les Torres.

Table 10. Visibility index of the Pep Ventura (R1) application case.

(I vis) Pep Ventura R1			
Angle	**Distance**	**Pylons**	**Cables**
0°	1.5 h	0.512	0.908

Table 11. Visibility index of the Avinguda Electricitat application case.

(I vis) Av. Electricitat R2			
Angle	**Distance**	**Pylons**	**Cables**
0°	1.5 h	1.000	1.000

Table 12. Visibility index of the Can Fatjó application case.

(I vis) Can Fatjó R3			
Angle	**Distance**	**Pylons**	**Cables**
0°	1.5 h	0.607	1.000

The *OAIV* was calculated by using (1). The values of the *OAIV* of the three transmission lines are shown in Table 13. The results of the visual impact assessment [1] and [25] were calculated on a scale from 0 to 27. In order to compare the results obtained using the novel methodology, the resulting *OAIV* was multiplied by 27. The comparison between the results of both methods is described in Table 14.

Table 13. The evaluation result of the proposed methodology.

Index	β	Indicator Color	Visibility of Pylons	Visibility of Wires	Quality	Fragility	27 *OAIV* [1] Total
R1	0.827	0.02	0.512	0.908	0.50	0.611	6.283
R2	0.827	0.02	1.000	1.000	0.25	0.792	5.616
R3	0.827	0.02	0.607	1.000	1.00	0.639	14.480

[1] The *OAIV* calculated in study [1] is on the scale from 0 to 27.

Table 14. Comparison of the impact ranking of the methodology used in [1] and the proposed methodology.

Index	The Methodology Used in [1] Result	Ranking	Proposed Methodology Result	Ranking
R1	6.24	2	6.283	2
R2	6.24	2	5.616	3
R3	18.72	1	14.480	1

It can be seen that the novel methodology provides a similar result in terms of the absolute value of the impact. That means that the expert opinion and the results of the presented methodology lead to similar results, and it validates the proposed methodology for the presented case study. However, the situations R1 and R2 are, in the methodology described in [1], evaluated equally, e.g., in the same position of the ranking, while the novel methodology provides a higher resolution and, therefore, better differentiation between both cases. The presented methodology shows a higher granularity and enables us to differentiate better between the two cases. The application of the methodology is not limited to the specific conditions in Spain, and it is applicable in other countries. The limitation of the methodology lies in the possibility of taking photographs at the distances and angles indicated.

6. Conclusions

The presented paper contributes to the improvement of the calculation of the overall aesthetic impact value (*OAIV*) for the assessment of the visual impact of high-voltage overhead lines. The proposed methodology is based on the systematic analysis of photographs taken of the impacted area to eliminate the subjective aspects of methods used in previous studies. The observer's position and angle have an impact on the results of the visual impact calculation, and therefore a sensibility analysis was performed. The analysis showed that photographs that were taken horizontally at a distance of 1.5 h, 2 h, and 2.5 h with an angle of 0° or 45° would not influence the result of the visibility index significantly. The methodology was applied to a study case near Barcelona, and results of the visual impact calculation show a higher resolution and better differentiation between the cases as compared with the previous methodologies used in the literature. The proposed methodology is not limited to the specific conditions in Spain and can be applied in the international context. It can improve the overall assessment of overhead lines and present an objective evaluation of their impact. This could improve the management of impacted areas by better public acceptance by using a scientific method.

Author Contributions: Conceptualization and methodology, A.S., O.B.-A., and J.R.-D.; investigation, J.W.; fieldwork, J.A.-A.; validation, J.W. and J.A.-A.; writing—original draft preparation, A.S., J.W., and J.A.-A.; writing—review and editing, A.S. and O.B.-A., and J.R.-D. All authors have read and agreed to the published version of the manuscript.

Funding: This research received no external funding.

Institutional Review Board Statement: Not applicable.

Informed Consent Statement: Not applicable.

Data Availability Statement: Not applicable.

Conflicts of Interest: The authors declare no conflict of interest.

References

1. Sumper, A.; Boix-Aragonés, O.; Villafáfila-Robles, R.; Ramírez-Pisco, R. Methodology for the assessment of the impact of existing high voltage lines. *Energy Policy* **2010**, *38*, 6036–6644. [CrossRef]
2. Sumper, A.; Ferreira, P. A New Open Access Journal. *Electricity* **2020**, *1*, 4. [CrossRef]
3. Franken, M.; Schrief, A.B.; Barrios, H.; Puffer, R.; Moser, A. Co-optimization of multi-stage transmission expansion planning and grid operation. *Electr. Power Syst. Res.* **2020**, *189*, 106686. [CrossRef]
4. Horst, D.V.d. NIMBY or not? Exploring the relevance of location and the politics of voiced opinions in renewable energy siting controversies. *Energy Policy* **2007**, *35*, 2705–2714. [CrossRef]
5. Vajjhala, S.P.; Fischbeck, P.S. Quantifying siting difficulty: A case study of US transmission line siting. *Energy Policy* **2007**, *35*, 650–671. [CrossRef]
6. Marshall, R.; Baxter, R. Strategic Routeing and Environmental Impact Assessment for Overhead Electrical Transmission Lines. *J. Environ. Plan. Manag.* **2002**, *45*, 747–764. [CrossRef]
7. Rahman, K.A. Mediation and Mediator Skills: A Critical Appraisal. Available online: https://ssrn.com/abstract=3231684 (accessed on 1 February 2012).
8. Cloquell-Ballester, V.-A.; Cloquell-Ballester, V.-A.; Monterde-Díaz, R.; San-tamarina-Siurana, M.-C. Indicators validation for the improvement of environmental and social impact quantitative assessment. *Environ. Impact Assess. Rev.* **2006**, *26*, 79–105. [CrossRef]
9. Salci, S. A framework for assessing economic and environmental benefits from transmission line rehabilitation investments. *Electr. J.* **2018**, *31*, 75–81. [CrossRef]
10. Bickel, P.; Friedrich, R. ExternE: Externalities of energy. *Eur. Commun.* **2005**. Available online: https://op.europa.eu/en/publication-detail/-/publication/b2b86b52-4f18-4b4e-a134-b1c81ad8a1b2 (accessed on 29 July 2021).
11. Doukas, H.; Karakosta, C.; Flamos, A.; Psarras, J. Electric Power Transmission: An overview of associated burdens. *Int. J. Energy Res.* **2011**, *11*, 970–988. [CrossRef]
12. Nassauer, J.I. Landscape as medium and method for synthesis in urban ecological design. *Landsc. Urban Plan.* **2012**, *106*, 221–229. [CrossRef]
13. Torres-Sibille, A.D.C.; Cloquell-Ballester, V.-A.; Cloquell-Ballester, V.-A.; Ramírez, M.Á.A. Aesthetic impact assessment of solar power plants: An objective and a subjective approach. *Renew. Sustain. Energy Rev.* **2009**, *13*, 986–999. [CrossRef]
14. Torres-Sibille, A.D.C.; Coloquel-Ballester, V.-A.; Coloquell-Ballester, V.-A.; Darton, R. Development and validation of a multi-criteria indicator for the assessment of objective aesthetic impact of wind farms. *Renew. Sustain. Energy Rev.* **2009**, *13*, 40–66. [CrossRef]
15. Hadrian, D.R.; Bishop, I.D.; Mitcheltree, R. Automated Mapping of visual impacts in utility corridors. *Landsc. Urbaaestheticng* **1988**, *16*, 261–282. [CrossRef]
16. Luken, J.O.; Hinton, A.C.; Baker, D.G. Forest edges associated with power-line corridors and implications for corridor siting. *Landsc. Urban Plan.* **1991**, *20*, 315–324. [CrossRef]
17. Tempesta, T.; Vecchiato, D.; Girardi, P. The Landscape benefits of the burial of high voltage power lines: A study in rural areas of Italy. *Landsc. Urban Plan.* **2014**, *126*, 53–64. [CrossRef]
18. Frontuto, V.; Corsi, A.; Novelli, S.; Gullino, P.; Larcher, F. The visual impact of agricultural sheds on rural landscapes: The willingness to pay for mitigation solutions and treatment effects. *Land Use Policy* **2020**, *91*, 104337. [CrossRef]
19. Wagner, J. Methodology of Environmental Impact Assessment for Evaluating Existing Overhead Lines. Master's Thesis, Universitat Politècnica de Catalunya, Barcelona, Spain, 2011.
20. Antrop, M. Why landscapes of the past are important for the future. *Landsc. Urban Plan.* **2005**, *70*, 21–34. [CrossRef]
21. Senetra, A.; Wasilewicz-Pszczółkowska, M.; Grzelak-Kostulska, E. The European landscape convention as a Tool for the protection, management and planning of landscapes. *Barom. Reg.* **2015**, *3*, 81–88.
22. Schripke, U.; Zoderer, B.M.; Tappeiner, U.; Tasser, E. Effects of past landscape changes on aesthetic landscape values in the European Alps. *Landsc. Urban Plan.* **2021**, *212*, 104109. [CrossRef]
23. Sevenant, M.; Antrop, M. Cognitive attributes and aesthetic preferences in assessment and differentiation of landscapes. *J. Environ. Manag.* **2008**, *90*, 2889–2899. [CrossRef] [PubMed]
24. Amat, J. Metodologia d'avaluació de l'impacte Visual de les línies d'alta Tensió Existents. Master's Thesis, Univeristat Politècnica de Catalunya, Barcelona, Spain, 2014.
25. CITCEA-UPC. *Estudi Sobre l'impacte de les Linies de Tensió Superior o Igual a 66 kV en les àrees Urbanes de Rubí*; CITCEA-UPC: Barcelona, Spain, 2004.

electricity

MDPI

Article

Adapted Stochastic PV Hosting Capacity Approach for Electric Vehicle Charging Considering Undervoltage

Enock Mulenga *, Math H. J. Bollen and Nicholas Etherden

Department of Engineering Sciences and Mathematics, Luleå University of Technology, 93187 Skellefteå, Sweden; math.bollen@ltu.se (M.H.J.B.); nicholas.etherden@ltu.se (N.E.)
* Correspondence: enock.mulenga@ltu.se

Abstract: This paper presents a stochastic approach to single-phase and three-phase EV charge hosting capacity for distribution networks. The method includes the two types of uncertainties, aleatory and epistemic, and is developed from an equivalent method that was applied to solar PV hosting capacity estimation. The method is applied to two existing low-voltage networks in Northern Sweden, with six and 83 customers. The lowest background voltage and highest consumption per customer are obtained from measurements. It is shown that both have a big impact on the hosting capacity. The hosting capacity also depends strongly on the charging size, within the range of charging size expected in the near future. The large range in hosting capacity found from this study—between 0% and 100% of customers can simultaneously charge their EV car—means that such hosting capacity studies are needed for each individual distribution network. The highest hosting capacity for the illustrative distribution networks was obtained for the 3.7 kW single-phase and 11 kW three-phase EV charging power.

Keywords: hosting capacity; Monte Carlo methods; stochastic; uncertainty; undervoltage; electric vehicle

Citation: Mulenga, E.; Bollen, M.H.J.; Etherden, N. Adapted Stochastic PV Hosting Capacity Approach for Electric Vehicle Charging Considering Undervoltage. *Electricity* **2021**, 2, 387–402. https://doi.org/10.3390/electricity2030023

Academic Editor: Pavlos S. Georgilakis

Received: 14 August 2021
Accepted: 13 September 2021
Published: 17 September 2021

1. Introduction

The transportation sector accounts for the emission of greenhouse gases responsible for global warming due to the use of fossil fuels. The increasing use of electric vehicles (EVs) reduces the emission of carbon dioxide when the electricity comes from clean and renewable energy sources [1–4]. The sale of EVs increased by 43% globally and 137% in the European Union in 2020 [5]. During December 2020, almost half of the new passenger cars registered in Sweden were fully electric or plug-in hybrid [6]. The increase in the share of electric vehicles will increase EV charging in distribution networks.

This increase will result in an expected increase in energy and peak power consumption [2,7]. The former is a challenge for the power generation side, while the latter is a challenge for the power distribution side. The increase in peak power consumption and other changes in consumption patterns will depend strongly on the charging pattern, which remains one of the unknowns [1,8]. The largest impact on the power consumption will occur when there is simultaneous charging of multiple EVs.

An alternative approach, independent of the actual charging pattern, is to estimate how much charging is possible as a function of days, weeks, and years. This amount of charging is referred to as the hosting capacity (HC). The hosting capacity is generally defined as the amount of new power consumption or generation that does not risk other customers' reliability or power quality [9–11]. This paper applies the hosting capacity approach to new power consumption, specifically simultaneous EV charging in distribution networks.

EV charging can be single-phase or three-phase. Larger penetration of EV charging can affect the distribution networks in different ways [12,13]. An increase in power demand beyond the thermal capacity of cables, lines, or transformers will cause an overload [2,14], which can result in accelerated ageing, component failure, or customers' supply fuse to blow. Other impacts caused by single-phase or three-phase chargers include harmonics, voltage unbalance, undervoltage, flicker or fast voltage fluctuations, and overvoltage in rare cases of single-phase chargers [2,13,14].

There are similarities concerning grid impact between EV charging and solar photovoltaic (PV) units. Both solar PV and EV charging were not initially planned by the distribution network operators (DSOs). Designing distribution grids that can cope with large amounts of solar PV and EV charging is technically possible, but such grids will be significantly more expensive than existing grids. Both solar PV and EV charging show fast and largely unpredictable growth, especially at a local level. The rapid growth and uncertainty result in several challenges to DSOs [12–15]. There is a need to determine the ability of the distribution network to cope with further growth [12,15], in this case growth in consumption due to EV charging. All impacts caused by EV charging on the distribution network should be considered to estimate the hosting capacity [15].

EVs can be charged at the home (residential charging), workplace, or public charging stations [16,17]. The charging behaviour will be a determining factor for the impact of EV charging on the distribution grid. The most common time of charging for residential customers is before work, around 8 am, and typically around 6 pm after arriving home from work [18]. This increased charging of EVs at home will impact the distribution network [19].

The uncertainties in EV charging have been addressed in hosting capacity studies by using a stochastic or time-series method. A Monte Carlo based simulation for EV charging assessment is proposed because of the stochastic elements it contains [20]. A Monte Carlo based stochastic approach was used for solar PV hosting capacity estimation in [21]. The method proposed and used in [21] is adapted for EV charging in this paper. The resulting approach is applied to obtain the hosting capacity for EV charging of a number of existing low-voltage networks in Sweden. The contributions and innovations of the study presented in this paper are as follows:

- A stochastic method is proposed for estimating the hosting capacity of the distribution grid for EV charging.
- The development of models to quantify the uncertainties associated with hosting capacity estimation for EV charging.
- The proposed method is non-specific and is applicable for estimating the EV charge hosting capacity for any time of the day, week, and year. None or only limited information of the charging pattern is needed to estimate the hosting capacity.
- The planning risk taken is one of the inputs to the method. This allows the method to be used in the trade-off between the risk of insufficient investment and the risk of overinvestment (stranded assets).
- Application of the method to a number of existing distribution networks.

The paper is divided into six sections. Section 1 introduces the paper, and Section 2 presents the state of the art in stochastic hosting capacity estimations. Section 3 presents the adaptation of the PV-HC stochastic method for EV charging estimation. Section 4 presents the results, and their discussion is given in Section 5. Finally, the conclusion is presented in Section 6.

2. Review on EV Stochastic Hosting Capacity

The hosting capacity estimation for EV can be performed using a deterministic, stochastic, or time-series method [19]. Simple deterministic methods can be used to assess the impact of either mean or maximum power consumption [14]. In addition, stochastic and time-series models can be used to include some of the uncertainties associated with EV charging and its grid impact [3,14,19].

Several EV charge hosting capacity methods using stochastic methods have been proposed, including some of the uncertainties [3,7,10,12,16,17,19–22]. There are two types of uncertainties, aleatory, and epistemic uncertainties [22]. Being aware of the difference between these is important; A detailed description of these two types of uncertainties and their application to solar PV hosting capacity is given in [21].

A study was undertaken to quantify power quality problems due to EV charging in [3]. A 3.7 kW EV was used with a Gaussian model for the harmonic currents. In [14], a method was implemented using survey and measurement data. The voltage and congestion indicators were applied for the charging cycle, mostly occurring between 6 pm and 10 pm. The hosting capacity for EVs was evaluated by considering the random distribution of EVs among households in [23]. Single-phase charging power of 3.7 kW and 7.4 kW was used for the study. Experimental results were applied in [19] with smart charging aspects, considering phenomena including undervoltage. The stochastic analysis is recommended to correctly quantify the likelihood and severity of EV charging in distribution networks [19]. The approach in [24,25] quantifies the risk of overloading by modelling the EV and customer loading with Poisson and Gaussian models. The presence of solar PV was considered in that study as well. In [9], combined EV charging and solar PV is considered with smart charging. It was shown that there is a small positive correlation between EV charging and production from solar PV. A stochastic model for EV charging was developed in [26]. In [27], a kernel distribution is applied to destination surveys. It was shown that the charging location was the most critical variable and that EV hosting capacity is very much restricted by the minimum voltage [27]. Charging time was modelled with a uniform distribution and occurring between 6 pm and 9 pm in [27]. None of the above-mentioned publications identifies the aleatory and epistemic uncertainties. Characterisation of the probability distribution function from measurements was also not addressed in any of the previous studies. Moreover, none of the studies develop a stochastic approach for EV charge hosting capacity from a successful approach for solar PV. These gaps motivated the approach to estimating EV charge hosting capacity presented in this paper.

Brief Description of Aleatory and Epistemic Uncertainties

The aleatory uncertainties emerge from the variables' natural randomness and built-in variability. Information on aleatory uncertainties can be obtained by statistical analysis of measurement data. The uncertainty of the variables can be characterised by means of probabilities or a probability distribution function [21,28]. Those can in turn be used as input in stochastic studies where aleatory uncertainties affect the outcome.

The epistemic uncertainties emerge from the lack of knowledge or information on a variable. The modelling of the variable is performed with either interval analysis or possibilistic and evidence theory. Gathering more information or built-up knowledge on the variable can reduce the variability [28]. Obtaining the data for the epistemic uncertainty may take a long time and obtaining statistical information from measurements is often not possible.

The application of the two types of uncertainties begins with recognizing them and their influence. By distinguishing between the aleatory from the epistemic, appropriate models can be developed and used for the stochastic hosting capacity approach.

In the context of EV charging, aleatory uncertainties are the voltage and consumption before the connection of EV charging. Examples of epistemic uncertainties are the number of customers that will charge their car simultaneously, their charging power, and the phase to which the charger is connected.

3. Stochastic EV Charge Hosting Capacity Approach

The hosting capacity method used in this paper is adapted from the stochastic approach developed and applied in [21]. In [21], the approach was used to estimate the hosting capacity of distribution networks for solar PV. The stochastic model used to estimate the hosting capacity in [18] used a probability distribution of the highest voltage during the hours of the day and year with high production from solar PV. This paper's approach instead uses a probability distribution of the lowest voltage resulting during those hours that EV charging is most likely to occur.

In [21], the method evaluated the combination of epistemic and aleatory uncertainties during the time-of-day (ToD) from 10 am to 2 pm and the time-of-year (ToY) with the highest solar PV power production. The obtained voltage rise leading to an overvoltage was due to solar power production during the sunny hours of the day and the year. EV charging can take place at any time of the day, week, or year. Contrary to solar PV, there is no defined ToD and ToY with EV charging. The approach developed for solar PV in [21] contributes to the probability distribution of a voltage rise compared to the "highest background voltage". The "background voltage" is the voltage with the customer in a low-voltage distribution network for zero local consumption and zero local production [29,30]. The background voltage used in the stochastic hosting capacity method for solar PV is the highest background that can be expected during the sunny hours of the day and year. The method adapted and applied to EVs causes a decrease compared to the lowest background voltage. When applied to EV charge hosting capacity, the 'background voltage' to be used in the stochastic hosting capacity method is the lowest value during the time of highest consumption, and when EV charging is expected to take place [31].

Both methods, the one in this paper and [20], include uncertainties in estimating the hosting capacity. Both epistemic and aleatory uncertainties associated with EV charging are considered. The probability distribution of the worst-case voltage due to a magnitude drop resulting from single-phase or three-phase EV charging is obtained.

3.1. Overall Stochastic Approach

The approach applied in this paper, treating aleatory and epistemic uncertainties differently as in [21], is used to obtain the worst-case undervoltage values as a probability distribution function (pdf). The worst-case undervoltage values are the minimum values of the voltage calculations described in Section 3.2.

The fundamental assumption for the aleatory uncertainties entails that the probability distribution functions (pdf) must be considered when there is high EV charging. High charging occurs when many customers with EVs are charging. The likelihood of an undervoltage occurring is highest for a combination of low background voltage and high consumption.

The severe impact of epistemic uncertainties on the hosting capacity is underlined by how many customers will purchase EVs, their charging location, and their charging pattern. None of these is known beforehand. The possible locations and interval range of EV charging for customers are evaluated in this paper. The overall approach is summarised in the flow chart shown in Figure 1.

3.2. Lowest Background Voltage and Undervoltage

The voltage magnitude in three-phase low-voltage networks with a contribution of solar PV was given in [21]. The customers' lowest consumption and solar PV injection were used in [21] to obtain the voltage rise. Using the equations in [21], Equation (1) is formulated to estimate the voltage at location a due to a customer power consumption and EV charging at location b [21].

$$U(\mathrm{a}) = U_0(\mathrm{a}) + \left[Z_{tf}(\mathrm{a},\mathrm{b}) \times (-I_{cons}(\mathrm{b}) - I_{EV}(\mathrm{b}))\right] \quad (1)$$

where $U_0(\mathrm{a})$ is the lowest background voltage, $Z_{tf}(\mathrm{a},\mathrm{b})$ is the transfer impedance, $I_{cons}(\mathrm{b})$ is the current at the customer during the highest consumption, and $I_{EV}(\mathrm{b})$ is the current due to EV charging.

In Equation (1), the highest customer consumption and EV charging power are added. It is also important to determine the lowest voltage occurring at the customer. In a distribution network with multiple customers, simultaneous EV charging at more than one location is possible. All the contributions of EV charging are superimposed, resulting in Equation (2).

$$U(\mathrm{a}) = U_0(\mathrm{a}) + \sum_{\mathrm{b}=1}^{\mathrm{N_{cust}}} Z_{tf}(\mathrm{a},\mathrm{b}) \times (-I_{cons}(\mathrm{b}) - I_{EV}(\mathrm{b})) \quad (2)$$

Equation (2) is used in this study to obtain the probability distribution of the lowest voltage due to *EV* charging. The obtained voltage, applying Equation (2) which is an extension of Equation (1), is the worst-case undervoltage distribution for the customers.

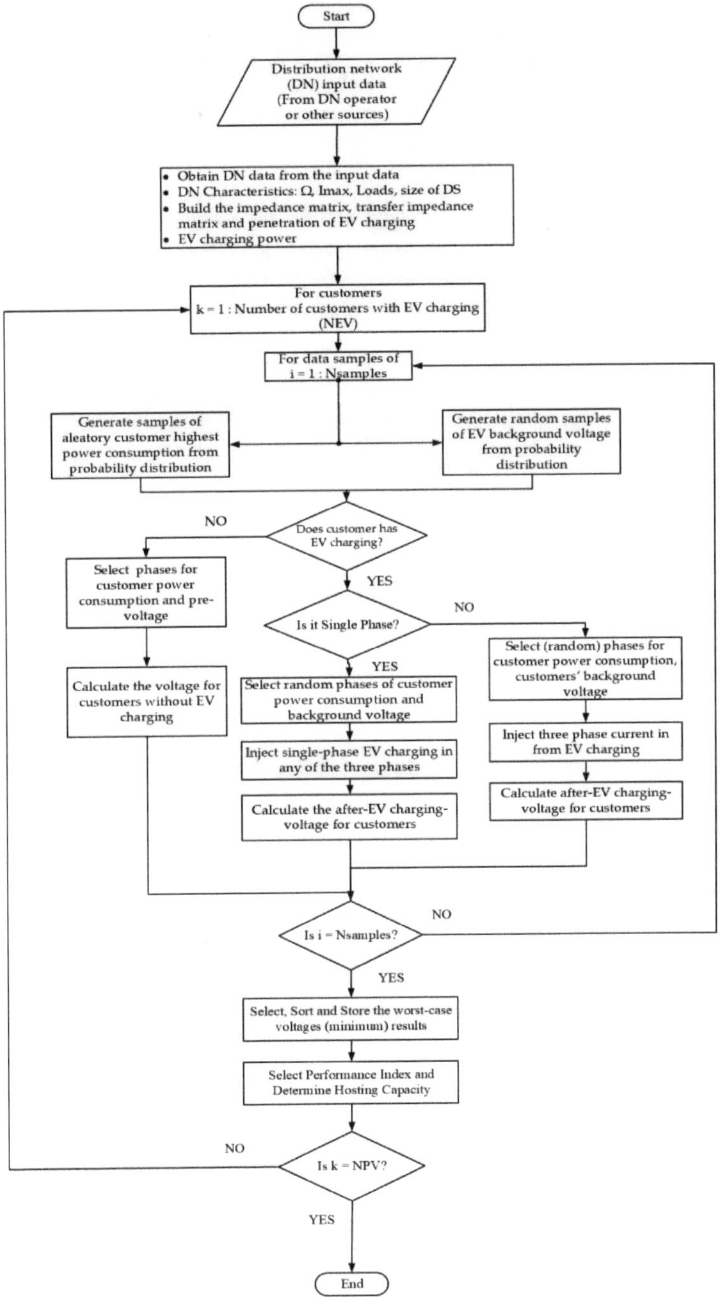

Figure 1. The flow chart for the stochastic hosting capacity approach for EV charging in a distribution network.

3.3. Uncertainties

In this paper, the adapted stochastic approach similar to [21] considers both aleatory and epistemic uncertainties.

The aleatory uncertainties considered are the background voltage and highest customer power consumption. They are modelled with their probability distribution functions (pdf). The difference with those applied in [21] is what matters most and how they are obtained. The background voltage obtained is the distribution of the lowest value during the time of highest consumption. The distribution of such values is characterised and a goodness-fit-applied to obtain the pdf, in a similar way as in [26].

The epistemic uncertainties considered are phase-type and connection (single-phase or three-phase), EV charger size, customer location with EV charging, and the number of customers charging simultaneously. Models are applied to estimate future occurrences. The data needed are obtained based on the possibilities, interval, or evidence of the occurrence. The possible future cases, possibilities, and interval ranges that can occur are applied in the stochastic approach.

3.4. EV Charging Power Size

The hosting capacity is first estimated for the charging power of 3.7 kW single-phase and 11 kW three-phase, corresponding to a 16-A fuse. The 3.7 kW and 11 kW are mentioned in [32] as the most popular charging power sizes and are also used in [9,32].

Other single-phase charging power sizes of 4.6, 5.75, 8, and 9.2 kW have been studied. The three-phase charging power sizes of 13.8, 17.3, 24.2, and 27.6 kW are applied too. The two sets of single-phase and three-phase charging power sizes correspond to 20, 25, 35, and 40-A fuse.

3.5. Study Distribution Networks

Two illustrative distribution networks have been used in this paper, with 6 and 83 customers. A 100 kVA, 10/0.4 kV, and Dyn11 (vector group) transformer with a 4% impedance supplies power to the 6-customer network [33]. The 83 customers are supplied by a 500 kVA, 10/0.4 kV, Dyn11 transformer with a 4.9% impedance [21].

The customers in both distribution networks are supplied with three-phase cables, including the 10-mm^2 service cable between the last cable cabinet and the customer. The service cable can supply 13 kW single-phase and 38 kW three-phase power.

3.6. Applied Highest Consumption and Lowest Background Voltage

The approach requires the input of the highest customer consumption and lowest background voltages. Measurements and DSO given data on consumption used in [29,34] have been applied in this paper. Measurements from 8 distribution transformers were used to obtain a distribution of the highest consumption per customer: transformers with 4–8 customers for 100 kVA and 76–94 customers for 500 kVA. The obtained probability distribution of the highest consumption per customer is shown in Figure 2.

The lower value obtained for both transformer sizes is 1 kW per phase. The upper value is higher for the 100-kVA transformer. The lower and upper limits of each transformer size are applied as input data for the hosting capacity estimation. The highest consumption per customer obtained and applied for the 6-customer network is in the range of 1–2.2 kW. It is 1–1.5 kW for the 83-customer distribution network. An approach is applied to the voltage measurements for the distribution network in Northern Sweden, also used in [29]. The daily lowest 10-min voltage measurement is obtained for a distribution network with 6 customers and one with 83 customers. These are voltages as measured at the low-voltage side of the distribution transformer. The results for 365 days of measurements during 2017 are shown in Figure 3.

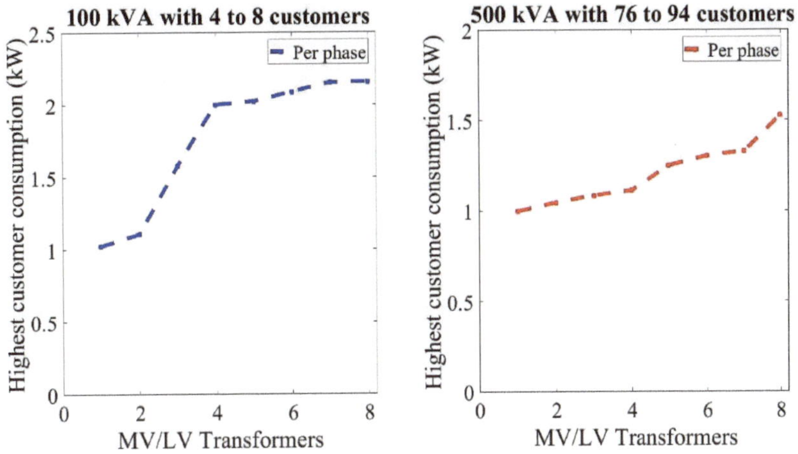

Figure 2. The highest consumption per customer per phase for 100 kVA and 500 kVA rated transformers.

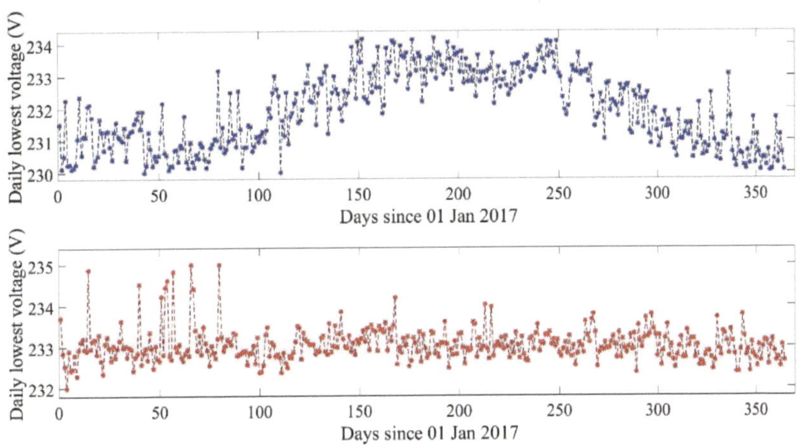

Figure 3. The measured background voltage for the 6-customers (100 kVA, **top**) and 83-customers (500 kVA, **bottom**) distribution network for 365 days of 2017.

The daily lowest voltages for the 6-customer distribution network show higher values during the summer and lower values in winter. The 83-customer network shows a few higher values in winter. In addition, the lowest values for an additional 31 transformers were obtained. The method used in [26] for the characterization of probability distribution functions was also applied to all the obtained measurements. The generalised extreme value (GEV) distribution was the one that most often fitted best to the background voltage that should be used for EV charging. The same was observed for the background voltage used for solar PV in [29]. The obtained lowest background voltage for the 6-customer distribution network is in the range of 230–234 V and the range of 232–235 V for the 83-customer distribution network. The parameters for the GED distribution used for the background voltage in the hosting-capacity estimation of the networks presented in Section 3.5 are shown in Table 1.

Table 1. The background voltage input parameter for the characterised GEV pdf for the 6-customers and 83-customers distribution networks was applied to estimate EV charge hosting capacity.

Distribution Network	6-Customers	83-Customers
k	−0.3228	−0.0309
Sigma (σ)	1.1491	0.3250
mu (μ)	231.55	232.88

In Table 1, k is the shape parameter of the GEV distribution. The value of k also describes any of the three types (Type 1: Gumbel, Type 2: Fréchet, or Type 3: Weibull) of a GEV distribution. The GEV in Table 1 is a Weibull distribution function. Sigma (σ) describes the scale parameter (standard deviation), and mu (μ) describes the location parameter of the GEV distribution (mean). The general representation of the GEV with location parameter (μ), scale parameter (σ), and the shape parameter ($k \neq 0$) is given by Equation (3).

$$y\,(x) = \left(\frac{1}{\sigma}\right) \exp\left(-\left(1 + k\left(\frac{x-\mu}{\sigma}\right)^{-\frac{1}{k}}\right)\left(1 + k\left(\frac{x-\mu}{\sigma}\right)^{-1-\frac{1}{k}}\right)\right) \tag{3}$$

The GEV is described as a Gumbel distribution for a zero value of k, Fréchet for k greater than zero, and Weibull for k less than zero [35].

3.7. Hosting Capacity Calculation, Limit and Performance Index

The EV hosting capacity in this paper refers to the number of customers connected to an LV distribution network that can charge their EV simultaneously without causing the voltage magnitude to go below the undervoltage limit. The hosting capacity is exceeded when at least one of the customers' voltages is below the limit.

The stochastic approach initially results in a probability distribution of each customer's worst-case voltage magnitude. To estimate the hosting capacity, a performance index and a limit are needed. The 10th percentile of the worst-case voltage distribution is used in this paper as a performance index, and 90% of the nominal voltage is used as a limit.

Distribution network operators often consider planning risk in one way or another for their networks. What is typically not explicitly known is the percentage risk they take, which, in mathematical terms, would be a percentile. The 10th percentile voltage values are used in this paper to estimate the hosting capacity for EV. This corresponds to what could be called a "planning risk" of 10%. An illustration of the approach described in Sections 3.1–3.6 is shown in Figures 4 and 5 for a 6-customer distribution network. The implementation is done with the following:

- The 6-customer distribution network input data and source impedance.
- The obtained background voltage of 230–234 V obtained in Figure 3 and GEV pdf was applied (k = −0.3228, sigma = 1.1491, mu = 231.55).
- The EV charging power of 4.6 kW single-phase [31].
- The highest power consumption per phase of 1–2.2 kW and its pdf (uniform).
- The interval and the possible number of customers with EV chargers applied. The interval range is from 1 to 6 customers (the total number of customers).
- All possible combinations and locations from 1 customer to 6 customers are assessed.
- The customers can install the charger in any of the three phases.
- Monte-Carlo is applied (100,000 to 1,000,000). The probability distribution of the customers' voltage magnitude is obtained.

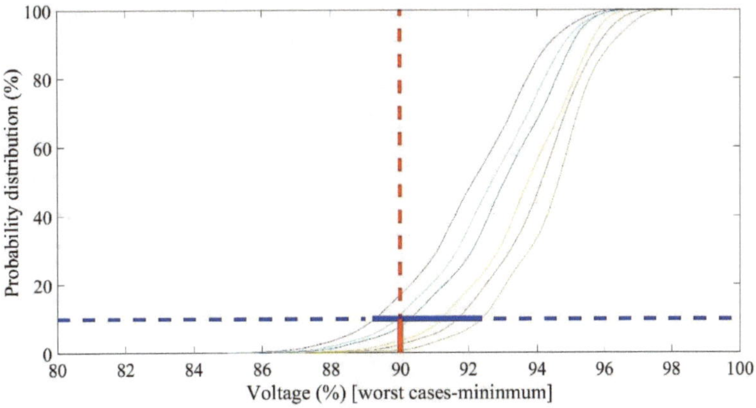

Figure 4. The customers' voltage probability distribution functions for the 6-customers distribution network. Each of the curves in the figure represents the voltage distribution of a single customer.

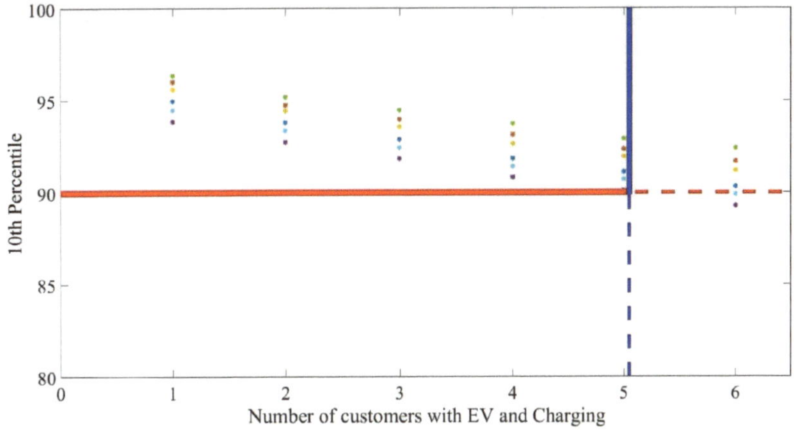

Figure 5. The 10th percentile voltage magnitude for the six customers is obtained from Figure 4 (colored dots) and the 90% undervoltage limit (solid red line).

The probability distributions for the voltage with the customers in the 6-customer network with EV chargers are shown in Figure 4.

In Figure 4, the horizontal blue line (solid and dotted) crossing the distributions indicates the 10th percentile. The vertical red line crossing the distributions indicates the probability that the voltage is less than the 90% undervoltage limit. Estimating the hosting capacity involves repeating the calculation of Figure 4 from one customer to the maximum (six). The 10th percentiles of the voltage (the performance index used), as obtained from Figure 4, and the limit are used to obtain Figure 5.

The results in Figure 5 include the possibilities of having 1 customer charging an EV up to all the customers (6) charging simultaneously. The blue vertical lines (solid and dotted) in Figure 5 show the highest number of customers charging simultaneously without any of the voltages falling below the undervoltage limit.

When one customer is charging, the 10th percentile values of the customers are above 90%. The index does not violate the limit. The increase in the number of customers charging shows the 10th percentile approaching 90%. The undervoltage limit is violated for 6-customer charging. At most, 5 customers can charge without the limit being violated. The hosting capacity in this case is 5 customers (83% of the total number of customers). The estimation procedure has been repeated for different distribution networks and different charging powers. The results are shown in Section 4.

4. EV Hosting Capacity Results

The approach in Section 3 and the steps outlined were applied to the six-customer and 83-customer distribution networks described in Section 3.

In the first step, the hosting capacity for single-phase EV simultaneous charging was assessed. In the second step, the assessment was performed for three-phase EV charging. Furthermore, the influence of the background voltage, highest customer power consumption, and the planning risk on the hosting capacity were also evaluated. The results obtained for the stochastic hosting capacity results applied to EVs are given in this section.

4.1. Single-Phase Hosting Capacity

The capacity of the six-customer and 83-customer distribution networks for single-phase EV charging of different power is shown in Figure 6.

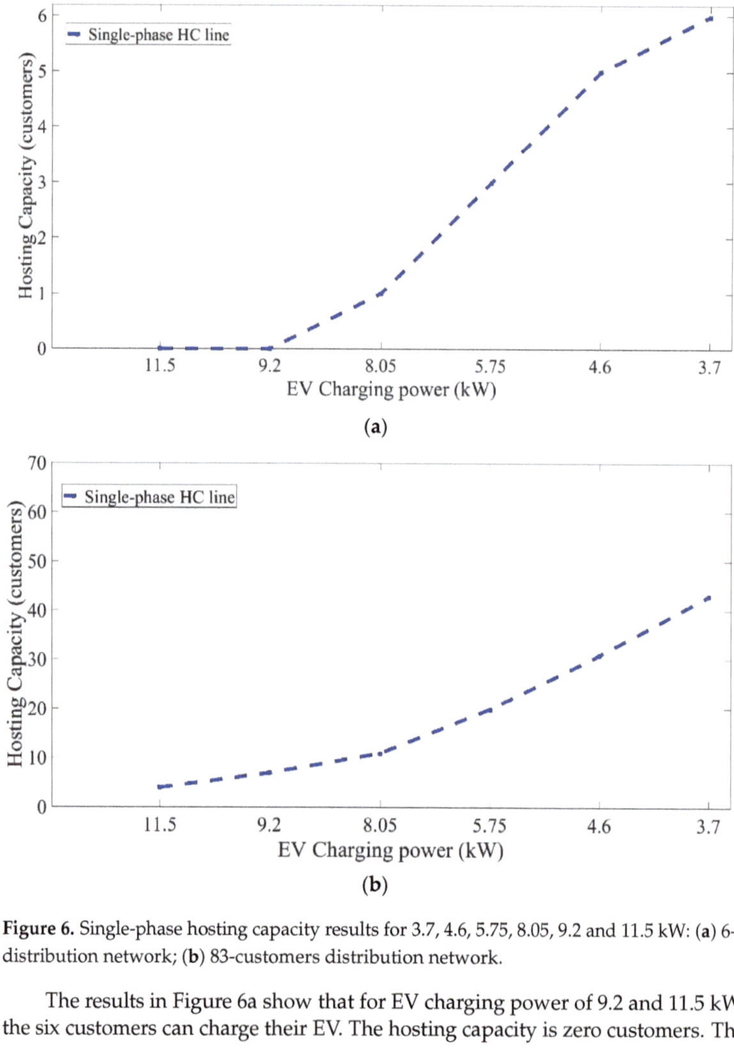

Figure 6. Single-phase hosting capacity results for 3.7, 4.6, 5.75, 8.05, 9.2 and 11.5 kW: (**a**) 6-customers distribution network; (**b**) 83-customers distribution network.

The results in Figure 6a show that for EV charging power of 9.2 and 11.5 kW, none of the six customers can charge their EV. The hosting capacity is zero customers. The hosting capacity is one customer for 8.05 kW charging power and becomes six customers when the charging power is 3.7 kW.

In Figure 6b, only a few customers can charge simultaneously with 11.5 kW. The hosting capacity is 5% (four customers). The hosting capacity increases to 8% at 9.2 kW charging power. It is 52% for the charging power of 3.7 kW. Over half the customers can have single-phase EV chargers with 3.7 kW.

4.2. Three-Phase Hosting Capacity

The three-phase hosting capacity for the six-customer and 83-customer distribution networks considering different EV charging powers are shown in Figure 7.

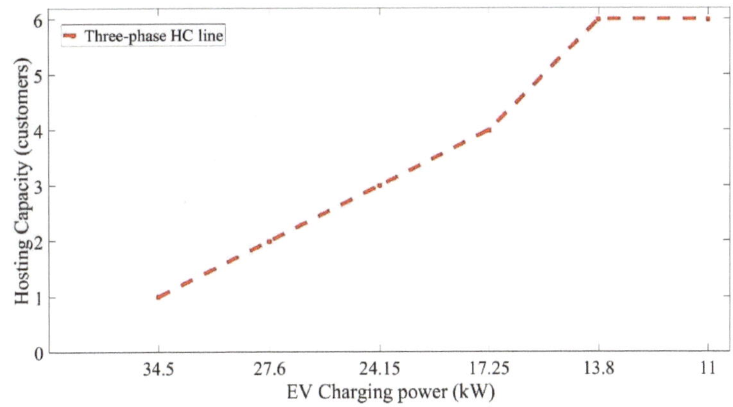

(**a**)

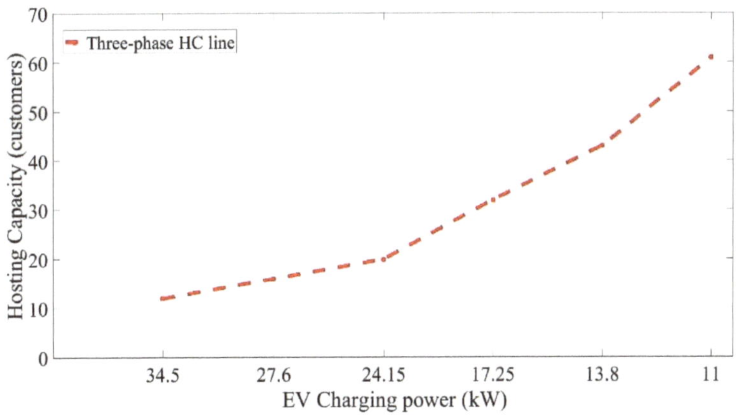

(**b**)

Figure 7. Three-phase hosting capacity results for 11, 13.8, 17.25, 24.15, 27.6 and 34.5 kW: (**a**) 6-customers distribution network; (**b**) 83-customers distribution network.

The results in Figure 7 show a similar pattern for single-phase charging, whereby more customers can simultaneously charge for smaller charging power per customer. The current per phase is the same for the single-phase and three-phase charging powers considered, but the hosting capacity is generally higher for three-phase charging. However, there is no clear relation between the two. Separate calculations are needed for single-phase and three-phase charging. The three-phase hosting capacity is one-sixth for the six-customer distribution network, at 34.5 kW of charging power needed. It is 14% (12 customers) for the 83-customer distribution network. In the six-customer distribution network, all customers can charge their EV simultaneously for 13.8 kW and below power needs. The hosting capacity for the 83-customer distribution network is 73% for the 11 kW three-phase charging power need.

4.3. *Influence of Lowest Background Voltage*

The obtained measured background voltage has been changed by 1.25%, 2.5%, 3.75%, and 5% of the nominal voltage. The GEV distribution function, which characterised measurements, was fitted to the increased voltage measurements. The input lowest background voltages applied to assess their influence on the hosting capacity are given in Table 2.

Table 2. The change in background voltage input parameters showing the standard deviation and mean for the GEV with the same shape (k = −0.0309).

Change in Background Voltage U (%)	GEV Sigma	GEV mu
−Δ5	0.3087	221.24
−3.75	0.3128	224.15
−2.5	0.3169	227.06
−Δ1.25	0.3209	229.97
0 (base case)	0.3250	232.88
+1.25	0.3291	235.79
+2.5	0.3331	238.70
+3.75	0.3372	241.61
+5	0.3412	244.52

The stochastic hosting capacity assessment, performed for the different background voltages shown in Table 2, resulted in Figure 8.

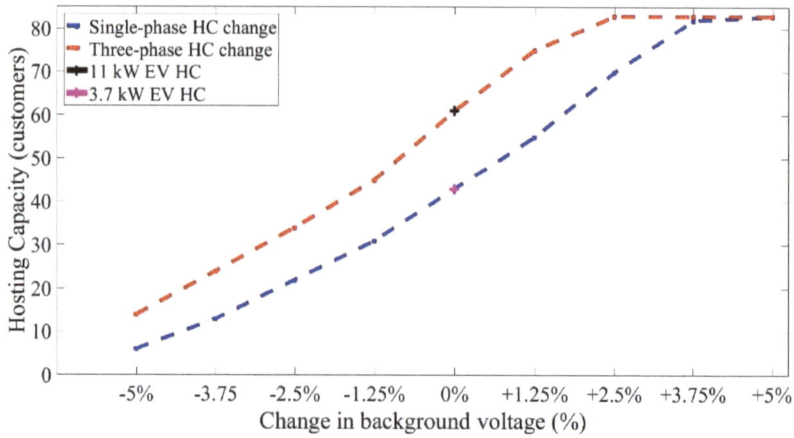

Figure 8. EV hosting capacity change for single-phase (3.7 kW) and three-phase (11 kW) charging with change in the lowest background voltage from +5% to −5%.

Again, the behaviour is similar for single-phase and three-phase charging, with the hosting capacity being generally higher for three-phase charging. Compared to the reference case, a decrease of 5% in the lowest background voltage results in a large decrease in hosting capacity. A similar increase in background voltage results in the hosting capacity reaching 100% of customers for both single-phase and three-phase charging.

4.4. *Influence of Customer Highest Consumption*

The highest customer consumption is changed for the 83-customer distribution network. The consumption ranges of 0–0.5, 0.5–1, 1.5–2, and 2–2.5 kW are applied. A stochastic approach is applied for each consumption range in addition to the reference case of 1–1.5 kW for single-phase and three-phase EV charging. The results are shown in Figure 9.

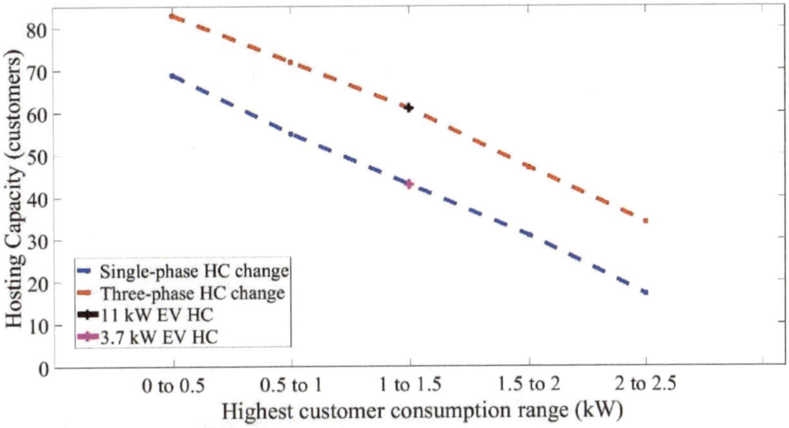

Figure 9. The 83-customers distribution network three-phase hosting capacity for EV charging power 11, 13.8, 17.25, 24.15, 27.6 and 34.5 kW.

The customers' highest consumption also has a significant impact on the hosting capacity. The smallest consumption range of 0–500 W in Figure 9 caused an increase in the hosting capacity by 36% for three-phase and 65% for single-phase EV charging. As the highest consumption is increased, there is a corresponding decrease in the hosting capacity. The hosting capacity decreases by 69% for single-phase EV charging and 44% for three-phase EV charging.

4.5. Influence of Planning Risk

The study applied before used the 10th percentile of the lowest voltage as a performance indicator. This can be interpreted as a 10% planning risk. The effect of selecting the 1st, 5th, 15th, or 20th percentiles on the hosting capacity for EV charging was studied. The results for single-phase (3.7 kW) and three-phase (11 kW) charging power is shown in Figure 10.

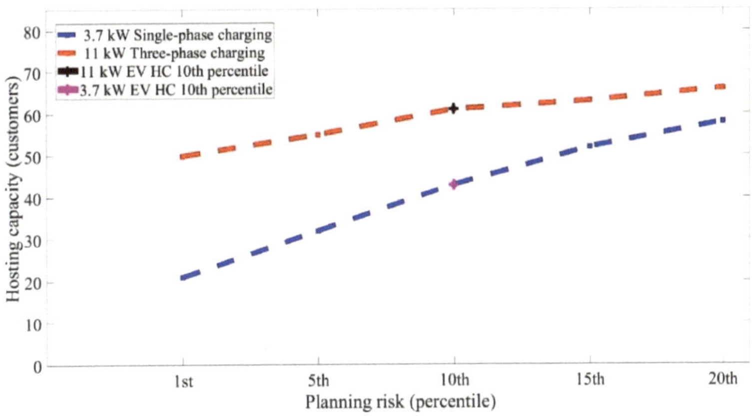

Figure 10. The planning risk (percentile) impact on the EV charge hosting capacity for the 83-customers distribution network.

The 10th percentile in Figure 10 is the reference case. The figure clearly shows that stricter planning risk decreases the hosting capacity. There is an 18% and 49% decrease in the EV hosting capacity for three-phase and single-phase when the planning is reduced from 10% to 1%. Higher planning risk increases hosting capacity, but the effect is lower than for lower-risk, especially three-phase charging.

The planning risk has a bigger influence on the single-phase hosting capacity (blue line).

5. Discussion

The EV charge hosting capacity in Section 4 estimates how many simultaneous customers can charge in a distribution network, considering undervoltage as the limiting factor. The method is applicable for any period of interest for the DSO and can bring out the bottlenecks for EV charging that could instill investment for future growth. The method presented can be utilized for making planning decisions regarding estimated hosting capacity without detailed knowledge of the charging patterns of the customers. DSOs typically take a certain risk when planning distribution networks. The method can also be used with varying planning risks shown in this paper, to make a trade-off between the risk of bad voltage quality (insufficient investment) and stranded assets (too much investment). The planning risk is essential, and efforts are needed to ascertain what the DSOs apply or a range of values to streamline the approach.

The presented results show that, for both example networks, the hosting capacity is higher for three-phase than for single-phase charging. It was also found to vary a lot for the range of the charging powers expected to appear in the coming years. There is a need to extend the studies to more distribution networks to verify the observations in this paper. The adapted PV hosting capacity method has been applied to two illustrative distribution networks. More distribution networks should be studied to obtain more general trends.

The method can be applied as an extension for calculating the hosting capacity as a function of time of day, week, or year. There is a need to apply the method for such other applications. In that way, the best periods for charging can be identified and used in designing smart charging mechanisms.

There are some challenges with the method and uncertainties in the output (not to be mixed up with the uncertainties in input). The results are stochastic, and their interpretation can often lead to uncertainties. It was also shown that both background voltage and consumption have a big impact on the hosting capacity. However, these are often not known and require detailed measurements. Data collection becomes important in order to estimate the hosting capacity accurately.

Fast charging in public with electric buses and trucks has not been assessed in this study. These are expected to be connected to the MV network and will have their main impact there. Further studies are needed for fast (public) charging, charging of electric buses and electric trucks. However, similar methods as applied in this paper can be applied for such studies.

The approach in this paper has considered the undervoltage phenomenon. Stochastic methods for estimating the hosting capacity of transformer and feeder overload are also needed. Their addition to the method proposed in this paper can inform the DSO of the two phenomena that are most likely to limit EV charging penetration in distribution networks.

6. Conclusions

A stochastic approach has been proposed and applied to evaluate distribution networks' hosting capacity for EV charging. An important application of the approach is relevant to future planning and investment decisions in distribution networks. The approach is non-specific and can be used for any time of day, week, or year, without detailed knowledge of the charging patterns.

The results obtained by applying the method to two existing low-voltage networks showed that the EV charge hosting capacity is sensitive to the lowest background voltage, highest power consumption, and planning risk. It is recommended to apply the method to more distribution networks, including medium-voltage networks, and to commence data collection to obtain input for stochastic hosting capacity studies.

Author Contributions: Conceptualization, E.M.; methodology, E.M.; software, E.M.; validation, E.M., M.H.J.B. and N.E.; formal analysis, E.M., M.H.J.B. and N.E.; investigation, E.M.; resources, M.H.J.B. and N.E.; data curation, E.M.; writing—original draft preparation, E.M.; writing—review and editing, E.M., M.H.J.B. and N.E.; visualization, E.M.; supervision, M.H.J.B. and N.E.; project administration, M.H.J.B.; funding acquisition, M.H.J.B. All authors have read and agreed to the published version of the manuscript.

Funding: This research was funded by the Swedish Energy Agency, Umeå Energi AB and Skellefteå Kraft Elnät AB.

Institutional Review Board Statement: Not Applicable.

Informed Consent Statement: Not Applicable.

Data Availability Statement: The data used in this paper is provided within and is also referenced.

Conflicts of Interest: The authors declare no conflict of interest.

References

1. Patil, H.; Kalkhambkar, V.N. Grid Integration of Electric Vehicles for Economic Benefits: A Review. *J. Mod. Power Syst. Clean Energy* **2021**, *9*, 13–26. [CrossRef]
2. ElNozahy, M.S.; Salama, M.M.A. A Comprehensive Study of the Impacts of PHEVs on Residential Distribution Networks. *IEEE Trans. Sustain. Energy* **2014**, *5*, 332–342. [CrossRef]
3. Leou, R.; Teng, J.; Lu, H.; Lan, B.; Chen, H.; Hsieh, T.; Su, C. Stochastic analysis of electric transportation charging impacts on power quality of distribution systems. *IET Gener. Transm. Distrib.* **2018**, *12*, 2725–2734. [CrossRef]
4. Vita, V.; Koumides, P. Electric Vehicles and Distribution Networks: Analysis on Vehicle to Grid and Renewable Energy Sources Integration. In Proceedings of the 2019 11th Electrical Engineering Faculty Conference (BulEF), Varna, Bulgaria, 11–14 September 2019; pp. 1–4.
5. Shahan, Z. Sweden EV Sales. Available online: https://cleantechnica.com/tag/sweden-ev-sales/ (accessed on 12 April 2021).
6. Holland, M. Sweden Hits Record 50% Electric Vehicle Share In December! Available online: https://cleantechnica.com/2021/01/05/sweden-hits-record-50-electric-vehicle-share-in-december/? (accessed on 12 April 2021).
7. Ping, J.; Yan, Z.; Chen, S. A Two-stage Autonomous EV Charging Coordination Method Enabled by Blockchain. *J. Mod. Power Syst. Clean Energy* **2021**, *9*, 104–113. [CrossRef]
8. Shepero, M.; Munkhammar, J.; Widen, J.; Bishop, J.D.K.; Bostrom, T. Modeling of photovoltaic power generation and electric vehicles charging on city-scale: A review. *Renew Sust. Energ. Rev.* **2018**, *89*, 61–71. [CrossRef]
9. Fachrizal, R.; Ramadhani, U.H.; Munkhammar, J.; Widén, J. Combined PV–EV hosting capacity assessment for a residential LV distribution grid with smart EV charging and PV curtailment. *Sustain. Energy Grids Netw.* **2021**, *26*, 100445. [CrossRef]
10. Bollen, M.H.J.; Häger, M. Power quality: Interactions between distributed energy resources, the grid, and other customers. In Proceedings of the 1st International Conference on Renewable Energy Sources and Distributed Energy Resources, Brussels, Belgium, 1–3 December 2004.
11. Bourgain, G. *Integrating Distributed Energy Resources in Today's Electrical Energy System*; Book Presenting the Results of the EU-DEEP Project; ExpandDER; European Commission: Saint Denis le Plaine, France, 2009.
12. Xin, T.; Chunlin, G.; Xiaoyan, Y.; Chenchen, C. Research on characteristics of electric vehicle charging load and distribution network supportability. In Proceedings of the IEEE PES Asia-Pacific Power and Energy Engineering Conference (APPEEC), Xi'an, China, 25–28 October 2016; pp. 1539–1542.
13. Ashfaq, M.; Butt, O.; Selvaraj, J.; Rahim, N. Assessment of electric vehicle charging infrastructure and its impact on the electric grid: A review. *Int. J. Green Energy* **2021**, *18*, 657–686. [CrossRef]
14. Leou, R.C.; Su, C.L.; Lu, C.N. Stochastic Analyses of Electric Vehicle Charging Impacts on Distribution Network. *IEEE Trans. Power Syst.* **2014**, *29*, 1055–1063. [CrossRef]
15. Zeng, B.; Zhu, Z.W.; Zhu, X.; Qin, L.; Liu, J.M.; Zhang, M. Estimation of Distribution System Capability for Accommodating Electric Vehicles. *J. Electr. Syst.* **2020**, *16*, 131–141.
16. Mishra, S.; Verma, S.; Chowdhury, S.; Gaur, A.; Mohapatra, S.; Dwivedi, G.; Verma, P. A Comprehensive Review on Developments in Electric Vehicle Charging Station Infrastructure and Present Scenario of India. *Sustainability* **2021**, *13*, 2396. [CrossRef]
17. Jones, C.B.; Lave, M.; Vining, W.; Garcia, B.M. Uncontrolled Electric Vehicle Charging Impacts on Distribution Electric Power Systems with Primarily Residential, Commercial or Industrial Loads. *Energies* **2021**, *14*, 1688. [CrossRef]
18. Quiros-Tortos, J.; Ochoa, L.F.; Butler, T. How Electric Vehicles and the Grid Work Together: Lessons Learned from One of the Largest Electric Vehicle Trials in the World. *IEEE Power Energy Mag.* **2018**, *16*, 64–76. [CrossRef]

19. Zou, Y.; Zhao, J.; Gao, X.; Chen, Y.; Tohidi, A. Experimental results of electric vehicles effects on low voltage grids. *J. Clean. Prod.* **2020**, *255*, 120270. [CrossRef]
20. Betancur, D.; Duarte, L.F.; Revollo, J.; Restrepo, C.; Díez, A.E.; Isaac, I.A.; López, G.J.; González, J.W. Methodology to Evaluate the Impact of Electric Vehicles on Electrical Networks Using Monte Carlo. *Energies* **2021**, *14*, 1300. [CrossRef]
21. Mulenga, E.; Bollen, M.H.J.; Etherden, N. Solar PV stochastic hosting capacity in distribution networks considering aleatory and epistemic uncertainties. *Int. J. Electr. Power Energy Syst.* **2021**, *130*, 106928. [CrossRef]
22. Awadallah, S.K.E.; Milanović, J.V. Quantification of aleatory and epistemic uncertainty in bulk power system reliability evaluation. In Proceedings of the IEEE Grenoble Conference, Grenoble, France, 16–20 June 2013; pp. 1–6.
23. Cundeva, S.; Mateska, A.K.; Bollen, M.H.J. Hosting capacity of LV residential grid for uncoordinated EV charging. In Proceedings of the 18th International Conference on Harmonics and Quality of Power (ICHQP), Ljubljana, Slovenia, 13–16 May 2018; pp. 1–5.
24. Palomino, A.; Parvania, M. Probabilistic Impact Analysis of Residential Electric Vehicle Charging on Distribution Transformers. In Proceedings of the 2018 North American Power Symposium (NAPS), Fargo, ND, USA, 9–11 September 2018; pp. 1–6.
25. Palomino, A.; Parvania, M. Data-Driven Risk Analysis of Joint Electric Vehicle and Solar Operation in Distribution Networks. *IEEE Open Access J. Power Energy* **2020**, *7*, 141–150. [CrossRef]
26. Hiesl, A.; Ramsebner, J.; Haas, R. Modelling Stochastic Electricity Demand of Electric Vehicles Based on Traffic Surveys—The Case of Austria. *Energies* **2021**, *14*, 1577. [CrossRef]
27. Barbosa, T.; Andrade, J.; Torquato, R.; Freitas, W.; Trindade, F.C.L. Use of EV hosting capacity for management of low-voltage distribution systems. *IET Gener. Transm. Distrib.* **2020**, *14*, 2620–2629. [CrossRef]
28. Billinton, R.; Huang, D. Aleatory and Epistemic Uncertainty Considerations in Power System Reliability Evaluation. In Proceedings of the 10th International Conference on Probablistic Methods Applied to Power Systems, Rincon, PR, USA, 25–29 May 2008; pp. 1–8.
29. Mulenga, E.; Bollen, M.H.J.; Etherden, N. Distribution networks measured background voltage variations, probability distributions characterization and Solar PV hosting capacity estimations. *Electr. Power Syst. Res.* **2021**, *192*, 106979. [CrossRef]
30. Energiforsk. Overvoltage from Single Phase Solar Panels. Available online: https://energiforskmedia.blob.core.windows.net/media/24707/oeverspanning-fran-enfasanslutna-solpaneler-energiforskrapport-2018-506.pdf (accessed on 15 December 2020).
31. Bollen, M.H.J.; Ronnberg, S.K. Hosting Capacity of the Power Grid for Renewable Electricity Production and New Large Consumption Equipment. *Energies* **2017**, *10*, 1325. [CrossRef]
32. Johansson, S.; Persson, J.; Lazarou, S.; Theocharis, A. Investigation of the Impact of Large-Scale Integration of Electric Vehicles for a Swedish Distribution Network. *Energies* **2019**, *12*, 4717. [CrossRef]
33. Schwanz, D.; Moller, F.; Ronnberg, S.K.; Meyer, J.; Bollen, M.H.J. Stochastic Assessment of Voltage Unbalance Due to Single-Phase-Connected Solar Power. *IEEE Trans. Power Deliv.* **2017**, *32*, 852–861. [CrossRef]
34. Mulenga, E. Impacts of Integrating Solar PV Power to An Existing Grid. Case Studies of Mölndal and Orust Energy Distribution (10/0.4 kV and 130/10 kV) Grids. Master's Thesis, Chalmers University of Technology, Gothenburg, Sweden, 2015.
35. Coles, S. *An Introduction to Statistical Modeling of Extreme Values*; Springer: London, UK, 2013.

electricity

MDPI

Article

Development of Planning and Operation Guidelines for Strategic Grid Planning of Urban Low-Voltage Grids with a New Supply Task

Patrick Wintzek [1,*], Shawki Alsayed Ali [1], Markus Zdrallek [1], Julian Monscheidt [2], Ben Gemsjäger [2] and Adam Slupinski [2]

1 Institute of Power Systems Engineering, University of Wuppertal, 42119 Wuppertal, Germany; shawki.ali@uni-wuppertal.de (S.A.A.); zdrallek@uni-wuppertal.de (M.Z.)
2 Siemens AG, Power Technologies International, 91058 Erlangen, Germany; julian.monscheidt@siemens.com (J.M.); ben.gemsjaeger@siemens.com (B.G.); adam.slupinski@siemens.com (A.S.)
* Correspondence: patrick.wintzek@uni-wuppertal.de

Citation: Wintzek, P.; Ali, S.A.; Zdrallek, M.; Monscheidt, J.; Gemsjäger, B.; Slupinski, A. Development of Planning and Operation Guidelines for Strategic Grid Planning of Urban Low-Voltage Grids with a New Supply Task. *Electricity* **2021**, 2, 614–652. https://doi.org/10.3390/electricity2040035

Academic Editor: Andreas Sumper

Received: 28 October 2021
Accepted: 9 December 2021
Published: 16 December 2021

Abstract: In contrast to rural distribution grids, which are mostly "feed-in oriented" in terms of electrical power, urban distribution grids are "load oriented", as the number of customer connections and density of loads in urban areas is significantly higher than in rural areas. Taking into account the progressive electrification of the transport and heating sector, it is necessary to assess the required grid optimization or expansion measures from a conventional, as well as an innovative point of view. This is necessary in order to be able to contain the enormous investment volumes needed for transforming the energy system and aligning the infrastructures to their future requirements in time. Therefore, this article first explains the methodological approach of allocating scenarios of the development of electric mobility and heat pumps to analyzed grids. The article continues with describing which power values need to be applied and which conventional and innovative planning measures are available for avoiding voltage band violations and equipment overloads within the framework of strategic grid planning. Subsequently, the results of grid planning studies are outlined and evaluated with an assessment model that evaluates capital as well as operational costs. On this basis, planning and operation guidelines for urban low-voltage grids are derived. The main result is that low-voltage grids can accommodate charging infrastructure for electric mobility, as well as heat pumps to a certain degree. In addition, it is concluded that conventional planning measures are not completely avoidable, but can be partially avoided or deferred through dynamic load management.

Keywords: charging infrastructure; grid planning; heat pumps; load management; low-voltage; planning and operation guidelines

1. Introduction

Technological progress is changing the energy industry, which is currently characterized by the increasing electrification of the mobility and heating sectors. Distribution system operators (DSOs) in particular are faced with the task of integrating more and more charging infrastructure (CI) for electric vehicles and electric heat pumps (HPs) for supplying heat to residential buildings in the future. For grid planning of low-voltage (LV) grids, DSOs usually apply planning and operation guidelines (POGs). The guidelines offer the advantage that not each grid has to be planned individually; general planning principles can be applied for the majority of grids using, for instance, predefined standard grid equipment. This standard equipment is kept in stock, e.g., a sufficient number of distribution transformers (DTs) or LV lines for the LV level, in order to be utilized quickly in the event of a fault. However, due to ever-increasing electrical loads, the planning guidelines need to be adapted in general and the currently available power classes of DTs and LV lines must be reviewed and adjusted specifically. Therefore, based on extensive LV grid planning, new POGs for urban LV grids are derived in the context of this article.

1.1. Literature Review and Novelty

POGs are not fundamentally new. In most cases, each DSO defines its own, company-specific POGs and reviews or updates them in fixed time intervals, which can even be longer than a decade. However, due to the increasing electrification of the mobility and heating sector, existing POGs must be reviewed in a timely manner, as the current standard equipment may no longer be sufficient. In publications, POGs have already been updated and in some cases supplemented, but to the best of our current knowledge, no extensive grid planning has, to date, been carried out with various grid plans supporting these POGs.

Therefore, this article carries out strategic grid planning with diverse power development scenarios for representative LV grids, and then evaluates the results. Subsequently, POGs are derived based on the evaluations. The diverse power assumptions offer the advantage that publications [1–7] have not yet considered high charging capacities in LV grids for public charging points (PuCPs) with a simultaneous consideration of private charging points (PrCPs). This article also analyses the impact of different load management systems for private and/or public CI as well as HPs on grid planning. Furthermore, a sensitivity analysis of the underlying costs for a grid-serving dynamic load management (DLM) is performed and analyzed. These aspects were not examined at this level of detail in previous publications, such as [8–10], either.

1.2. Structure and Objective

In this article, first the procedure for strategic grid planning is explained in Section 2. This includes both the basic planning steps and the handling of new loads (CI and HPs). For the latter, the development scenarios are selected, and the method for allocation of CI and HPs at grid level is explained in order to interpret the results accordingly. Section 3 explains the general grid conditions under which the grid planning is performed. In addition to the basic power assumptions, further planning parameters are determined, such as the operating points (OPs), the planning perspectives with their associated simultaneity factors (SFs), and the technical limits for the grid operation. Section 4 explains conventional and innovative planning measures that are available for applying in the LV level in order to avoid the expected limit violations. To assess individual planning studies, all grid planning variants are evaluated in Section 5 based on an assessment model consisting of a primary and a secondary assessment model. In Section 6, the POGs are derived based on the results of the primary assessment model and then explained in detail. This article concludes with a discussion of the results.

2. Strategic Grid Planning

First, this Section explains the basics of strategic grid planning and the steps necessary for fulfilling the respective planning objective. Based on this, new requirements for grid planning with regard to new loads are presented. These new requirements need to be taken into account in the future as part of the energy transition.

2.1. Basic Planning Steps

Starting with the definition of basic planning objectives, it is important to note that these must comply with the current laws, regulations, standards, ordinances, and directives, as well as the commonly acknowledged rules of technology and technical guidelines. These general framework conditions can be supplemented by company-specific requirements, which must first be identified. It is possible to define the standards, that need to be taken into account more strictly, such as standard DIN EN 50160 [11] with regard to the voltage band, if this is necessary from the point of view of grid engineering. In addition, specifications must be made concerning to the integration of new loads. Once all the relevant conditions are available, the corresponding information and data must be obtained and processed. It should be noted that the greater this information density, the less robust the respective grid must be in the event of deviations from these forecasts. Therefore, it is important to derive information that is not available with suitable models, if necessary, or to approach it in some other way. On the basis of this information, suitable conventional and innovative planning measures are selected, used to develop various target grid planning studies, subsequently compared, and finally evaluated on the basis of various criteria. The result is an optimal target grid planning which is used as the basis for the implementation [12–16].

2.2. *New Loads in Urban Low-Voltage Grids*

CI and HPs play an increasingly important role in the dimensioning of urban LV grids. Furthermore, according to [16], decentralized energy conversion systems such as photovoltaic systems (PVSs) are particularly relevant for grid planning of rural and suburban areas.

2.2.1. Charging Infrastructure for Electric Vehicles

Figure 1 shows different development scenarios for electric vehicles (EVs) in Germany. To create a corridor, a conservative (cons) scenario (Q) from the lower development range and a progressive (prog) scenario (R) from the upper development range are used for further grid planning. The corresponding sources are listed in Table 1.

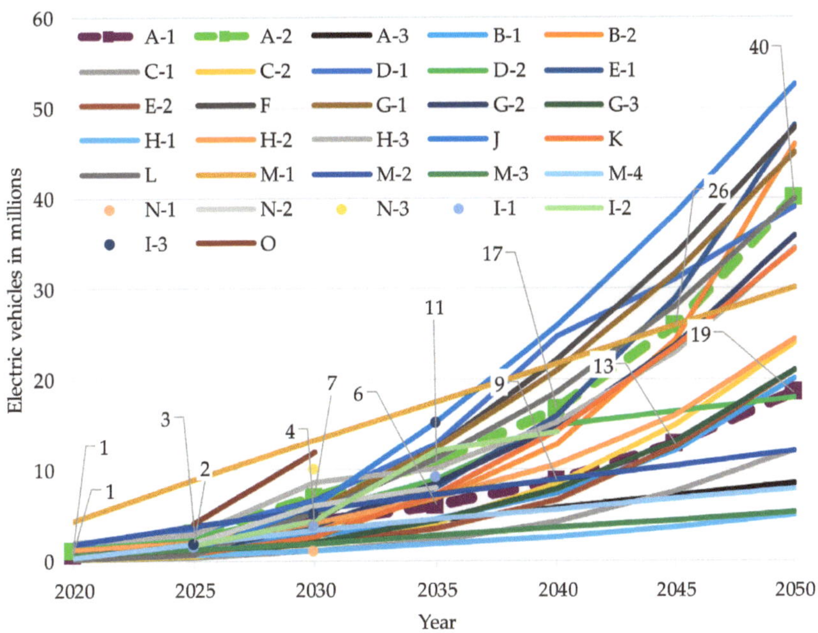

Figure 1. Ramp-up trajectories for EV development scenarios in Germany based on [17].

Table 1. Researched scenarios for the development of electric vehicles.

Scenario	Based on Source	Scenario	Based on Source
A-1/A-2/A-3	[18]	B-1/B-2	[19]
C-1/C-2	[20]	D-1/D-2	[21]
E-1/E-2	[22]	F	[23]
G-1/G-2/G-3	[24]	H-1/H-2/H-3	[25]
J	[26]	K	[27]
L	[28]	M-1/M-2/M-3/M-4	[29]
N-1/N-2/N-3	[30]	I-1/I-2/I-3	[31]
O	[32]		

According to the apportionment methodology in [17], the values of the scenarios for Germany are apportioned to city level using various factors and weightings. To apportion the number of EVs to the LV level, commercial market and geodata from [33,34] are used. For the distribution at the respective LV grids, an iterative Saint-Laguë technique [35] is employed for private EVs. Commercial EVs and commuter EVs are allocated based on building types. The methods apply data available at street level regarding the building and income structures. A more detailed description of the apportionment methodology can be found in [17].

2.2.2. Electric Heat Pumps

In accordance with the apportionment methodology for electric mobility, different scenarios for the development of HPs can be taken from Figure 2 with the sources listed in Table 2. Scenarios G (cons) and H (prog) are selected for the grid planning.

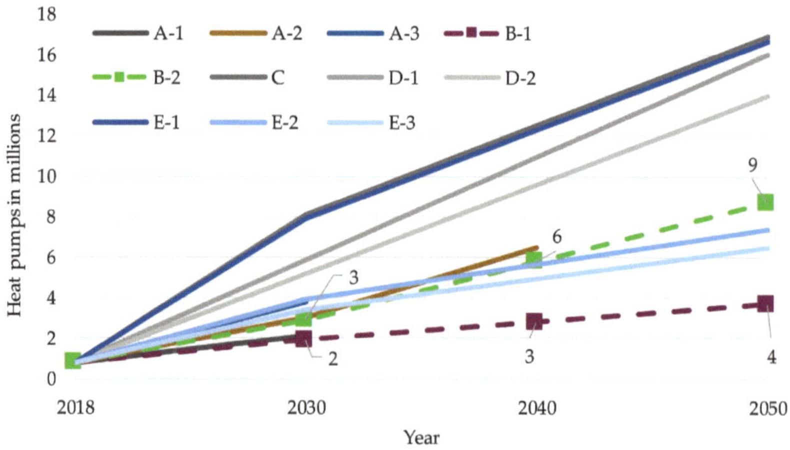

Figure 2. Ramp-up trajectories of heat pump development scenarios for Germany based on [17].

Table 2. Studies used as a basis for the development of heat pumps.

Scenario	Based on Source	Scenario	Based on Source
A-1/A-2/A-3	[31]	B-1/B-2	[36]
C	[37]	D-1/D-2	[38]
E-1/E-2/E-3	[39]		

Additionally, analogous to EVs, the values of the scenarios are apportioned with various factors to the city level. A further distribution to the LV grids uses market and geodata from [33]. A more detailed description of the apportionment methodology can be found in [17].

2.2.3. Photovoltaic Systems

This article focuses on new loads. However, as PVSs are also relevant, especially at the LV level, scenarios of the installed photovoltaic capacity in Germany (analogous to Figures 1 and 2, see Figure 3 with sources stated in Table 3), in particular scenario (A) and scenario (E), are assessed. A more detailed description of the apportionment methodology can be found in [40] and is not part of the present analysis.

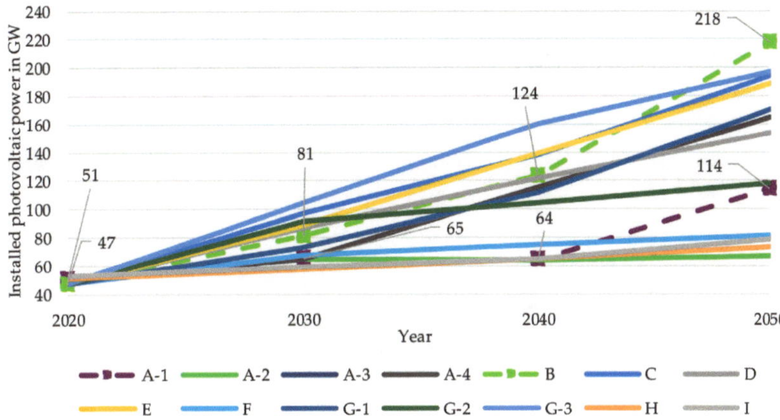

Figure 3. Ramp-up trajectories of photovoltaic power development scenarios for Germany based on [41].

Table 3. Studies used as a basis for the development of heat pumps.

Scenario	Based on Source	Scenario	Based on Source
A-1/A-2/A-3/A-4	[29]	B	[42]
C	[43]	D	[44]
E	[45]	F	[46]
G-1/G-2/G-3	[47]	H	[48]
I	[49]		

3. General Grid Conditions

In addition to the scenarios for new loads presented in the previous Section, it must be determined how these loads are taken into account in grid planning in terms of their power value. The first step is to specify the relevant OPs. These OPs are then valid for certain planning perspectives, for each of which different SFs are applied. After the LV grid has been modelled for the specified planning perspectives, the grid limit violations must be determined on the basis of the permissible limit values for the voltage band and the equipment loading.

3.1. Power Value Assumptions

For new loads, power values must be assumed for using in strategic grid planning. Therefore, possible charging capacities of CI are firstly analyzed. Here, a distinction is made between PrCPs and PuCPs, as the categories are assigned to different power classes. For example, PrCPs usually have 3.7 kW, 11 kW, or 22 kW, whereas PuCPs also cover 50 kW and 150 kW at LV level. Higher charging capacities or charging parks are generally connected to the medium-voltage (MV) and high-voltage (HV) levels [50].

For HPs, the assumption of suitable power values is somewhat more challenging, as each building has a different insulation standard and different heat requirements. As a result, each HP should be specifically designed. To take into account different configurations, three power values 3.0 kW, 6.5 kW, and 9.0 kW are assumed. HP variant 1 (HP-V1) represents a HP without an additional heating element as a minimum power value. HP-V3, on the other hand, takes into account additional heating elements with an average electrical output of 6.0 kW as a maximum power value. HP-V2 represents a combination of the two variants with further assumptions, resulting in 6.5 kW per HP. All three HP variants are applied to the conservative and progressive scenarios for EVs [50].

Table 4 shows a consolidation of power value assumptions for CPs with the corresponding assumed development over the years 2030, 2040, and 2050.

Table 4. Power value assumptions for private and public charging infrastructure over three years for grid planning, based on [50], note: Details of the distribution or number of the respective additions per year.

Private Charging Points	2030	2040	2050	Public Charging Points	2030	2040	2050
3.7 kW	10%	0%	0%	3.7 kW	0%	0%	0%
11 kW	60%	65%	65%	11 kW	5%	5%	5%
22 kW	30%	35%	35%	22 kW	75%	20%	20%
50 kW	0%	0%	0%	50 kW	15%	50%	50%
150 kW	0%	0%	0%	150 kW	5%	25%	25%

3.2. Operating Points

In general, LV grids are dimensioned for certain OPs, i.e., certain grid use cases. The two most common OPs are "peak generation" (PG, or OP-PG) and "peak load" (PL, or OP-PL) [14–16]. The OP-PG defines a summer day on which, for example, the highest feed-in from decentralized PVSs and minimum power consumption can be expected at midday. The OP-PL defines a winter day in the early evening on which there is no feed-in from PVSs with simultaneous maximum power demand from HPs, CI and conventional loads.

Alternatively, power time series can also be used for grid planning. However, as no power time series are available for all nine different new loads at the time of the analysis and these can be taken into account indirectly via SFs, the following two OPs are used:

- Peak load: It is assumed that the loads draw the maximum simultaneous power demand while grid feed-in is minimal [51].
- Peak generation: Here, it is assumed that the feed-in in the grid area is maximum, while load demand is minimum [51].

3.3. Planning Perspectives and Simultaneity Factors

Based on both OPs, different planning perspectives must be taken into account in order to dimension the equipment correctly. Otherwise, either over-dimensioning leads to unnecessary costs or under-dimensioning leads to overloading of the equipment. Therefore, the so-called SFs are used for grid planning. The SF is defined as the ratio of the maximum simultaneous sum reference to the sum of the maximum individual powers [51].

According to Figure 4, two planning perspectives are considered for the subsequent analyses. For dimensioning the respective DT, all loads in the grid are considered (left side). For the dimensioning of main feeders, all loads connected per feeder are considered with the corresponding SF. A feeder is defined as a line that is laid from the DT to the first load. SF calculations are carried out separately for conventional loads, CI and HPs. The resulting power is then cumulated, yielding the total load. However, it must be noted with these two planning perspectives that, depending on the circumstances, there is an over- or under-dimensioning of the ends of the lines, as the SF is not determined and modelled with node precision. However, as the feeder, which is important for the fundamental supply of many end consumers, is correctly dimensioned, this procedure represents a trade-off between the dimensioning of relevant equipment close to the feeder and the avoidance of over- or under-dimensioning of distant lines.

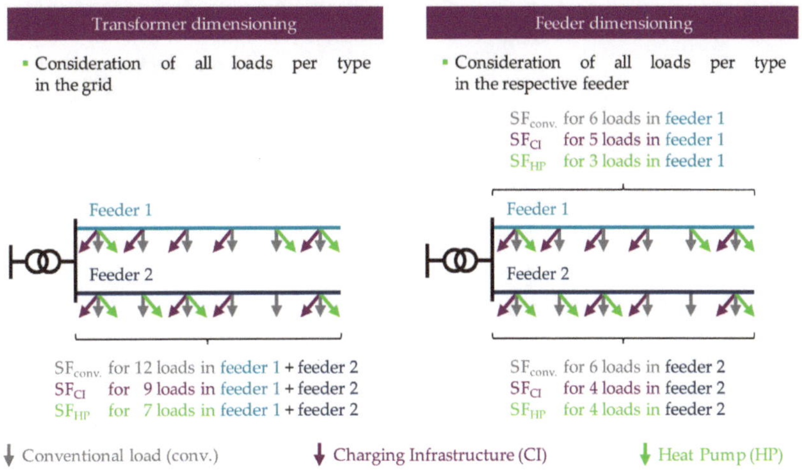

Figure 4. Planning perspectives taking into account the respective simultaneity factor (SF).

3.3.1. Simultaneity Factors for Charging Points

Due to the diverse power values for charging points (CPs), there are different ways to calculate SFs. Therefore, four different calculations (C1 to C4) with Equations (1)–(4) are presented below [17].

$$C1: P_{CP_i} = i \cdot SF_{P_i, n_{CP_i}} \tag{1}$$

$$C2: P_{CP_i} = i \cdot SF_{P_i, \sum n_{CP}} \tag{2}$$

$$C3: P_{CP_i} = i \cdot SF_{P_\varnothing, n_{CP_i}} \tag{3}$$

$$C4: P_{CP_i} = i \cdot SF_{P_\varnothing, \sum n_{CP}} \tag{4}$$

where I = charging power type; P_{CPi} = charging power per charging point type; P_i = charging power per type; n_{CPi} = number of charging points per charging power; $\sum n_{CP}$ = number of all charging points; and $P_\varnothing$ = average charging power based on the distribution in the respective grid.

- Calculation 1: The respective charging power multiplied with the SF for the respective charging power for the number of CPs for this charging power (several SFs per grid or feeder).
- Calculation 2: The respective charging power multiplied with the SF for the respective charging power for the number of all CPs and all charging powers (several SFs per grid or feeder).
- Calculation 3: The respective charging power multiplied with the SF for the average charging power based on the respective distribution for the number of CPs for this charging power (several SFs per grid or feeder).
- Calculation 4: The respective charging power multiplied with the SF for the average charging power based on the respective distribution for the number of all CPs and all charging powers (one SF per grid or feeder).

The calculation results are shown for three examples in Figure 5. It can be seen that the calculation methods C1 and C3 may result in an overestimation of simultaneity as the different charging powers are considered separately from each other. C4 follows an aggregated approach in which the charging powers are combined in an SF based on an average charging power per feeder, which is a practicable approach for grid planning and results in a lower SF. C4 is therefore used in the analyses here. Similarly, it is assumed that EVs can use either PrCPs or PuCPs for charging.

Figure 6 shows the SFs for CPs which are used for C4. The SF curves of the main five charging powers are shown in colors. The figure also shows the SF for charging powers between 3.7 and 22 kVA in 1 kVA steps, which are later on required for C4.

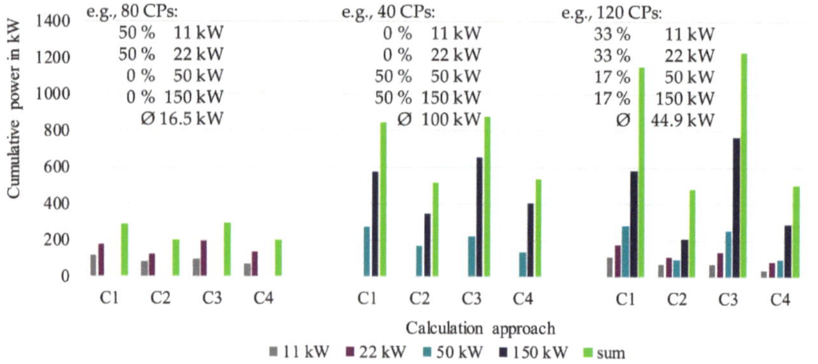

Figure 5. Example of cumulative charging capacities with a different calculation of simultaneities for charging points (CPs).

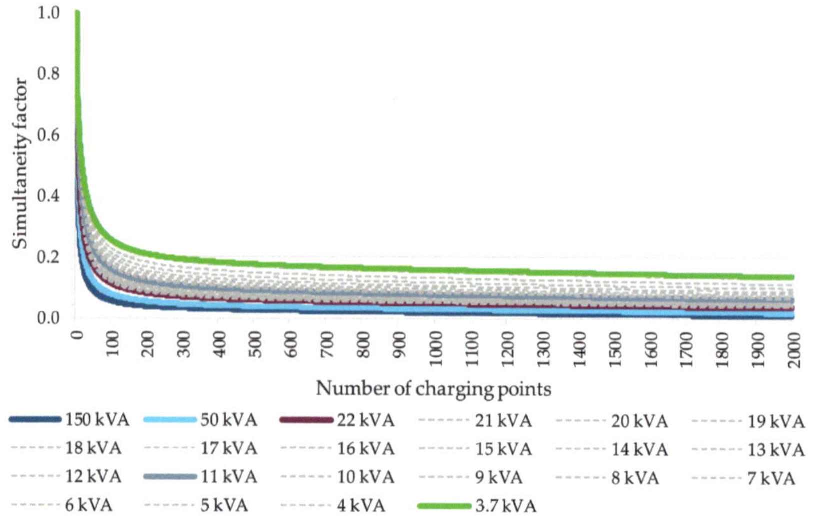

Figure 6. Simultaneity factors for electromobility based on [52].

3.3.2. Simultaneity Factors for Heat Pumps

Figure 7 shows the SFs for HPs. It is apparent that within a grid area, the simultaneity does not decrease as much with an increasing number of HPs as it does for the CI. The reason for this is that the same outdoor temperature is present almost everywhere in a grid area, so that the HPs normally operate simultaneously for heat generation.

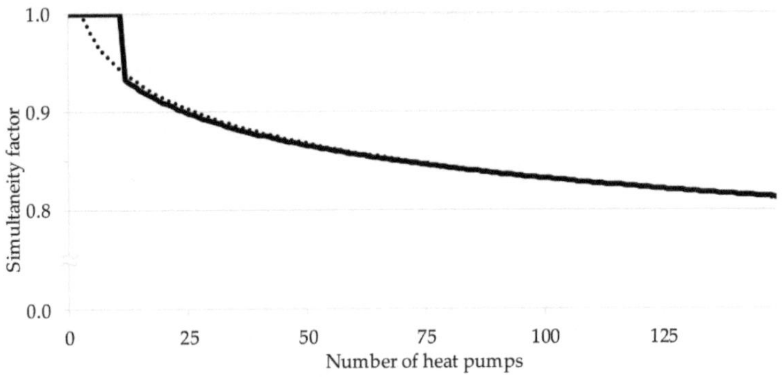

Figure 7. Simultaneity factors for heat pumps based on [53].

3.4. Limit Violations for Grid Planning

In order to identify limit violations after the grid modelling, the voltage band and the equipment loading capacity must first be defined.

Regarding the voltage band, DIN EN 50160 [11] must always be maintained. It specifies that slow voltage changes must not exceed $\Delta U_{\mathrm{max}}/U_{\mathrm{n}} = \pm 10\%$. Although there are loading specifications in VDE-AR-N 4100 [54], no specific requirements are made for the OP-PL. In contrast, VDE-AR-N 4105 [55] recommends that slow voltage changes caused by decentralized generation and storage facilities with a grid connection point on the LV level may not exceed 3% of the original voltage level without such generation plants and energy storages. However, this recommendation may be deviated from according to the DSOs specifications, e.g., if regulated DTs (RDTs) are used. This is particularly relevant for the OP-PG. Based on these assumptions, the available voltage band is divided according to Figure 8 and used for grid planning. On the LV level, a voltage drop of 5% of the nominal voltage at the OP-PL and a voltage increase of 3% at the OP-PG is allowed.

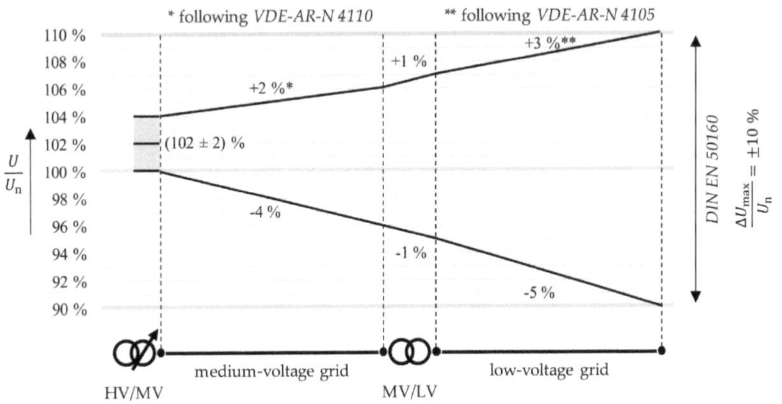

Assumed voltage values for the low-voltage analyses

Voltage value U/U_{n} at the operating point "peak load" on the low-voltage busbar:	95 %
Lowest voltage value U/U_{n} in the operating point "peak load" that is permissible in the grid:	90 %
Voltage value U/U_{n} at the operating point "peak generation" on the low-voltage busbar:	107 %
Highest voltage value U/U_{n} in the operating point "peak generation" that is permissible in the grid:	110 %

Figure 8. Assumed voltage band division.

As for equipment loading, DIN EN 60076-1 [56] specifies that the DTs may be operated with 100% of their rated apparent power. As for lines, DIN VDE 0276-1000 [57] specifies that they may be operated with the maximum permissible current capacity I_z (current carrying capacity). As line installations differ from grid to grid in the type of installation, as well as the accumulation and degree of loading, standard loading conditions are assumed for the derivation of general POGs. This ensures that a uniform system is used for all performed analyses.

4. Planning Measures

After identifying the limit value violations according to the previous Section, the violations can then be remedied with the following appropriate measures. The conventional measures correspond to the current state of the art, have been tried and tested, and are generally accepted. Innovative measures, on the other hand, are already being applied in isolated cases, yet do not represent the current state of the art as they cannot draw on years of experience.

4.1. Conventional Measures

Conventional measures are, in many cases, carried out without taking the direct influence on the other voltage levels into account, such as a subsequent adjustment of the voltage band. Apart from that, dimensioning takes place for a specific year in the future. Thus, for example, a transformer is not designed for its current power demand, but for a specific power demand in the future. This way, it will not have to be replaced over its lifespan, resulting in cost savings.

Conventional measures are used as a reference planning variant for the subsequent analyses and are explained in detail below.

4.1.1. Replacement or Reinforcement of Local Distribution Transformers

If, depending on the load development, the power of the installed DT is no longer sufficient, it must be replaced or reinforced by a second transformer. The latter is possible only if the corresponding space is available. Normally, an existing DT is replaced without changes to the substation. However, it must also be checked whether a new local substation is needed if a larger transformer is to be installed in a compact substation and there is not enough available space.

4.1.2. Tap Changer (Load or Voltage-Free Switchable Tap Changer)

In most cases, a tap changer can be used to adjust the voltage of conventional DTs. This is accomplished in the load-free state. For analyses, it is assumed that a tap changer is always available and that it provides a total of five taps, each with a voltage change of $\Delta U/U_n = \pm 2.5\%$. Depending on the OP, the voltage for the downstream LV grid can thus be raised or lowered by a total of 5.0% [58].

4.1.3. Replacement of Lines

For many DSOs, line replacements are subject to asset management and thus DSO-specific strategies. The approach chosen for the analyses is therefore that lines are replaced if they are overloaded and also have an old insulation type or no longer correspond to the standard line cross-section $q = 150$ mm^2.

4.1.4. Reinforcement of Lines

If a line with a modern insulation type or with the standard line cross-section $q = 150$ mm^2 is overloaded, it can be assumed in most cases that, in contrast to lines with older insulation types, the end of its useful life has not yet been reached and the asset therefore still has a value for the DSO. Therefore, in this case, it is not replaced in the analyses but reinforced. There are three different options for reinforcements, which are briefly explained below.

1. Minimum Reinforcement If, for example, a feeder in a radial grid is overloaded up to a certain load, it can be checked to which point a new feeder can be laid as the shortest route if the existing connection is severed at this connection point.
2. Maximum Reinforcement If the nearest line distribution cabinet (LDC) is only a few meters away in the case of a line separation, no joints are set up to the connection point as the shortest route, but a new feeder is laid up to the LDC. In the LDC, the other line can then be disconnected so that it can continue to be fed from the existing feeder.
3. Parallel Line with Redistribution of Loads A new feeder is laid. The loads are distributed between the two parallel lines so that both lines are subjected to similar loads.

Figure 9 shows the above-mentioned line and transformer measures. In principle, no line reduction factors are applied as it is assumed that the lines have been laid at a sufficient distance from each other and the OPs are not permanently in operation. For assumptions that deviate from this in practice, grid-specific derating factors must be taken into account.

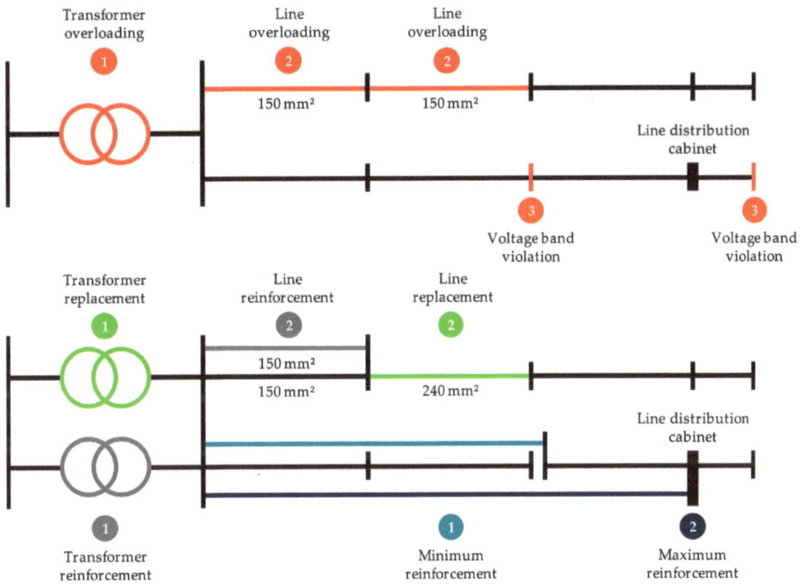

Figure 9. Conventional planning measures in the low-voltage level.

4.1.5. Separation of Grids

In addition to the transformer and line measures, the existing grid can also be split up. However, corresponding properties for new local substations must be available for this, as local DTs are placed in the respective load center based on optimal voltage distribution and equipment loading. As the respective ownership structures are not available for all grids, this measure is not considered within the framework of the uniform grid planning.

4.1.6. Topology Change

The last conventional measure is to change the grid topology. For example, a radial grid is changed to a meshed grid or vice versa. The greatest difficulty with topology changes is that the existing protection concepts may also have to be affected. For these reasons, this measure is also not considered for further analysis.

4.2. Innovative Measures

Independent of the conventional measures applied in practice, there are also a number of innovative measures that are already applied on a small scale, such as RDT, or are currently being researched and tested in selected grids, such as DLM. The relevant innovative measures are shown in Table 5 with effects on all voltage levels (as well as the conventional measures) and are briefly explained below with their significance for the LV level.

Table 5. Overview of the effects of innovative equipment and technologies on the various voltage levels when used in or for the low-voltage level as well as for complementary conventional measures based on [41].

Measure Relocation/Assembly	Influence of the Measure on the Grid Parameters					
	Voltage (U/U_n)			Loading (I/I_z)		
	LV	MV	HV	LV	MV	HV
Lines [A,B]	↕	-	-	↓ [D]	-	-
Distribution transformer with tap changer *	↕	-	-	↓ [E]	-	-
Voltage regulation at the HV/MV-substation	↕	↕	-	-	-	-
Regulated distribution transformer	↕	-	-	↓ [E]	-	-
Line voltage regulator [B]	↕	-	-	-	-	-
Grid-serving energy storage [B]	↕	-	-	↕ [F]	↕ [F]	↕ [F]
Reactive power management [C]	↕	↕	-	↕ [F]	↕ [F]	↕ [F]
Dynamic load management [C]	↑	↑	-	↓ [F]	↓ [F]	↓ [F]
Grid-serving energy storage	↕	-	-	↕ [F]	↕ [F]	↕ [F]

[A] exchange of equipment. [B] new equipment. [C] for low-voltage connected charging infrastructure and heat pumps. [D] line utilization(s). [E] transformer utilization(s) with higher dimensioning. [F] line utilization(s) and transformer utilization(s). "-" means no or negligible influence. "↑" means increase. "↓" means decrease. "↕" both increase and decrease. * Note: Within the framework of LV grid planning, the tap changer with two steps is always used first, before conventional line measures or innovative planning measures (except load management) are applied.

4.2.1. Voltage Regulation at the HV/MV-Substation

Voltage regulation at the HV/MV-substation (VRS) is the permanent voltage adjustment at the MV busbar by changing the tap position of the HV/MV-transformer(s). Depending on the supply task of the MV and the downstream LV grids, the voltage can thus be increased or reduced. According to Figure 8, the setpoint value is U_{target}/U_n = 102% and the control tolerance is $\Delta U/U_n = \pm 2.0\%$. Within the scope of grid planning, it is assumed that the setpoint can be adjusted in six steps of $\Delta U/U_n = \pm 0.5\%$ each. If limit value violations persist despite VRS, conventional planning measures are carried out.

4.2.2. Regulated Distribution Transformer

An RDT is a transformer in a local substation that is equipped with a switchable actuator, a so-called on-load tap changer, and a control unit. In contrast to conventional DTs with a tap changer, voltage adjustments can take place under load and thus lead to a better decoupling of the MV and LV levels with regard to voltage maintenance and therefore to a more flexible division of the available voltage band [59].

The modelling considers an RDT that has nine taps including neutral position, each with a voltage step of $\Delta U/U_n = \pm 3\%$, so that a total band of $\Delta U/U_n = \pm 12.0\%$ is possible. If limit value violations persist despite RDT, conventional planning measures are carried out.

4.2.3. Dynamic Load Management

A DLM accesses CI and HPs, considering them controllable consumers according to the DSOs' specifications. This happens without customer-side load management to ensure safe grid operation. The active power is controlled or regulated depending on the current grid status in terms of voltage and current.

The basic mode of operation of the employed DLM is shown in Figure 10. Limit violations are first identified for each planning perspective. These can be voltage band violations and/or equipment overloads. To ensure that the end consumer does not feel any loss of comfort, HPs are switched off before CPs, as the HPs usually have a heat storage to bypass the so-called blocking periods. Blocking periods are defined as periods in which an electrical system is temporarily, automatically and actively disconnected (switched off/blocked) from the distribution grid by the DSO and is not (fully) available to be connected by the consumer during this time [60]. If limit violations persist, the charging power of the CPs are reduced gradually to a minimum of 3.7 kW. If limit violations are still present despite the use of a DLM, additional conventional planning measures are carried out [61].

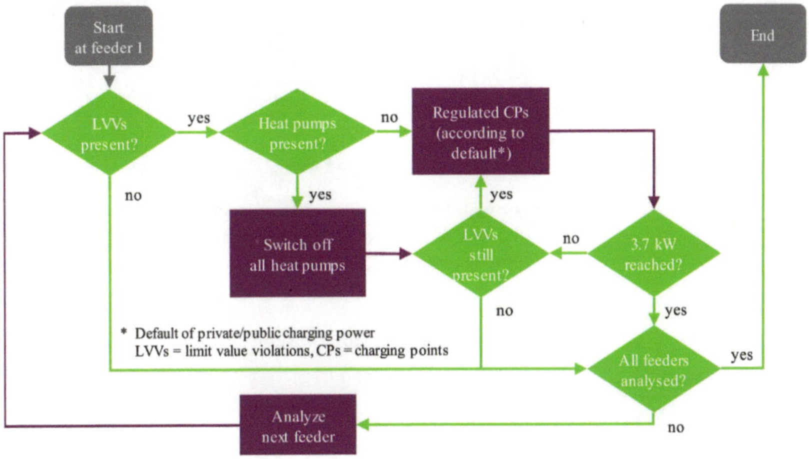

Figure 10. Basic operating principle of dynamic load management.

Within the framework of the DLM, three different variants are considered as use cases according to Table 6. DLM-V1 regulates HPs and PrCPs, DLM-V2 only PrCPs, and DLM-V3 only PuCPs. It should be noted that DLM-V1 and DLM-V3 are rather theoretical scenarios. On the one hand, HPs usually switch back on with full power after the blocking period is over, and on the other hand, there are currently no large-scale efforts to regulate PuCPs. Nevertheless, the three DLM variants span a relevant corridor for grid planning and offer tendencies for the grid-serving use of a DLM with the inclusion of various loads [61].

Table 6. Control of the loads within the DLM variants (DLM-V1, DLM-V2, and DLM-V3) [61].

Controllable Loads	DLM-V1	DLM-V2	DLM-V3
3.7 kW PrCPs	No	No	No
11 kW PrCPs	Yes	Yes	No
22 kW PrCPs	Yes	Yes	No
11 kW PuCPs	No	No	Yes
22 kW PuCPs	No	No	Yes
50 kW PuCPs	No	No	Yes
150 kW PuCPs	No	No	Yes
HPs	Yes	No	No

In addition to the proposed DLM variants, a sensitivity analysis is carried out for the measurement, information, and communication technology (MICT) to be used and is taken into account in the evaluation as part of the economic efficiency analysis.

Figure 11 therefore shows the measuring equipment to be used for radial grids (a.1 and a.2) and meshed grids (b.1 and b.2). Each measuring device can take up to four measurements.

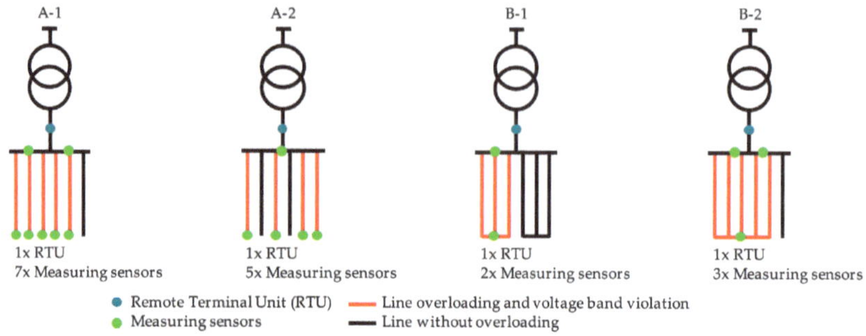

Figure 11. Examples of consideration of a grid automation system for different topologies with (**A-1**) radial grid with five overloaded feeders, (**A-2**) radial grid with four overloaded feeders, (**B-1**) meshed grid with one overloaded mesh and three feeders, and (**B-2**) meshed grid with one overloaded mesh and five feeders.

The following four variants are calculated for the economic efficiency analysis:

- Full equipment (F): Basic amount (remote terminal unit)
  + feeder measurement (current)
  + worst node measurement (voltage)
- Reduced measuring sensors (M): Basic amount (remote terminal unit)
  + feeder measurement (current)
- Basic amount (B): Basic amount (remote terminal unit)
- No costs (0): no MICT

The basic amount represents the remote terminal unit (RTU) as well as the hardware and software integration into the local substation. Feeder measurements are current measurements and worst node measurements are voltage measurements. The latter are not included in (M), as it is assumed that these values will be made available in the future via existing smart metering systems at the end consumers.

4.2.4. Grid-Serving Energy Storages

A grid-serving energy storage (ES) can act as a source or load in order to decrease or remedy limit violations. Its position and dimensioning depends on the type and extent of the identified violations. For this purpose, the necessary active power is determined and used as the basis for the ES dimensioning.

4.2.5. Measures Not Considered

In Table 5, all identified measures were presented for reasons of complete equipment and technology research. Similarly, all innovative planning measures were taken into account in the grid planning. However, for certain operating equipment and technologies, it was foreseeable at an early stage that they would have no relevance for grid planning. Hence, corresponding planning measures, described below, are no longer taken into account.

Line Voltage Regulator

Line voltage regulators (LVRs) were not used for the LV grid studies here for two basic reasons. First: for meshed LV grids that were also part of the assessment sample aside from grids with radial topology, LVRs have only very limited applications. Second: for radial grids, mostly line overloads could be identified in almost all feeders depending on the scenario and could not be eliminated with an LVR. Consequently, the use of an LVR was in any case more expensive than conventional grid expansion.

Reactive Power Management

Reactive power management was applied for the LV level. As the adjustment of the power factor $\cos(\varphi)$ in the majority of cases made it possible to eliminate voltage band violations at certain nodes, however, equipment overloads were intensified and, in some cases, increased, reactive power management is not considered further.

5. Assessment Model

To compare all planning variants, they must be evaluated using a uniform approach. According to the development in [41], an assessment model is used for this purpose. This model consists of a primary and a secondary assessment model.

In both models, only equipment is taken into account that was newly introduced into the grid compared to the base year 2021. Thus, no maintenance, renewal, and equipment costs for the existing grid are considered in the assessment. The basic structure of the assessment model is shown in Figure 12 and is explained below.

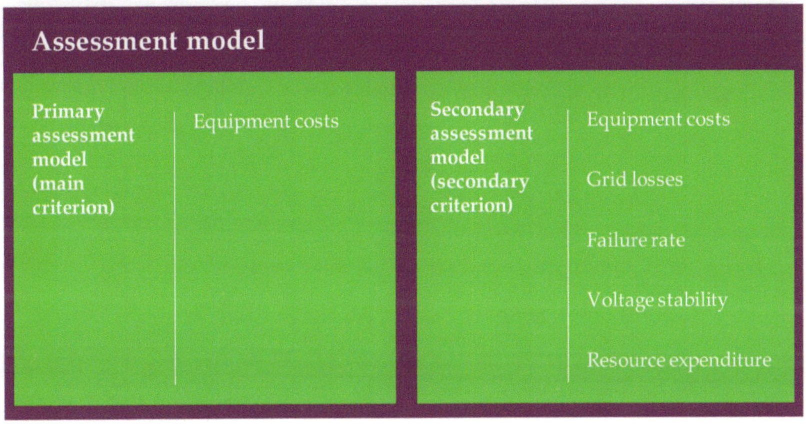

Figure 12. Assessment model with different assessment criteria.

5.1. Primary Assessment Model

The primary assessment model is used to derive the POGs. The costs (Appendix A) for equipment are used as the main criterion, consisting of both capital expenditures (CapEx) and operational expenditures (OpEx) and the residual values for the period between year 2021 and year 2050. The resulting total costs are calculated using the net present value method. Figure 13 illustrates the basic principle of discounting, which makes it possible to objectively compare different planning variants on the basis of net present values despite different investment dates. The equipment costs are taken into account annually and also discounted to the year 2021. For the DSOs, the resulting total costs are in most cases a decisive factor, as they intend to plan the grid in a cost-optimal, and thus economical, way as much as possible.

5.2. Secondary Assessment Model

If DSOs want to focus on additional parameters rather than solely on costs, a secondary assessment model can be used that takes four other criteria into account and applies different weightings.

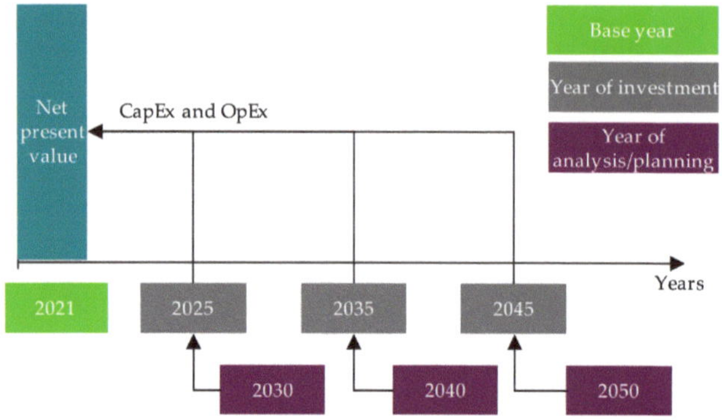

Figure 13. Method of the primary assessment model based on the determination of the net present value.

5.2.1. Secondary Criteria and Weightings

Equipment costs, which represent the investment costs in euros, are again the basic criterion of the secondary assessment model. These are supplemented by the grid losses as an additional secondary criterion and are described as the increase in annual energy in relation to the original grid as percentage. They represent a technical criterion that evaluates the efficiency of the grid and must be borne permanently by the DSO as equipment costs. The attractiveness of a planning variant decreases with increasing grid loss energy in the respective grid.

The third secondary criterion is the failure rate based on a simplified reliability calculation which indirectly estimates the maintenance effort. The failure rate is calculated using Equation (5).

$$H = \sum n_k \cdot H_k \quad (5)$$

where H = failure rate per grid in $1/a$; n_k = number of affected assets in pieces or meters per equipment type; and H_k = failure rate per equipment type.

The failure rate H per grid for the respective quantity structure n_k in pieces or meters per equipment type in the grid is determined with the respective failure rate H_k per equipment type. The mean value of the years 2013 [62], 2014 [63], 2016 [64], and 2018 [65] of stochastic failures for the LV level is used for calculation.

The fourth secondary criterion is defined as voltage stability and represents the robustness of a planning variant to a change in the supply task, where $\Delta U/U_n$ corresponds to the largest voltage drop in percent from the grid interconnection to the upstream grid to the furthest point in the grid without occurring voltage band violations.

The fifth secondary criterion reflects the extent of resource expenditures, i.e., the expenditure for construction activities. Here, the length of a line's route, in which several lines can be located, is determined. This criterion is relevant as, especially in urban LV grids, construction work causes noise and road closures, so that the attractiveness of a planning variant decreases with the length of the necessary construction work.

To be able to carry out different assessments with the five secondary criteria, five different weightings are introduced with the addition of a sixth weighting, whose percentage distribution can be seen in Table 7. In the weighting "Equally weighted", all criteria have the same importance in determining the optimal planning variant. In the weighting "Cost-oriented", the equipment costs play the greatest role in identifying an optimal planning variant. For the weighting "Grid resilience", the failure rate and the voltage stability against unpredictable grid conditions have the greatest importance. In a "Technically oriented" weighting, technical aspects of grid operation are given greater relevance than non-technical criteria. In the "Resource-saving" weighting, a resource-saving measure is rated highly in terms of both grid losses and resource expenditures. The weighting

"Use of primary equipment" focuses on the equipment costs and the resource expenditure, aiming at evaluating each planning variant with regard to the share of required primary equipment.

Table 7. Six different weightings for the secondary criteria.

Secondary Criterion	Equally Weighted	Cost-Oriented	Grid Resilience	Technically Oriented	Resource-Saving	Use of Primary Equipment
Equipment costs	20%	60%	10%	5%	10%	35%
Grid losses	20%	10%	10%	30%	35%	10%
Failure rate	20%	10%	35%	30%	10%	10%
Voltage stability	20%	10%	35%	30%	10%	10%
Resource expenditure	20%	10%	10%	5%	35%	35%

5.2.2. Scoring System

Methodically, the input parameters per planning measure are first determined, as they can be taken from Table 8 for a hypothetical example.

Table 8. Exemplary input parameters of six planning variants (PV) for the secondary assessment model.

Secondary Criterion	PV1	PV2	PV3	PV4	PV5	PV6	"Worst"	"Best"
Equipment costs [A]	12,500 EUR	17,750 EUR	15,000 EUR	35,500 EUR	4500 EUR	42,000 EUR	42,000 EUR	4500 EUR
Grid losses [B]	7.1%	4.5%	2.3%	1.0%	2.8%	0.1%	7.1%	0.1%
Failure rate [C]	0.087 $\frac{1}{a}$	0.080 $\frac{1}{a}$	0.092 $\frac{1}{a}$	0.082 $\frac{1}{a}$	0.094 $\frac{1}{a}$	0.083 $\frac{1}{a}$	0.094 $\frac{1}{a}$	0.080 $\frac{1}{a}$
Voltage stability [D]	4.9%	3.5%	0.8%	3.3%	4.7%	1.2%	4.9%	0.8%
Resource expenditure [E]	550 m	425 m	410 m	260 m	375 m	75 m	550 m	75 m

[A] Results of the primary assessment model. [B] Grid losses in year 2050. [C] Total failure rate in year 2050. [D] Voltage value of the feed-in minus voltage loss value in percent. [E] Resource expenditure (length) in meters on which the results of the primary assessment model are based.

Subsequently, the value range per secondary criterion is identified. Here, either the highest value can correspond to the "worst" result (e.g., equipment costs) or the lowest value to the "best" result (e.g., grid losses). Afterwards, the score ("baselining") per secondary criterion (SC) and planning variant (PV) is determined in the form of a point system according to Equation (6):

$$Score_{SC_n,PV_m} = Top_{Score} - \frac{Value_{SC_n,PV_m}}{Worst\ Value\ SC_n\ over\ all\ PVs} \cdot Top_{Score} \quad (6)$$

The Top_{Score} is defined for all assessments and used for all secondary criteria.

The respective planning variants are then calculated with the weighting factors (WF) to an overall result $Score_\Sigma$ according to Equation (7):

$$Score_\Sigma = WF_{SC_1} \cdot Score_{SC_1} + WF_{SC_2} \cdot Score_{SC_2} + WF_{SC_3} \cdot Score_{SC_3} + WF_{SC_4} \cdot Score_{SC_4} + WF_{SC_5} \cdot Score_{SC_5} \quad (7)$$

Finally, the evaluation of the planning alternatives is carried out depending on the overall result. Here, the planning measure with the highest $Score_\Sigma$ represents the optimal overall result for the respective weighting.

The scoring result of the "baselining", based on the input parameters, can be found in Table 9. On the one hand, it can be seen that the "worst" planning measure per secondary criterion receives no points (e.g., PV1 voltage stability). Furthermore, a value close to the optimum shows that the respective score is approaching the top score and the top score is almost reached (e.g., PV1 grid losses). The values of the other planning measures are calculated according to the equation.

Table 9. Exemplary "baselining" for six planning variants (PV) in the secondary assessment model with a defined Top_{Score} of 6 (without taking weightings into account).

Secondary Criterion	PV1	PV2	PV3	PV4	PV5	PV6	"Max"
Equipment costs	4.21	3.46	3.86	0.93	5.36	0.00	5.36
Grid losses	0.00	2.20	4.06	5.15	3.63	5.92	5.92
Failure rate	0.45	0.89	0.13	0.80	0.00	0.70	0.89
Voltage stability	0.00	1.71	5.08	2.02	0.24	3.86	5.08
Resource expenditure	0.00	1.36	1.53	3.16	1.91	5.18	5.18
Sum	4.66	9.63	14.65	12.07	11.14	15.66	15.66

Figure 14 shows the respective total scores ($Score_{\Sigma}$) of the planning measures per weighting. The example of planning variant PV5 shows that it performs best in the weighting in "Cost-oriented" due to the high score for equipment costs, among other things. Within the weighting "Resource saving", on the other hand, PV5 has the most points for equipment costs, but PV6 has significantly more points for grid losses and resource expenditures, which is why PV6 can be identified here as the optimal planning variant.

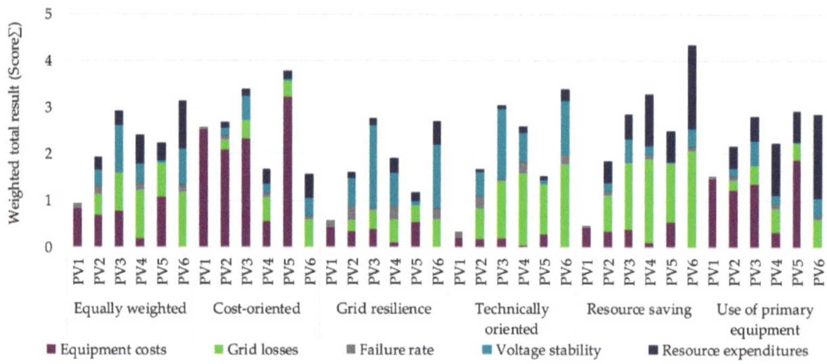

Figure 14. Exemplary overall results of different weightings of the secondary assessment model for six planning variants (PV).

6. New Planning and Operation Guidelines for Urban Low-Voltage Grids

For the LV level, there is generally a reactive and short-term need for planning measures in the case of acute problems, which can largely be defined in standardized POGs, as this involves a large number of similar planning and construction measures. The challenge here is usually the variety of necessary measures over time and the reconciliation with the equipment available at the DSO and its subcontractors. These measures are driven, especially in the area of CI and HPs, by the corresponding political subsidies and are regulated by the use of public transport routes for local general supply via route utilization contracts within the framework of concession awards.

6.1. Load Development

In order to derive POGs, grids are first required to which the methodology described in Section 2 can be applied. For this purpose, 20 representative LV grids (Appendix B) have been selected from 4200 grids on the basis of [66], and are shown in Figure 15.

Based on the scenarios and apportionment methodology, Figure 16 shows the respective load development for all 20 LV grids from the perspective of the DTs. The currently installed transformer capacity is also indicated (horizontal black dotted lines). The grey bars represent the conventional household and commercial loads. Building on this, the private charging power is shown in purple and the public charging power in turquoise. The continuous green bar then shows the power for the 3.0-kW-HP variant. The bar on top of this (green left hatched line) represents the power increase that must be considered if the 6.5-kW-HP variant is adopted. As for the 9.0-kW-HP variant (green right hatched line), the same applies to the last green hatched bar accordingly. It can be seen that more suburban grids with a correspondingly higher proportion of HPs are reaching their limits from the perspective of the DTs (e.g., G11). In contrast, more inner-city grids show a higher robustness against the integration of new loads (e.g., G01), which are comparatively fewer than in suburban grids. It can be seen that a significantly higher utilization of the transformers by new loads can be assumed in the future, a factor that must be taken into account in the grid planning. As an alternative calculation of SFs for calculation method C4, the last blue dotted bar shows the additional power results when private and public CI is determined with calculation method 4 and on the assumption of a simultaneity factor of 1 separate to each other (C4s).

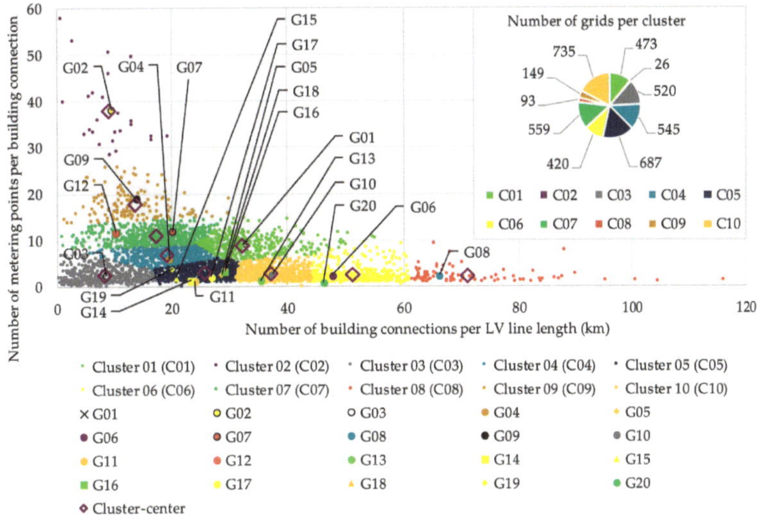

Figure 15. Results of a clustering for the low-voltage level and grid selection based on [66].

6.2. Technical-Economic Evaluation

Applying the relevant conventional and innovative planning variants for the LV grids described in Sections 4.1 and 4.2 results in the consolidated line measures shown in Figure 17 across all analyzed planning variants for each scenario, HP variant, and year. It can be seen that the line measures decrease from the VRS through the RDT to the ES. There are also differences within the DLM variants. As for the resulting costs, Figure 18 shows that conventional grid expansion is the most economical planning variant next to the VRS. One should keep in mind that, if full equipment is necessary for the DLM due to shorter useful lives of the MICT components, renewals are necessary at least once during the lifetime of conventional equipment of these same components. The use of RDT is also partly more expensive as it primarily resolves voltage band violations, so that further measures are necessary in the event of line overloads. ES is the most expensive planning measure. As an alternative cost representation, Figure 19 shows the respective saving potential compared to conventional planning representing the 0-%-line. It can be seen that the DLM-V1 with necessary full equipment is less expensive than the conventional grid expansion in approximately 23% of all planning variants. ES is also cheaper than conventional expansion, but only in very few planning variants.

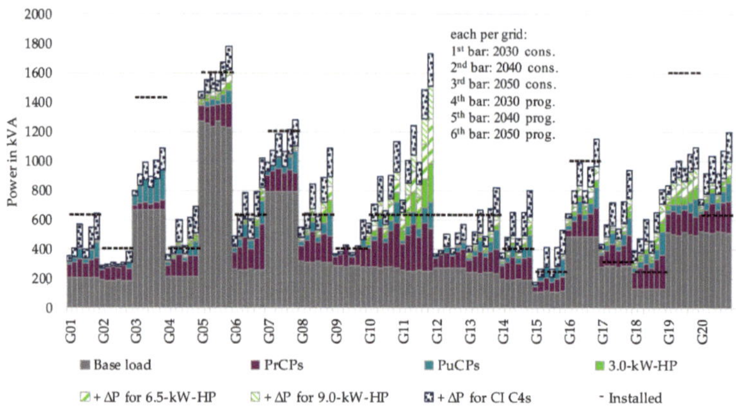

Figure 16. Load development and installed capacity from the perspective of local distribution transformers for 20 low-voltage grids.

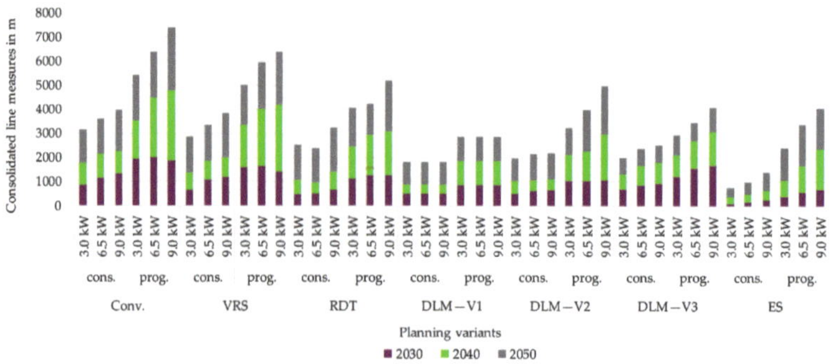

Figure 17. Necessary line measures consolidated across all conventional and innovative planning variants from the perspective of the feeders for 18 low-voltage grids.

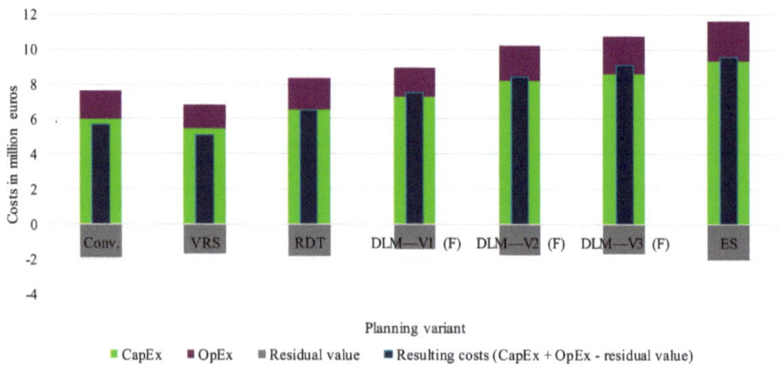

Figure 18. Resulting costs (CapEx + OpEx—residual values) consolidated across all conventional and innovative planning variants for 18 low-voltage grids.

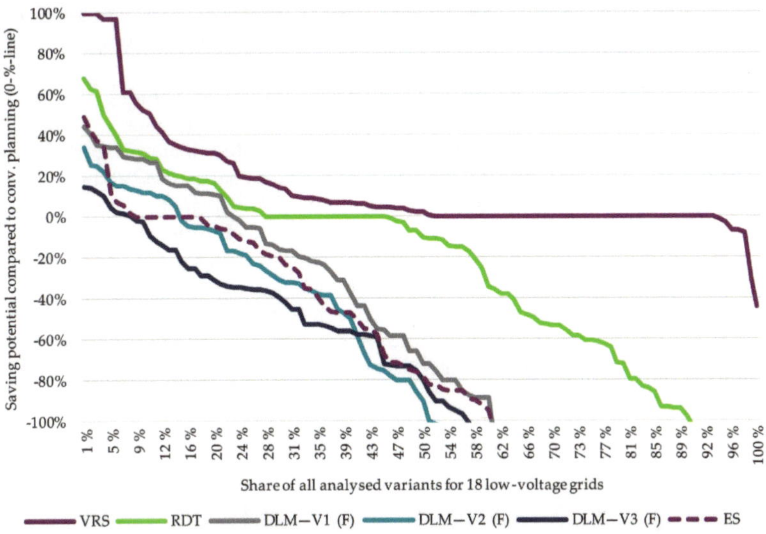

Figure 19. Savings potential related to the conventional planning variant as a reference variant consolidated across all innovative planning variants for 18 low-voltage grids, note: not shown further than −100%.

6.3. *Derivation of New Planning and Operation Guidelines*

Based on the findings of the individual grid planning variants as well as the corresponding technical-economic assessment from Section 5, seven new POGs for urban LV grids are defined (in italics) in Sections 6.4–6.10 and explained.

The first guideline provides power value assumptions for different (new) loads, which can be used for grid planning if no own reliable findings are available. Following these assumptions, the relevant OP for the design of urban LV grids is identified. Afterwards, recommendations for standard equipment are made for lines and DTs. Subsequently, innovative measures, primarily voltage regulating measures, are discussed with regard to their effects on grid planning. Based on this, the different DLM variants are thoroughly investigated. The last guideline then discusses the grid structures and where there is a need for grid reinforcement. Thus, all strategically important topics are dealt with within the framework of the principles for the LV level, which can be supplemented by DSO-specific POGs.

Note: The LV grids G02 and G12 can integrate the new loads without limit violations. so that some evaluations include only 18 LV grids instead of 20. Likewise, G10 has only one transformer overload, which is why only 17 instead of 20 grids are considered for the secondary assessment model in Section 6.12.

6.4. *First Planning and Operation Guideline*

For the dimensioning of local distribution transformers, an average effective power for private charging points per building connection of $P_{PrCP,DT,BC}$ = *[1.7; 2.9] kW or alternatively* $P_{PrCP,DT,MP}$ = *[0.4; 0.8] kW per metering point plus* $P_{conv,DT,OTFH}$ = *2.0 kW for conventional loads of one- and two-family detached houses or* $P_{conv,DT,MFH}$ = *1.0 kW per metering point for multi-family houses in the grid is recommended.*

For the dimensioning of the low-voltage feeders, $P_{PrCP,Feeder,BC}$ = *[8.2; 12.0] kW or alternatively* $P_{PrCP,Feeder,MP}$ = *[3.2; 4.3] kW per metering point plus* $P_{conv,Feeder,OTFH}$ = *2.7 kW for conventional loads of one- and two-family detached houses or* $P_{conv,Feeder,MFH}$ = *1.3 kW per metering point for multi-family houses are recommended.*

If additional electric heat pumps are to be considered, it is recommended to add an additional power of P_{HP} = *[3; 9] kW per heat pump, which is equally valid for local distribution transformers and low-voltage feeders* [41].

For the future development and planning of urban LV grids, it is important to plan with appropriate power value assumptions. The first POG therefore aims to provide the grid planner with power values that can be used for conventional loads, PrCPs and HPs. Taking into account the scenarios used and the apportionment methodology and SF calculation carried out (further explanations in [41]), the power values for PrCPs given in Figures 20–25 are developed for different planning perspectives (DT, feeder, building connection, metering point, and CP) for 20 LV grids. The public CI is not shown here, as it can be assumed that the DSO can plan them much better than the private CI. Due to the applied SF calculation method, it should be noted that public CI is indirectly taken into account. For the predominant share of the grids, the presented power value assumptions represent applicable values that reflect a balanced distribution of charging services.

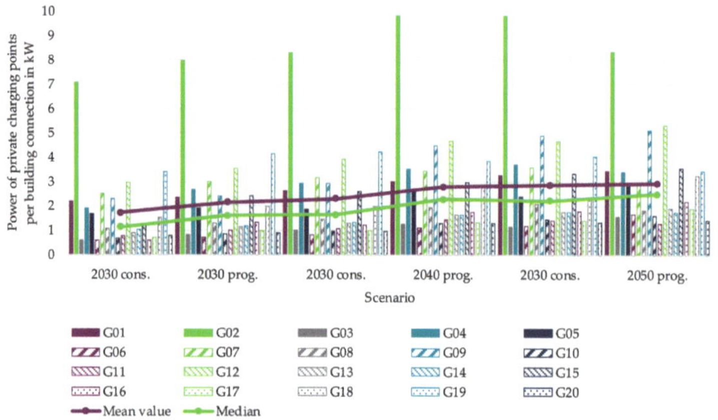

Figure 20. Average effective charging power of private charging points per building connection from the perspective of local distribution transformers for 20 low-voltage grids based on [41].

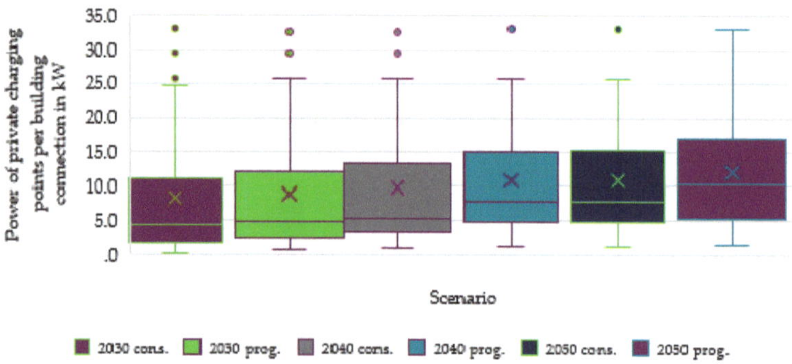

Figure 21. Average effective charging capacity of private charging points per building connection from the perspective of feeders (not including feeders without private charging infrastructure) for 20 low-voltage grids based on [41].

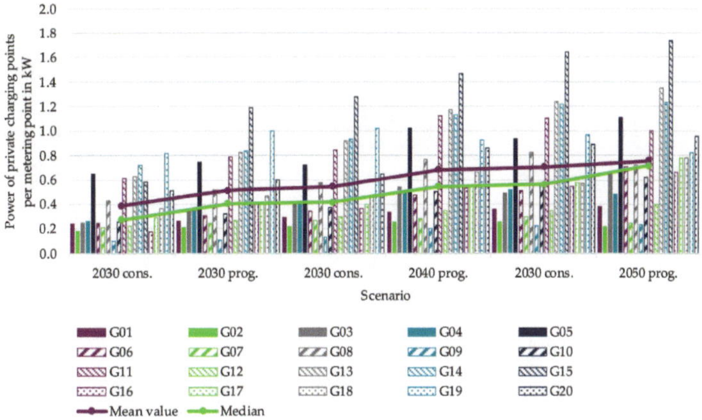

Figure 22. Average effective charging power of private charging points per metering point from the perspective of local distribution transformers for 20 low-voltage grids based on [41].

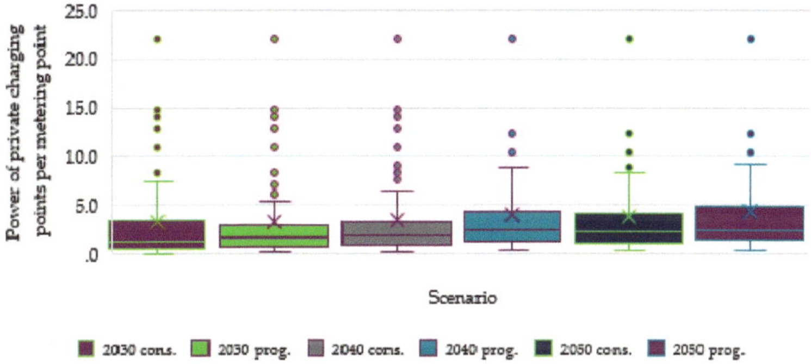

Figure 23. Average effective charging capacity of private charging points per metering point from the perspective of feeders (not including feeders without private charging infrastructure) for 20 low-voltage grids based on [41].

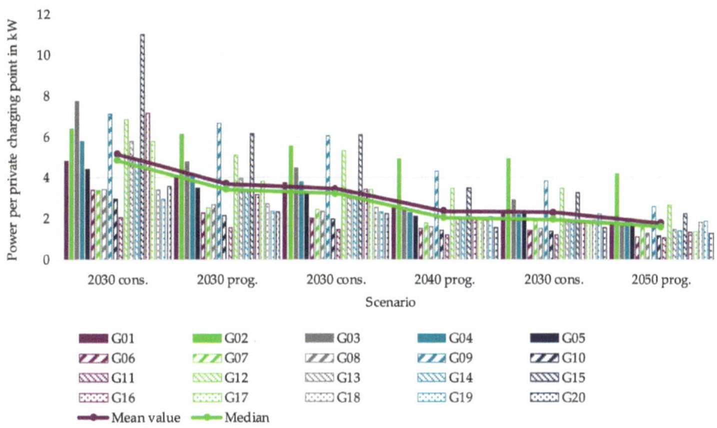

Figure 24. Average effective charging power per private charging point from the perspective of local distribution transformers for 20 low-voltage grids based on [41].

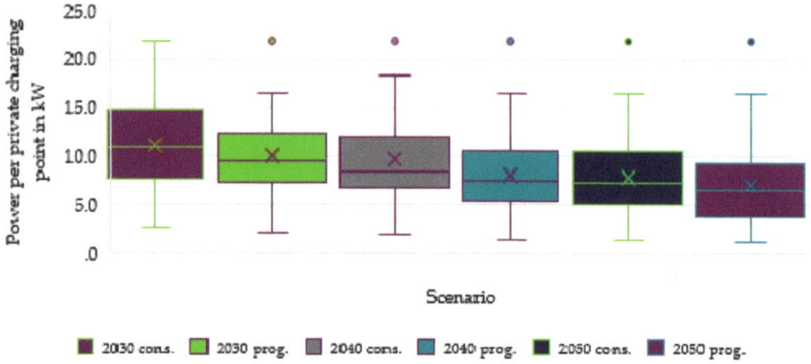

Figure 25. Average effective charging capacity per private charging point from the perspective of feeders (not including feeders without private charging infrastructure) for 20 low-voltage grids based on [41].

As the DSOs do not always have all grid parameters for each grid, the values necessary for strategic grid planning is displayed for different grid parameters. In particular, in analogy to Section 3.3 regarding the dimensioning of DTs and feeders, different SFs in the area of the CI are applied, which accordingly result in different power value assumptions. Furthermore, the power value assumptions currently represent the greatest uncertainty as, to date, only limited reliable knowledge is available for the grid-wide load behavior. Hence, the following six figures show different value assumptions for private CI depending on the considered grid parameter. Figure 20 shows both the detailed results and, among other values, the average values for private CI from the perspective of the DT based on the building connections per grid. Figure 21 shows corresponding values to Figure 20, however from the perspective of the feeders. Figures 22 and 23 represent, analog to Figures 20 and 21, respectively, the values for private CI but on the basis of the metering points, and Figures 24 and 25 per CP. These evaluations form the basis of the power value assumptions for private CI in Table 10.

Figures 26–28 show evaluations of conventional loads. Although they represent the smallest share of future impacts according to Figure 16, these must also be differentiated for different building types and taken into account in grid planning.

If HPs are to be taken into account, an additional value of 3 to 9 kW must be applied in the grid planning, which hardly differs between the planning perspectives DTs and feeders due to the SF shown in Figure 7 and can therefore be used for both planning perspectives.

An overview and summary of all power value assumptions can be seen in Table 10.

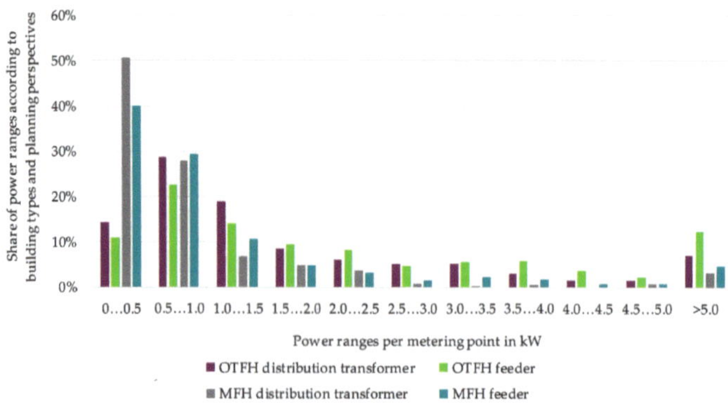

Figure 26. Performance ranges of conventional loads per building type and planning perspective based on [41].

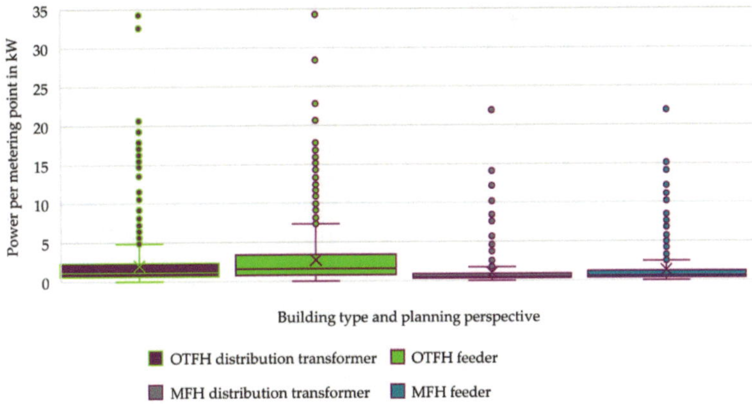

Figure 27. Distribution of conventional loads per building type and planning perspective based on [41].

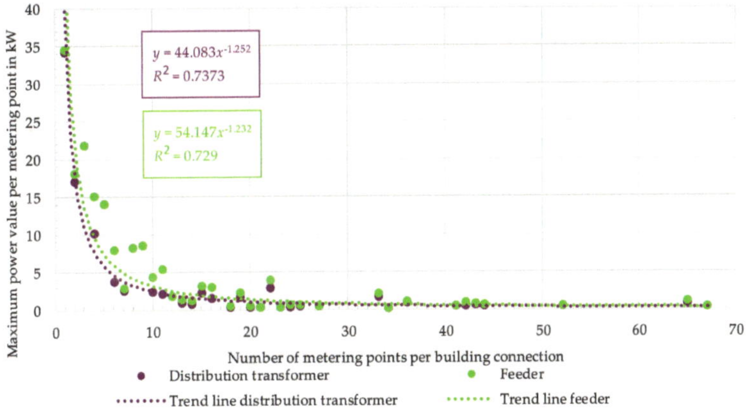

Figure 28. Maximum simultaneous conventional power consumption per household metering point based on [41].

Table 10. Power value assumptions for different loads in the low-voltage level.

Load	Local Distribution Transformer			Feeder		
	kW/BC [5]	kW/MP [6]	kW/PrCP *	kW/BC [5]	kW/MP [6]	kW/PrCP *
PrCP [1,A,B]	[1.7 [D]; 2.9 [E]]	[0.4 [D]; 0.8 [E]]	[5.1 [D]; 1.8 [E]]	[8.2 [D]; 12.0 [E]]	[3.2 [D]; 4.3 [E]]	[11.1 [D]; 7.1 [E]]
OTFH [2,B,C]	-	2.0	-	-	2.7	-
MFH [3,B,C]	-	1.0	-	-	1.3	-
HP [4]	[3; 9] (no differentiation for HPs, as SF [7] is almost 1)					

[1] Private charging point. [2] One- and two-family house (per metering point). [3] Multi-family house (per metering point). [4] Heat pump. [5] Building connection. [6] Metering point. [7] Simultaneity factor. [A] 3.7 kW, 11 kW, and 22 kW. [B] Arithmetic mean values per building connection, metering point, or charging points in the grid or feeder. [C] There is no distinction for electric water heating. [D] Conservative scenario in 2030. [E] Progressive scenario in 2050. * Note: The charging capacity per PrCP decreases over the years as the number of PrCPs increases. Building connections and metering points of OTFH and MFH, however, remain constant.

6.5. Second Planning and Operation Guideline

For urban low-voltage grids, the operating point "peak load" is the relevant operating point for grid dimensioning.

To determine the relevant OP for urban LV grids, it is necessary to investigate the extent to which limit value violations are caused by loads and feed-ins in both analyzed OPs. For this purpose, Figure 29 shows an evaluation for the conservative and progressive scenario with 9.0 kW HPs. Equipment overloads are not shown here, as equipment overloads do not occur in the OP-PG, neither from the perspective of the DT, nor from the perspective of the feeders. With regard to the voltage band being respected according to [11], it is clear that the OP-PL is more important for grid planning. Voltage band violations occur due to feed-ins (progressive scenario) in only four suburban grids. Voltage band violations, however, occur in 17 of 20 grids due to new loads, some of which are severe (progressive scenario). In addition, Figure 30 shows the year in which limit violations occur for the first time. Figures 29 and 30 should therefore be considered in combination, as they also show that the OP-PL is relevant for the planning of urban LV grids.

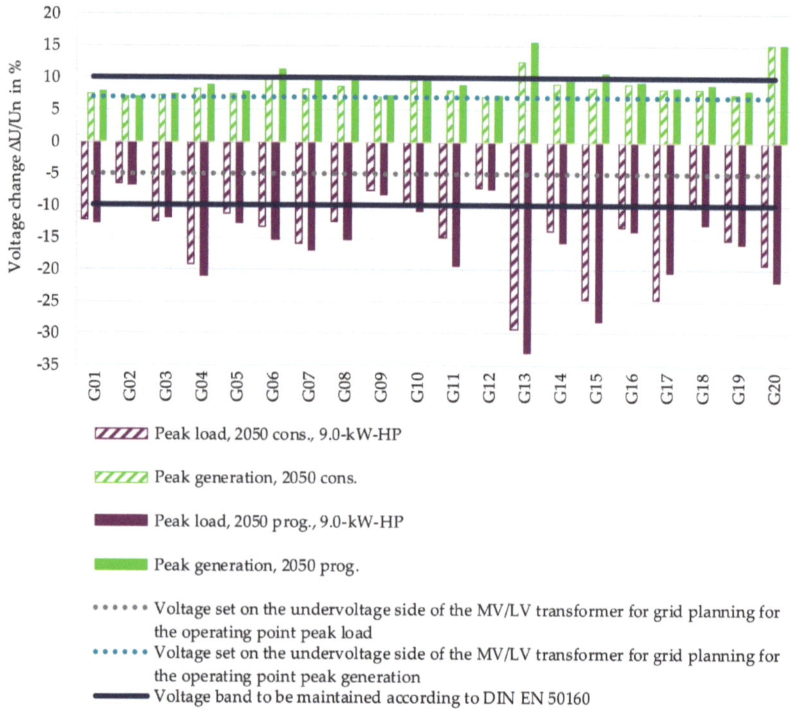

Figure 29. Maximum voltage change for the operating points "peak generation" and "peak load" for 20 low-voltage grids at the year 2050 in the progressive scenario with 9.0 kW heat pumps for the conservative (cons.) and progressive (prog.) scenario.

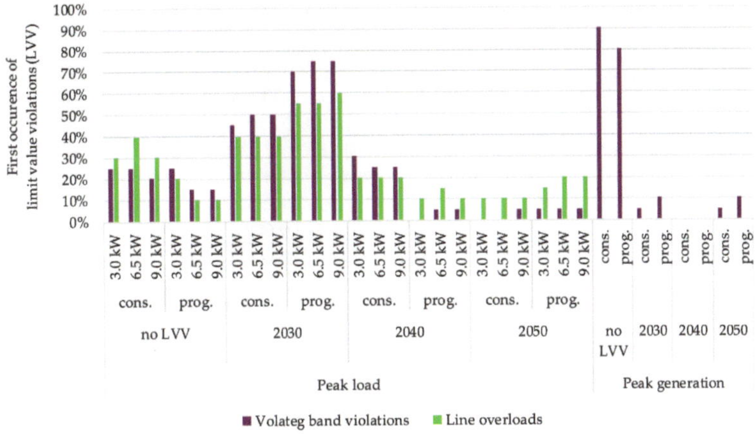

Figure 30. First occurrence of limit value violations for low-voltage lines for the operating points "peak generation" and "peak load" consolidated for 20 low-voltage grids based on [41].

6.6. Third Planning and Operation Guideline

It is recommended to supplement the current standard line cross-section q = 150 mm^2 (Al) with a second new standard line cross-section q = 240 mm^2 (Al) for urban low-voltage grids.

Standard line cross-sections are another essential component of POGs. They are kept in stock for short-term line measures, so that a restriction of line cross-sections is intended here in order to keep storage costs at a minimum. Figure 31 therefore shows an evaluation that essentially recommends the cross-sections q = 150 mm^2 (Aluminum, short: Al) and q = 240 mm^2 (Al). Among the different line measures given in Section 4.1, the cross-section q = 150 mm^2 (Al) is sufficient for line reinforcement in approximately 83% of all line measures. On the other hand, in the case of a line replacement, a cross-section of q = 240 mm^2 (Al) is necessary in a quarter of all the cases, which also covers q = 185 mm^2 (Al). In the case of a "forced" replacement, i.e., if no parallel q = 150 mm^2 (Al) line is installed to redistribute the loads, but rather is replaced by a higher dimensioned line, a line cross-section q = 240 mm^2 (Al) is already necessary in one third of all the cases.

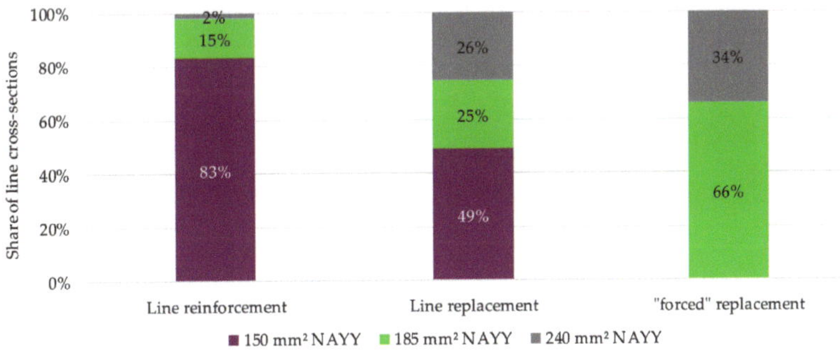

Figure 31. Shares of line measures by cross-section for 18 consolidated grids.

It is therefore recommended to introduce a further cross-section of q = 240 mm^2 (Al) in addition to the current standard line cross-section of q = 150 mm^2 (Al), which equally covers q = 185 mm^2 (Al).

6.7. Fourth Planning and Operation Guideline

Regarding standard transformer classes, it is recommended to increase the respective existing dimension by one power class in relation to the rated power.

In addition to lines, standard DTs, which are also kept in stock in order to react quickly as a DSO, must also be defined. Figure 32 shows the necessary power classes of DTs that result from the respective planning variants. It can be seen that the standard size of 630 kVA—used by most DSOs—is still significantly important to be kept in stock. However, it then replaces lower power classes such as 400 kVA and below. Furthermore, it can be seen that 800 kVA DTs and 1000 kVA DTs will also become necessary in some grids, despite the application of the DLM described in Section 4.2.3. It is therefore recommended to define a standard size of 800 kVA in addition to the current standard size of 630 kVA, which is henceforth to be defined as the smallest size. Alternatively, it can be examined whether two 630 kVA DTs can be used instead of one 1000 kVA DT if an 800 kVA DT is not sufficient.

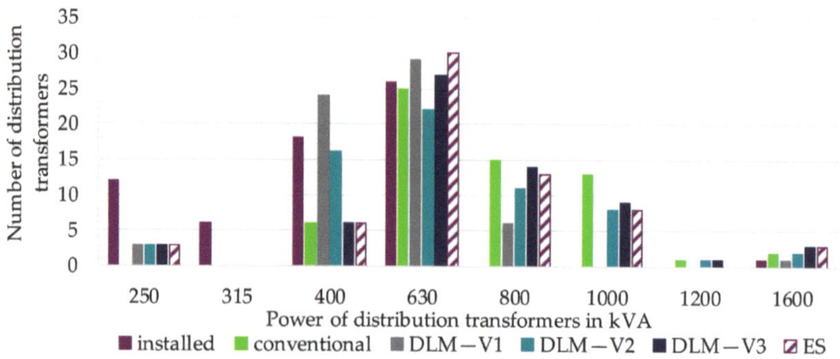

Figure 32. Necessary increases in the power classes of local distribution transformers for 18 consolidated grids in the presence of limit violations.

6.8. Fifth Planning and Operation Guideline

Innovative voltage regulation measures do not offer a long-term advantage for the elimination of limit violations, as in urban low-voltage grids, equipment overloads are the dimensioning factor for reinforcement measures [41].

Figure 30 has shown that in the years 2030 and 2040, voltage band violations can also occur without the presence of equipment overloads. These can be remedied almost free of charge by applying the tap changer. Only negligible costs are incurred by the operating personnel who have to adjust the tap on site. In the case of voltage band violations and equipment overloads, the same remedy applies. If voltage band violations persist, they can almost always be remedied in the same way equipment overloads are remedied.

As, according to the second POG, the OP-PL is relevant to the planning and it is shown in Figure 30 that line overloads are predominant, these usually cannot be remedied with innovative voltage regulating measures.

In addition to this fundamental observation, the RDT must be considered for another reason. According to [17,38,48], this is recommended when the OP-PG also becomes relevant in suburban grids.

6.9. Sixth Planning and Operation Guideline

Grid-serving load management can defer and partly avoid conventional grid expansion. If the acquisition of measurement, information and communication technology (MICT) is fully necessary, conventional planning measures are usually more cost-effective and are therefore recommended. However, if the MICT is already available or its deployment is independently planned and can be used for load management, load management is always significantly more cost-efficient and is therefore recommended [41].

Figure 17 already shows that, with a DLM, considerable line measures can be saved. On the other hand, it was evident in Figures 18 and 19 that a DLM is the economical solution in only very few cases when the MICT has to be built, and that conventional grid expansion is still necessary to a small extent in most cases. Therefore, a sensitivity analysis with further cost calculations can be taken from Figures 33 and 34 in relation to Section 4.2.3. This clearly shows that if existing MICT is used, or if it is already planned and can be used, a DLM is considerably more cost-effective than conventional grid expansion. In the case that either reduced measurement sensors or only a RTU is to be considered, DLM does become more economical in some variants, but at maximum in 49% of the variants. This proportion is still too low to qualify for a new recommendation. However, if the necessary MICT equipment is already available in the respective grids, DLM can be recommended, as it is cheaper in at least 88% of the planning variants and even in approximately 50% of all variants, in some cases significantly more than 50% in DLM-V2.

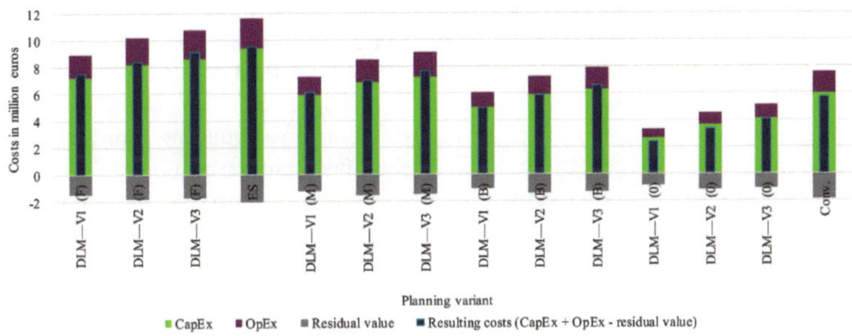

Figure 33. Resulting costs (CapEx + OpEx—residual values) consolidated over all analyzed variants for 18 low-voltage grids in different variants of dynamic load management based on [41].

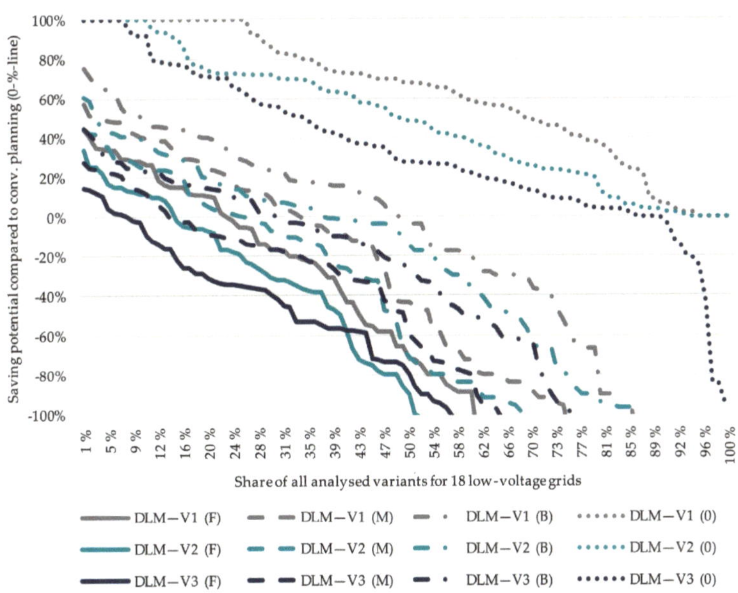

Figure 34. Savings potential of different variants of dynamic load management related to conventional reference variant based on the conventional reference variant for 18 low-voltage grids based on [41]; note: not shown further from −100%.

In principle, it should be noted that the regulation of private CI has a higher effectiveness, as this outweighs public CI in terms of numbers in the grids. The DLM is also more successful if in early years, such as in the year 2030, the base load is not yet so high due to new loads and thus limit violations are avoided.

Furthermore, it is recommended to include HPs in DLM in addition to the private CI, as the grid load can be temporarily reduced in any peak load time window through the intelligent use of blocking periods.

6.10. Seventh Planning and Operation Guideline

For inner-city low-voltage grids in which no building renovations or new constructions are planned and which have an average load density of more than 10 metering points per building connection, it can be assumed that no significant need for reinforcement of the low-voltage lines is required [41].

To interpret the results in Figure 35, it should first be noted that there were no underground garages or larger garage yards in any of the analyzed 20 LV grids. In particular, this leads to the fact that no limit violations were identified in G02 from cluster 2 and G12 from cluster 7. The grids have a load density of approximately 38 and 12 metering points per building connection, respectively. This is also the case in G09 from cluster 9, with approximately 19 metering points per building connection, where a negligible reinforcement due to public CI is identified. Similarly, in grid G07 from cluster 7, with approximately 12 metering points per building connection, the need for expansion is significantly lower compared to suburban areas with less than 10 metering points per building connection on average. Accordingly, provided that no building renovations or new buildings are planned, and no underground parking garages are available, it can be assumed that no significant expansion is required with an average load density of approximately 10 metering points or more per building connection.

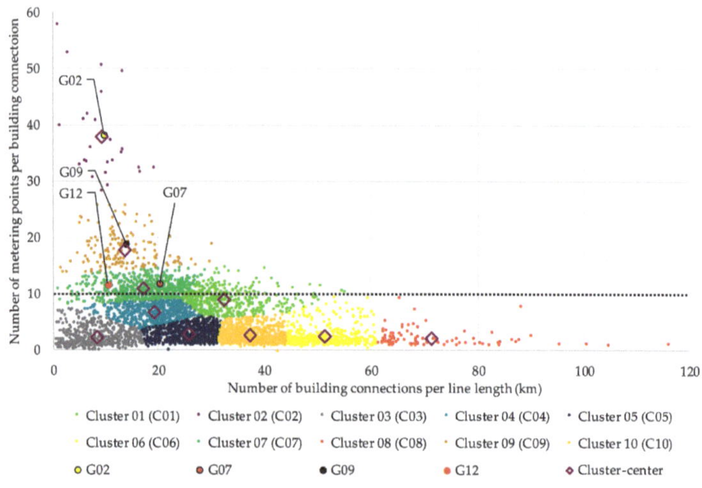

Figure 35. Analyzed low-voltage grids with more than 10 metering points per building connection in the cluster evaluation based on [41].

6.11. Overarching New Planning and Operation Guidelines over All Voltage Levels

From the new POGs for LV grids derived in Sections 6.4–6.10, the following three POGs in Sections 6.11.1–6.11.3 can be derived from the LV level, which basically apply to all voltage levels.

6.11.1. First Overarching POG concerning All Voltage Levels

In principle, a cross-voltage level consideration of HV, MV and LV grids should be aimed for.

The first POG concerning all voltage levels addresses the importance of an overarching and target-oriented consideration of the three voltage levels relevant for distribution grids. As the LV and MV grids, in particular, are coupled via still largely conventional DTs, a voltage increase in the MV level results in a voltage increase in the LV level. Thus, voltage band violations can be solved simultaneously by adjusting the permissible voltage band in both voltage levels. Vice versa, a DLM in the LV level results in a load reduction in the MV level and thus also eliminates not only equipment overloads in the LV level but also those in the MV level.

6.11.2. Second Overarching POG concerning All Voltage Levels

Equipment overloads are the driving factor behind the need for expansion of urban distribution grids, whereas voltage band violations are not.

The second POG concerning all voltage levels, in combination with the second POG from the LV level, addresses the fact that the new loads such as CI and HPs, which are mostly connected in the LV level, also have a considerable impact on the two upstream voltage levels in the distribution grid. According to Figure 29, Figure 30, and Figure 32, these power increases primarily result in equipment overloads. Voltage band violations, on the other hand, can be remedied almost free of charge in most cases, especially via tap positions in DTs, so that voltage band violations play a subordinate role.

6.11.3. Third Overarching POG concerning All Voltage Levels

Innovative technologies, such as grid-serving load management or energy storages, are the most economical solution only in some parts of the grids. In the remaining grids, conventional expansion is recommended.

The third POG concerning all voltage levels is to be understood in combination with the second POG over all voltage levels and the sixth POG in the LV level. Although the measures primarily driven by equipment overloads can in principle be remedied by a grid-serving DLM, conventional measures are still required in many cases in addition to the DLM. If the MICT then has to be installed, a DLM is usually significantly more expensive. However, it becomes particularly interesting where the necessary MICT can already be used or is already in the planning stage. Regardless of this purely economic consideration, a DLM can make sense wherever there are time or resource bottlenecks for the grid expansion, grid conversion or renewal requirements.

6.12. Decision Path for Strategic Grid Planning

Based on the new POGs for the LV level, a decision path (see Figure 36) is derived that takes into account both conventional and innovative planning measures. Based on the underlying limit value violations (voltage band violations or equipment overloads) for LV grids, decision-making for suitable planning measures is thus facilitated. It should be noted, however, that this is a highly simplified flowchart that cannot replace the POGs described in detail and must be supplemented by the specific requirements of the respective DSO.

6.13. Results from the Perspective of the Secondary Assessment Model

With regard to the secondary assessment model presented in Section 5.2, the results consolidated over 17 LV grids can be seen in Figures 37 and 38 for the progressive scenario with 9.0-kW-HPs and year 2050.

Figure 37 shows, in analogy to the primary assessment model, that conventional planning and VRS perform best when only the secondary criterion "Equipment costs" is used as a basis for evaluation. They both have a predominantly high rank between 5 and 7, whereas rank 7 is the best and rank 1 is the worst in terms of the underlying assessment. VRS and RDT score best regarding the secondary criterion "voltage stability" and DLM on the secondary criterion "Resource expenditure".

If the evaluations of the secondary criteria are combined in the respective weightings, Figure 38 shows a different picture depending on the weighting to be considered. If, for example, the weighting "Cost-oriented" is left out, conventional planning no longer performs as well, as other secondary criteria are assigned a higher relevance. Thus, in all other weightings, the RDT and DLM-V1 measures are in the lead. In summary, it can be said that when grid-related criteria are given a higher relevance, other planning measures come out ahead of conventional planning.

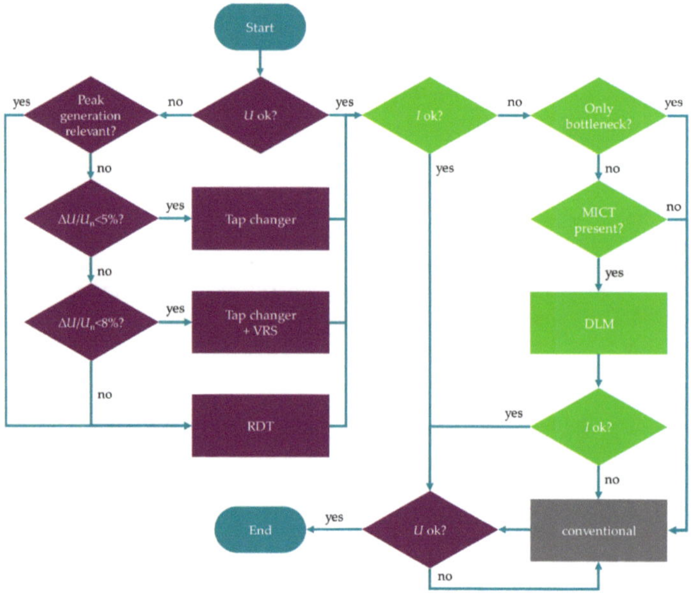

Figure 36. Decision path for strategic grid planning of urban low-voltage grids, with: "ok" = compliance with the specified limit values, bottleneck = short line section (length and/or small cross-section), $\Delta U/U_n$ = additional voltage difference beyond the lower voltage band according to [11].

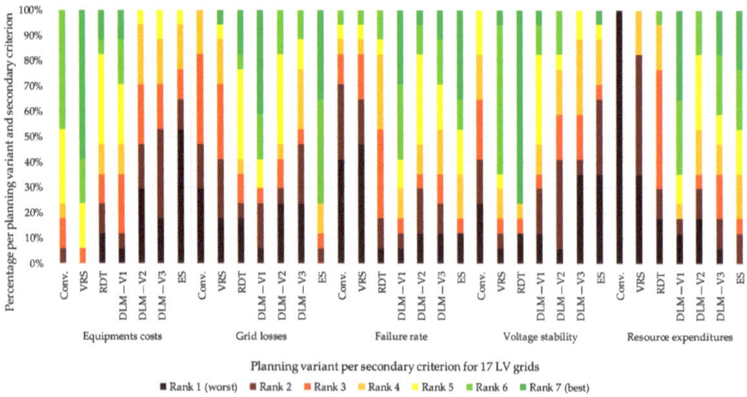

Figure 37. Percentage distribution of the ranking consolidated for 17 low-voltage grids in relation to the individual secondary criteria for the progressive scenario with 9.0-kW-HPs.

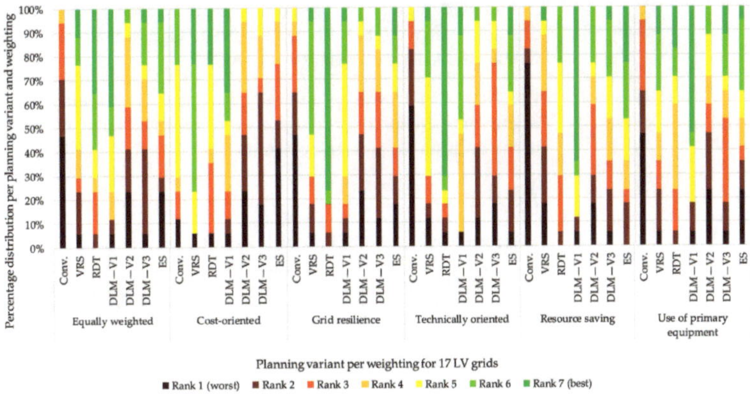

Figure 38. Percentage distribution of the ranking consolidated for 17 low-voltage grids in relation to the different weightings for the progressive scenario with 9.0-kW-HPs.

7. Discussion

Finally, it should be noted that urban LV grids have a certain capability for integrating new loads, depending on the area structure, so that the load growth driven by CI and HPs does not lead to limit violations in every case. The POGs derived for the LV level therefore enable DSOs to develop, standardize, or supplement DSO-specific POGs, taking into account conventional and innovative planning and operating variants for a cost-optimal reinforcement of the grids.

Furthermore, it should be noted that, in many cases and in the near future, voltage band violations can be partially or completely eliminated by conventional voltage regulation measures, such as tap changers, without additional investment costs. On the other hand, equipment overloads often have to be remedied by conventional means if measurement, information, and communication technology is not yet available for the use of load management, even though load management can defer and partially prevent planning measures in many variants.

If costs are not to be the only criterion to be considered, the secondary assessment model has shown that other innovative planning measures can also be beneficial in contrast to conventional planning measures. In any case, the results show that target grid planning for urban LV grids and the revision or expansion of the company's own planning and operation guidelines are urgently needed.

Author Contributions: Conceptualization, P.W., S.A.A., J.M. and B.G.; methodology, P.W. and S.A.A.; validation, M.Z. and A.S.; formal analysis, J.M.; resources, M.Z.; data curation, P.W. and S.A.A.; writing—original draft preparation, P.W.; writing—review and editing, S.A.A. and J.M.; supervision, M.Z.; project administration, M.Z.; and funding acquisition, M.Z. All authors have read and agreed to the published version of the manuscript.

Funding: This research is a result of the project "PuBStadt" which was funded by the German Federal Ministry for Economic Affairs and Energy following the decision of the "German Bundestag", "grant number 0350038A".

Supported by:

Federal Ministry
for Economic Affairs
and Energy

on the basis of a decision
by the German Bundestag

Acknowledgments: The used data were provided by the six German DSOs Stromnetz Berlin GmbH, Stuttgart Netze GmbH, Rheinische NETZGesellschaft mbH, SachsenNetze GmbH, Enercity Netz GmbH and Erlangener Stadtwerke AG.

Conflicts of Interest: The authors declare no conflict of interest. The funders had no role in the design of the study; in the collection, analyses, or interpretation of data; in the writing of the manuscript, or in the decision to publish the results.

Appendix A

Table A1. Cost assumptions for low-voltage equipment based on [41].

Cost Position	Parameter	Value	Unit
NAYY cables	Service life	45	a
	Operating costs	2.5	% [1]
	Cost increase	0.5	%/a
NAYY 150 single [2]	Cables and installation	150	Euro/m
NAYY 150 parallel [3]	Cables and installation	+20	Euro/m
NAYY 185 single [2]	Cables and installation	175	Euro/m
NAYY 185 parallel [3]	Cables and installation	+40	Euro/m
NAYY 240 single [2]	Cables and installation	200	Euro/m
NAYY 240 parallel [3]	Cables and installation	+60	Euro/m
Distribution transformer (DT)	Service life	40	a
	Operating costs	2.5	% [1]
1.630-kVA-DT	Costs	10,000	Euro/Piece
1.800-kVA-DT	Costs	12,500	Euro/Piece
1000-kVA-DT	Costs	15,000	Euro/Piece
1200-kVA-DT	Costs	18,500	Euro/Piece
1600-kVA-DT	Costs	25,000	Euro/Piece
Regulated DT (RDT)	Service life	40	a
	Operating costs	2.5	% [1]
1.630-kVA-RDT	Costs	21,500	Euro/Piece
1.800-kVA-RDT	Costs	24,200	Euro/Piece
1000-kVA-RDT	Costs	27,500	Euro/Piece
1200-kVA-RDT	Costs	30,000	Euro/Piece
1600-kVA-RDT	Costs	36,000	Euro/Piece
Remote terminal unit	Basic amount	9500	Euro/Piece
	Service life	15	a
	Operating costs	2.5	% [1]
Measuring sensors (up to 4)	Service life	15	a
	Operating costs	3500	Euro/Piece
Line cabinet distribution	Service life	40	a
	Operating costs	2.5	% [1]
	Installation	5000	Euro/Piece
Line voltage regulator	Service life	30	a
	Operating costs	2.5	% [1]
	Installation costs	1000	Euro/Piece
	Capacity costs	78	Euro/kVA
Grid-serving energy storage	Service life	16	a
	Operating costs	2.5	% [1]
	Basic amount	16,500	Euro/Piece
	Capacity costs (2 h)	550	Euro/kW

[1] Percent of capital expenditures per anno. [2] These lump-sum cost figures include, among other things, costs for construction work, joints, etc. [3] For the cost calculation, with parallel cables of different cross-sections, the larger cross-section is used as the basis for the cost and the smaller cross-section is used for the additional cost of the parallel cable (cables and installation).

Appendix B

Table A2. Grid structure parameters of the low-voltage grids (values of the scenarios: first line in each case conservative, second line in each case progressive, first column in each case 2030, second column in each case 2040, third column in each case 2050) based on [41].

Grid Cluster	Installed Transformer Capacity in kVA	Power Line Length in m	Building Connections in Pieces	Metering Points in Pieces	Feeders/Meshes in Pieces	Charging Points (Private + Public) in Pieces			Heat Pumps in Pieces			Photovoltaic Systems in Pieces		
G01 C01	630	1233	40	353	4	20	31	66	1	1	2	3	3	5
						26	55	94	1	2	2	4	5	8
G02 C02	400	1042	10	382	5	12	16	21	0	0	0	1	1	1
						14	21	24	0	0	0	1	1	2
G03 C03	1430	4757	41	94	13	8	18	32	0	0	0	2	2	4
						15	39	74	0	0	0	3	4	8
G04 C04	400	2043	40	245	5	15	33	75	2	3	4	4	4	7
						29	69	89	3	7	7	5	7	13
G05 C05	1600	2058	59	150	9	24	39	68	5	7	8	8	8	15
						36	85	118	7	11	17	10	16	28
G06 C06	630	4017	192	448	4	37	84	180	2	2	4	14	14	25
						68	158	327	2	7	10	18	27	47
G07 C07	1200	2173	44	521	14	37	65	97	0	0	1	2	2	4
						59	100	112	0	3	4	3	4	8
G08 C08	630	1493	99	226	6	35	67	149	3	5	10	13	13	23
						54	136	175	6	17	31	16	25	44
G09 C09	400	2097	29	486	7	9	13	33	0	0	0	2	2	3
						10	27	55	0	0	0	2	4	7
G10 C10	630	5054	189	489	6	46	99	214	9	11	14	22	22	39
						79	187	303	11	18	26	28	43	75
G11 C05	630	9412	226	289	2	93	180	286	24	37	50	18	18	35
						160	299	320	38	79	106	23	35	61
G12 C07	630	2529	26	306	2	11	18	35	0	0	1	1	1	2
						18	35	53	0	2	2	2	3	5
G13 C10	630	2486	88	111	4	15	39	87	5	5	6	10	10	18
						29	77	137	5	8	11	13	19	34
G14 C05	400	3892	91	119	8	21	44	99	4	5	8	10	10	18
						37	87	142	5	9	15	13	20	35
G15 C05	250	1299	27	51	3	4	14	35	0	0	0	3	3	5
						13	30	59	0	1	4	4	6	10
G16 C05	1000	3003	88	287	10	9	36	103	6	9	12	7	7	13
						43	92	178	9	14	18	9	14	24

Table A2. *Cont.*

Grid Cluster	Installed Transformer Capacity in kVA	Power Line Length in m	Building Connections in Pieces	Metering Points in Pieces	Feeders/Meshes in Pieces	Charging Points (Private + Public) in Pieces			Heat Pumps in Pieces			Photovoltaic Systems in Pieces		
G17 C05	315	4010	108	231	6	15	36	92	3	6	6	9	9	15
						32	84	178	6	8	16	1	16	29
G18 C05	250	2458	72	301	1	36	63	101	6	7	10	10	10	18
						58	115	152	7	14	23	13	19	34
G19 C05	1600	1832	37	140	8	47	75	78	12	19	24	7	7	13
						71	81	89	20	26	26	9	14	24
G20 C06	630	4991	231	166	4	35	74	146	5	7	8	77	77	77
						65	139	193	7	11	18	77	77	77

Table A3. Classification of building and urban structure based on location and building types for 20 low-voltage grids (with: OTFH = One- and two-family detached house, MFH = multi-family house, CB = commercial buildings, S = Suburban, U = Urban, I = Inner-city, X = high proportion, O = low proportion, - = negligible or not present) based on [41].

Structure	G01	G02	G03	G04	G05	G06	G07	G08	G09	G10	G11	G12	G13	G14	G15	G16	G17	G18	G19	G20
OTFH	-	-	-	O	X	X	-	X	-	X	X	-	X	X	X	X	X	X	X	X
MFH	X	X	-	X	O	O	X	O	X	O	-	X	-	-	-	X	O	X	X	O
CB	-	-	X	-	-	-	O	-	O	-	-	O	-	-	-	-	-	-	-	-
S				X	X	X		X		X	X		X	X	X	X	X	X	X	X
U	X		X						X											
I		X					X					X								

References

1. Meinecke, S.; Klettke, A.; Sarajlic, D.; Dickert, J.; Hable, M.; Fischer, F.; Braun, M.; Moser, A.; Rehtanz, C. General Planning and Operational Principles in German Distribution Systems Used for SimBench. In Proceedings of the CIRED 2019 Conference, Madrid, Spain, 3–6 June 2019; p. 139.
2. Müller, M.; Schulze, Y. Future grid load with bidirectional electric vehicles at home. In Proceedings of the International ETG Congress, Wuppertal, Germany, 18–19 May 2021.
3. Möller, C.; Kotthaus, C.; Zdrallek, M.; Schweiger, F. Comparison of Incentive Models for Grid-supporting Flexibility Usage of Private Charging Infrastructure. In Proceedings of the International ETG Congress, Wuppertal, Germany, 18–19 May 2021.
4. Wruk, J.; Cibis, K.; Resch, M.; Sæle, S.; Zdrallek, M. Optimized strategic planning of future Norwegian low-voltage networks with a genetic algorithm applying empirical electric vehicle charging data. *Electricity* **2021**, 2, 91–109. [CrossRef]
5. Ramaswamy, P.C.; Chardonnet, C.; Rapoport, S.; Czajkowski, C.; Sanchez, R.R.; Arriola, I.G.; Bulto, G.O. Impact of electric vehicles on distribution network operation: Real world case studies. In Proceedings of the CIRED Workshop, Helsinki, Finland, 14–15 June 2016; p. 415.
6. Stoeckl, G.; Witzmann, R.; Eckstein, J. Analyzing the Capacity of Low Voltage Grids for Electric Vehicles. In Proceedings of the IEEE Electrical Power and Energy Conference, Winnipeg, MB, Canada, 3–5 October 2011; ISBN 978-1-4577-0404-8.
7. Kreutmayr, S.; Storch, D.J.; Niederle, S.; Steinhart, C.J.; Gutzmann, C.; Finkel, M.; Witzmann, R. Time-Dependent and Location-Based Analysis of Power Consumption at Public Charging Stations in Urban Areas. In Proceedings of the CIRED 2021 Conference, Geneva, Switzerland, 20–23 September 2021; p. 82.

8. Propst, A.; Siegel, M.; Braun, M.; Tenbohlen, S. Impacts of electric mobility on distribution grids and possible solution through load management. In Proceedings of the 21st International Conference on Electricity Distribution, Frankfurt, Germany, 6–9 June 2011; p. 126.
9. Schoen, A.; Ulffers, J.; Maschke, H.; Junge, E.; Bott, C.; Thurner, L.; Braun, M. Considering Control Approaches for Electric Vehicle Charging in Grid Planning. In Proceedings of the International ETG Congress, Wuppertal, Germany, 18–19 May 2021.
10. Petrou, K.; Quiros-Tortos, J.; Ochoa, L.F. Controlling Electric Vehicle Charging Points for Congestion Management of UK LV Networks. In Proceedings of the IEEE/PES Innovative Smart Grid Technologies ISGT 2015 Conference, Washington, DC, USA, 17–20 February 2015. [CrossRef]
11. DKE Deutsche Kommission Elektrotechnik Elektronik Informationstechnik in DIN und VDE. *DIN EN 50160:2020-11—Voltage Characteristics of Electricity Supplied by Public Electricity Networks*; German Version EN 50160:2010 + Cor.:2010 + A1:2015 + A2:2019 + A3:2019; Beuth Verlag GmbH: Berlin, Germany, 2020.
12. Harnisch, S.; Steffens, P.; Thies, H.; Monscheidt, J.; Münch, L.; Böse, C.; Gemsjäger, B.; Zdrallek, M. *Planungs- und Betriebsgrundsätze für Ländliche Verteilungsnetze—Leitfaden zur Ausrichtung der Netze an ihren Zukünftigen Anforderungen*; Bergische Universität: Wuppertal, Germany, 2016; Volume 8.
13. Sallam, A.A.; Malik, O.P. *Electric Distribution Systems*, 2nd ed.; Wiley-IEEE Press: Hoboken, NJ, USA, 2018; ISBN 978-1-119-50933-2.
14. Riedlinger, T.; Wintzek, P.; Wierzba, B.B.; Zdrallek, M. Cross-sectoral planning of energy grids in development areas with different heating concepts. In Proceedings of the Power and Energy Student Summit (PESS), online, 5–7 October 2020; ISBN 978-3-8007-5337-6.
15. Steffens, P.; Harnisch, S.; Zdrallek, M.; Monscheidt, J.; Münch, L.; Böse, C. New Planning Guidelines for Rural Distribution Grids. In Proceedings of the International ETG Congress 2017, Bonn, Germany, 28–29 November 2017.
16. Harnisch, S.; Steffens, P.; Thies, H.H.; Zdrallek, M.; Lehde, B. New Planning Principles for Low Voltage Networks with a High Share of Decentralised Generation. In Proceedings of the CIRED Workshop, Helsinki, Finland, 14–15 June 2016; p. 0086.
17. Wintzek, P.; Ali, S.A.; Kotthaus, K.; Wruk, J.; Zdrallek, M.; Monscheidt, J.; Gemsjäger, B.; Slupinski, A. Application and Evaluation of Load Management Systems in Urban Low-Voltage Grid Planning. *World Electr. Veh. J.* **2021**, *12*, 91. [CrossRef]
18. Zimmerman, L. *Progress Report 2018—Market Ramp-Up Phase*; German National Platform for Electric Mobility NPE: Berlin, Germany, 2018.
19. Bundesministerium für Umwelt, Naturschutz und nukleare Sicherheit BMU. *Langfristszenarien und Strategien für den Ausbau der Erneuerbaren Energien in Deutschland bei Berücksichtigung der Entwicklung in Europa und Global*; Bundesministerium für Umwelt, Naturschutz und nukleare Sicherheit BMU: Bonn, Germany, 2012.
20. Shell Deutschland Oil GmbH. *Shell Passenger Car Scenarios for Germany to 2040*; Shell Deutschland Oil GmbH: Hamburg, Germany, 2014.
21. Öko-Institut e.V. *eMobil 2050*; Öko-Institut e.V.: Berlin, Germany, 2014.
22. Institut für Mobilitätsforschung ifmo. *Die Zukunft der Mobilität*; Institut für Mobilitätsforschung Ifmo: Munich, Germany, 2015.
23. PricewaterhouseCoopers PwC. *Autofacts*; PricewaterhouseCoopers PwC: Frankfurt, Germany, 2016.
24. Center of Automotive Management CAM. *Marktentwicklung von Elektrofahrzeugen für das Jahr 2030*; Center of Automotive Management CAM: Bergisch Gladbach, Germany, 2017.
25. Fraunhofer-Institut für System-und Innovationsforschung ISI. *Entwicklung der Regionalen Stromnachfrage und Lastprofile*; Fraunhofer-Institut für System-und Innovationsforschung ISI: Karlsruhe, Germany, 2016.
26. Transmission System Operators in Germany. *Netzentwicklungsplan Strom 2035, Version 2021*; Zweiter Entwurf, Übertragungsnetzbetreiber CC-BY-4.0; 50Hertz Transmission GmbH, Amprion GmbH, TenneT TSO GmbH, TransnetBW GmbH: Berlin, Germany, 2021.
27. The Boston Consulting Group BCG. *The Electric Car Tipping Point*; The Boston Consulting Group BCG: Munich, Germany, 2018.
28. BloombergNEF. Electric Vehicle Outlook 2018. 2018. Available online: https://europeanlithium.com/electric-vehicle-outlook-2018-bloomberg/ (accessed on 23 June 2021).
29. RBC Capital Markets. Electric Vehicle Forecast through 2050 & Primer. 2018. Available online: http://www.fullertreacymoney.com/system/data/files/PDFs/2018/May/14th/RBC%20Capital%20Markets_RBC%20Electric%20Vehicle%20Forecast%20Through%202050%20%20Primer_11May2018.pdf (accessed on 23 June 2021).
30. Deutsche Energie-Agentur dena. *dena-Leitstudie Integrierte Energiewende—Ergebnisbericht*; Deutsche Energie-Agentur dena: Berlin, Germany, 2018.
31. Transmission System Operators in Germany. *Netzentwicklungsplan Strom 2030, Version 2019*; Zweiter Entwurf, Übertragungsnetzbetreiber; 50Hertz Transmission GmbH, Amprion GmbH, TenneT TSO GmbH, TransnetBW GmbH: Berlin, Germany, 2019.
32. Nationale Plattform Zukunft der Mobilität. *Roadmap—Markthochläufe alternativer Antriebe und Kraftstoffe aus Technologischer Perspektive*; Bundesministerium für Verkehr und Digitale Infrastruktur (BMVI): Berlin, Germany, 2021.
33. GfK SE Growth from Knowledge. *Population Structure Data*; GfK SE Growth from Knowledge: Nuremberg, Germany, 2018.
34. Federal Office of Statistics. *Census Database 2011*; Federal Office of Statistics: Wiesbaden, Germany, 2011.
35. Plurality and Majority Systems. Available online: www.britannica.com/topic/election-political-science/Plurality-and-majoritysystems (accessed on 6 February 2021).
36. Bundesverband Wärmepumpe BWP. *Branchenstudie 2018: Marktanalyse-Szenarien Handlungsempfehlungen*; Bundesverband Wärmepumpe BWP e.V.: Berlin, Germany, 2018.

37. Fraunhofer-Institut für Windenergie und Energiesystemtechnik IWES. *Wärmewende 2030*; Fraunhofer-Institut für Windenergie und Energiesystemtechnik IWES: Kassel, Germany, 2017.
38. The Boston Consulting Group BCG. *Klimapfade für Deutschland*; The Boston Consulting Group BCG: Munich, Germany, 2018.
39. Deutsche Energie-Agentur GmbH dena. *Gebäudestudie-Szenarien für eine Marktwirtschaftliche Klima- und Ressourcenschutzpolitik 2050 im Gebäudesektor*; Deutsche Energie-Agentur GmbH dena: Berlin, Germany, 2017.
40. Göbelsmann, F.; Wintzek, P.; Ali, S.A.; Zdrallek, M.; Monscheidt, J.; Gemsjäger, B.; Slupinski, A. Potentials of a cost-optimized application of regulated distribution transformers in urban low-voltage grid planning with a high share of charging infrastructure, heat pumps and photovoltaic systems. In Proceedings of the Power and Energy Student Summit (PESS), online, 25–26 November 2021.
41. Wintzek, P.; Ali, S.A.; Monscheidt, J.; Gemsjäger, B.; Slupinski, A.; Zdrallek, M. *Planungs- und Betriebsgrundsätze für Städtische Verteilnetze—Leitfaden zur Ausrichtung der Netze an Ihren Zukünftigen Anforderungen*; Zdrallek, M., Ed.; Neue Energie aus Wuppertal, Bergische Universität: Wuppertal, Germany, 2021.
42. German Environment Agency. *Rescue—Resource-Efficient Pathways to Greenhouse-Gas-Neutrality*; German Environment Agency: Dessau-Roßlau, Germany, 2020.
43. Fraunhofer ISI. *Langfristszenarien für die Transformation des Energiesystems in Deutschland*; Fraunhofer ISI: Karlsruhe, Germany, 2017.
44. World Wide Fund for Nature (WWF). *Zukunft Stromsystem II—Regionalisierung der erneuerbaren Stromerzeugung*; World Wide Fund for Nature (WWF): Berlin, Germany, 2018; ISBN 978-3-946211-22-8.
45. EWI Energy Research & Scenarios GmbH. *Energiemarkt 2030 und 2050—Der Beitrag von Gas- und Wärmeinfrastruktur zu Einer Effizienten CO2-Minderung*; EWI Energy Research & Scenarios GmbH: Düsseldorf, Germany, 2017.
46. Zentrum für Sonnenenergie- und Wasserstoff-Forschung Baden-Württemberg (ZSW). *Dynamische Simulation der Ausbauszenarien für Erneuerbare Stromversorgung in Baden-Württemberg bis 2050*; Zentrum für Sonnenenergie- und Wasserstoff-Forschung Baden-Württemberg (ZSW): Stuttgart, Germany, 2014.
47. Bundesnetzagentur. *Genehmigung des Szenariorahmens 2021–2035*; Bundesnetzagentur: Bonn, Germany, 2020.
48. Bundesnetzagentur. *Genehmigung in dem Verwaltungsverfahren Wegen der Genehmigung des Szenariorahmens für die Netzentwicklungsplanung und Offshore-Netzentwicklungsplanung gem. § 12a Abs. 3 EnWG.*; Bundesnetzagentur: Bonn, Germany, 2014.
49. CDU Deutschlands, CSU-Landesleitung und SPD. *Deutschlands Zukunft gestalten, Koalitionsvertrag zwischen CDU, CSU und SPD*; Union Betriebs-GmbH: Rheinbach, Germany, 2013.
50. Monscheidt, J.; Gemsjäger, B.; Slupinski, A.; Yan, X.; Ali, S.A.; Wintzek, P.; Zdrallek, M. An energy demand model for electric load forecast in urban distribution networks. In Proceedings of the CIRED Workshop, Berlin, Germany, 22–23 June 2020.
51. Wintzek, P.; Göbelsmann, F.; Ali, S.A.; Zdrallek, M.; Brinkmann, F.; Monscheidt, J.; Gemsjäger, B.; Slupinski, A. Relevant Operating Points for Future Grid Planning of Urban Low-Voltage Grids. In Proceedings of the CIRED 2021 Conference, Geneva, Switzerland, 20–23 September 2021; p. 0472.
52. Ali, S.A.; Wintzek, P.; Zdrallek, M.; Böse, C.; Monscheidt, J.; Gemsjäger, B.; Slupinski, A. Demand Factor Identification of Electric Vehicle Charging Points for Distribution System Planning. In Proceedings of the CIRED 2021 Conference, Geneva, Switzerland, 20–23 September 2021; p. 0470.
53. Rehtanz, C.; Greve, M.; Häger, U.; Hagemann, Z.; Kippelt, S.; Kittl, C.; Klouobert, M.-L.; Pohl, O.; Rewald, F.; Wagner, C. *Verteilnetzstudie für das Land Baden-Württemberg*; ef.Ruhr GmbH: Dortmund, Germany, 2017.
54. VDE Verband der Elektrotechnik Elektronik Informationstechnik e.V. *Technical Rules for the Connection and Operation of Customer Installations to the Low Voltage Network (TAR Low Voltage)*; VDE-AR-N 4100:2019-04; VDE-Verlag GmbH: Berlin, Germany, 2019.
55. VDE Verband der Elektrotechnik Elektronik Informationstechnik e.V. *Generators Connected to the Low-Voltage Distribution Network—Technical Requirements for the Connection to and Parallel Operation with Low-Voltage Distribution NETWORKS*; VDE-AR-N 4105:2018-11; VDE-Verlag GmbH: Berlin, Germany, 2018.
56. DKE Deutsche Kommission Elektrotechnik Elektronik Informationstechnik in DIN und VDE. *Power Transformers—Part 1: General (IEC 60076-1:2011)*, German version EN 60076-1:2011; Beuth Verlag GmbH: Berlin, Germany, 2012.
57. Deutsche Elektrotechnische Kommission im DIN und VDE (DKE). *Power Cables—Part 1000: Current-Carrying Capacity, General, Conversion Factors*; DIN VDE 0276-1000:1995-06; VDE-Verlag GmbH: Berlin, Germany, 1995.
58. DKE Deutsche Kommission Elektrotechnik Elektronik Informationstechnik in DIN und VDE. *Medium Power Transformers 50 Hz with Highest Voltage for Equipment not Exceeding 36 kV—Part 1: General Requirements*; German Version EN 50588-1:2017; VDE-Verlag GmbH: Berlin, Germany, 2019. [CrossRef]
59. FNN-Hinweis. *rONT—Einsatz in Netzplanung und Netzbetrieb*; Forum Netztechnik/Netzbetrieb im VDE: Berlin, Germany, 2016.
60. Becker, M.; Jung, D.; Wintzek, P.; Müller, T.; Zdrallek, M. Rechtliche Rahmenbedingungen zur Anwendung von Sperrzeiten bei Wärmepumpen für die Integration in ein Lastmanagement. *Z. Energ.* **2021**, *45*, 271–285. [CrossRef]
61. Wintzek, P.; Ali, S.A.; Talmond, F.; Zdrallek, M.; Röstel, T.; Monscheidt, J.; Gemsjäger, B.; Slupinski, A. Influence of a dynamic load management on the future grid planning of urban low-voltage grids. In Proceedings of the International ETG Congress, Wuppertal, Germany, 18–19 May 2021; ISBN 978-3-8007-5549-3.
62. FNN-Hinweis. *Störungs- und Verfügbarkeitsstatistik—Berichtsjahr 2013*; Forum Netztechnik/Netzbetrieb im VDE: Berlin, Germany, 2014.

63. FNN-Hinweis. *Störungs- und Verfügbarkeitsstatistik—Berichtsjahr 2014*; Forum Netztechnik/Netzbetrieb im VDE: Berlin, Germany, 2015.
64. FNN-Hinweis. *Störungs- und Verfügbarkeitsstatistik—Berichtsjahr 2016*; Forum Netztechnik/Netzbetrieb im VDE: Berlin, Germany, 2017.
65. FNN-Hinweis. *Störungs- und Verfügbarkeitsstatistik—Berichtsjahr 2018*; Forum Netztechnik/Netzbetrieb im VDE: Berlin, Germany, 2019.
66. Wintzek, P.; Ali, S.A.; Möhrke, F.; Zdrallek, M.; Asman, A.; Brinkmann, F.; Monscheidt, J.; Gemsjäger, B.; Slupinski, A. Identification of typical urban grid structures across voltage levels by clustering distribution grid big data. In Proceedings of the International ETG Congress, Wuppertal, Germany, 18–19 May 2021; ISBN 978-3-8007-5549-3.

electricity

MDPI

Article

Conception of High-Frequency Power Planar Transformer Prototypes Based on FabLab Platform

Simon Thomy, Xavier Margueron *, Jean-Sylvio Ngoua Teu Magambo, Reda Bakri and Philippe Le Moigne

Univ. Lille, Arts et Metiers Institute of Technology, Centrale Lille, Junia, ULR 2697 - L2EP, F-59000 Lille, France; simon.thomy@centralelille.fr (S.T.); jeansylvio@yahoo.fr (J.-S.N.T.M.); reda.bakri@centralelille.fr (R.B.); philippe.lemoigne@centralelille.fr (P.L.M.)

* Correspondence: xavier.margueron@centralelille.fr

Abstract: Conceiving planar magnetic components for power electronic converters is very constraining, especially in the case of prototype development. Indeed, such making requires skills, specific appliances as well as human time for setting up the machine tools and the fabrication process. With the emergence of Fabrication Laboratory (FabLab), conceiving of planar copper foil prototypes becomes more feasible in a shortened time process for engineers and researchers. This paper presents a methodology and process for conceiving power planar transformers with the help of machines and tools that can be found in the usual FabLab.

Keywords: HF planar transformer; fabrication laboratory; prototype manufacturing

Citation: Thomy, S.; Margueron, X.; Ngoua Teu Magambo, J.-S.; Bakri, R.; Le Moigne, P. Conception of High-Frequency Power Planar Transformer Prototypes Based on FabLab Platform. *Electricity* **2022**, 3, 1–15. https://doi.org/10.3390/electricity3010001

Academic Editors: Pavlos S. Georgilakis and Sérgio Cruz

Received: 14 October 2021
Accepted: 15 December 2021
Published: 21 December 2021

1. Introduction

For some years now, planar transformers have gradually been replacing traditional high-frequency (HF) wounded transformers in embedded systems and electric vehicles. Planar components present many advantages like efficiency, power density, small size and less weight as well as good thermal characteristics [1–3].

Planar transformers are usually made of a printed circuit board (PCB) or copper foil windings combined with low profile magnetic core. Even if PCB windings present many advantages in terms of industrialization (manufacturability, cost reduction, repeatability), making multi-layer PCB transformer prototypes is complex and requires specific machines, technology and engineers' time. Planar component prototypes are then difficult to conceive, expensive and need to be subcontracted to specialists. This drawback was not as important when developing HF wounded transformers. For power planar components, copper foil can also be an interesting solution for a transformer's windings. Regarding prototypes, with such technology combined with the emergence of Fabrication Laboratory (FabLab) almost everywhere [4], making planar transformer prototypes becomes more feasible in an acceptable time, with a do-it-yourself (DIY) conception philosophy.

To address this issue, a complete conception process has been developed in order to achieve planar transformer prototypes. The process is suitable for any FabLab and creates opportunities to quickly elaborate affordable components. Therefore, this paper presents the complete process for the conception of planar transformer prototypes based on tools that can be found in FabLab. Based on electrical specifications, a planar transformer is designed, conceived in a FabLab and tested with specific power electronic equipment.

The paper is organized as follows: In Section 2, an introduction to the world of FabLabs is undertaken. In Section 3, technical considerations are listed dealing with the planar transformer design specificities. Then, a complete design of the planar transformer is performed based on models from the literature and finite element analysis (FEA). In Section 4, the full process to make a copper foil planar transformer prototype is described. The focus is on the key points and difficulties encountered in the prototype's development. In Section 5, the planar prototype is tested and characterized to validate its functioning. Finally, the process is discussed and potential improvements are highlighted.

2. FabLab

The first FabLabs were opened in the 2000s. These laboratories are dedicated to open-source creation, co-creation and local production [5] based on modern equipment and software. In 2016, 490 FabLabs were referenced in 72 countries [5]. In 2018, this number reached 650 in 80 countries [6] and it is still increasing quickly. That has led to a worldwide network of fabrication laboratories where almost everything can be made everywhere [4]. FabLabs are made for prototyping, making, learning, meeting and contributing to the maker community [7]. A lot of universities have created their own FabLab to form a new generation of high-tech FabLab-oriented makers [8].

The emergence of FabLabs is linked to the development of new philosophy of conception as well as the growth of new practical tools such as 3-D printers; 3-D printing, also called additive manufacturing, consists of machines that are able to add material layer-by-layer to create an object from 3D CAD (computer aided design) models. This topical subject concerns a lot of application domains from micro-systems [9] to aerospace and automotive industries [10] as well as education [11]. In 2020, during the health crisis, the makers' community, FabLabs and 3D-printing have played an important role in manufacturing personal protective equipment and ventilator replacement parts that could not be satisfied by regular suppliers [12–14]. Regarding power electronics, 3-D printing technology is now applied to magnetics, dealing with 3-D printed air-core inductors [15], low power planar inductors [16], shaped profile windings [17], cool fins for planar inductors [18] and wireless power transfer system [19].

The FabLab (Figure 1) involved in this study hosts a laser-cutter, four types of 3-D printers, a milling machine, wood-cutter, embroiderer, sewing machine and other more classical tools that can be found in electronic labs and workshop manufacturing.

Figure 1. FabLab "Make" at Centrale Lille Institut.

3. Technological Considerations for the High-Frequency (HF) Planar Transformer Prototype

3.1. *Conceiving Planar Transformer: Printed Circuit Board (PCB) vs. Copper Foil*

Planar transformer windings can be made with PCB (rigid or flex technology), copper foil and more rarely with Litz wires [20–25]. PCB technology is the most widespread, offering many advantages like mass production, repeatability, reliability or low leakage inductance [26]. PCB planar transformers can be connected as stand-alone or embedded within a PCB-assembled converter to save space [27] or to increase power integration. On the other hand, the main drawback of PCB planar transformers is their parasitic capacitances [23] that can cause electro-magnetic interference (EMI) problems. Other limits deal with costs, especially in the case of prototypes, and technological limitations linked to the manufacturing capabilities.

With PCB technology, copper tracks are a real issue. Indeed, PCB copper thicknesses are limited to standard values: 17.5 μm, 35 μm, 70 μm, 105 μm, 210 μm and 435 μm. Due to these small values, paralleling layers is necessary to carry power electronic high current values. As a consequence, vias are needed to connect different layers of the same winding. Connections between layers is a known problem for multi-layer PCB. It can increase cost as well as track's equivalent resistance and can lead to hot spots, especially in HF.

PCB Flex technology is a solution to reduce vias in windings. A Flex PCB is adequate for 3D power electronics [28]. Even if it is more suitable for medium series and prototypes than traditional multi-layer PCB, Flex PCB copper tracks are not optimal in the case of HF power electronic magnetic components [21,22]. Such windings are more dedicated to low power HF transformers or specific devices like bendable transformers [29].

Regarding leakage inductances, distance between PCB layers and interleaving for reducing HF copper losses are factors that limit the leakage inductance values [26]. For some converters operating in soft-switching, it is interesting to increase the leakage value of a transformer to avoid the use of an additional inductance [30]. With a planar PCB transformer, this increase is limited due to PCB technological constraints and traditional available cores.

For an HF power planar transformer, copper foil windings appear like an interesting alternative to PCB windings. Some benefits can be highlighted:

- Copper track thicknesses are not limited. Bigger conductors can be selected in order to limit or at least reduce complex paralleling layers with respect to skin depth that is problematic. Moreover, primary and secondary thicknesses can easily be set to different values.
- Distance between layers can also be set to different values, stacking insulated layers. Insulation material can also be chosen differently according to layers.
- Leakage values can be increased by spacing primary and secondary windings or by introducing ferrite polymer composite (FPC) material between layers, like C350 for example [31].

As a major drawback, connections between layers need to be well thought out in the design step.

3.2. *Transformer Prototype Specifications*

The planar transformer developed in this study is defined for an aeronautic application for a More Electric Aircraft (MEA) [3]. It is designed for a 2 kW DC/DC Dual active bridge (DAB) power converter [32]. The circuit schematic is presented in Figure 2. The DAB electrical characteristics are listed in Table 1.

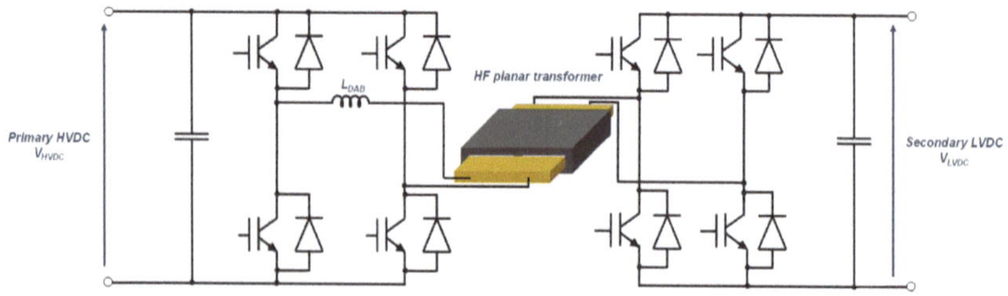

Figure 2. Dual active bridge (DAB) converter.

Table 1. Planar transformer electrical specifications.

Electrical Parameter	Value
Nominal Power	P_{nom} = 2 kW
Primary voltage (rms)	V_p = 400 V
Switching frequency	f_s = 100 kHz
Transformer ratio	η = 0.05
Magnetizing inductance	L_m = 0.5 mH

Due to the application, the prototype has to be lightweight. In terms of target design, a maximal temperature rise is fixed to ΔT = 120 °C, for a 20 °C ambient temperature and no cooling device (heat transfer by natural convection). The leakage inductance is not a key parameter for this design. Its value can be low. Indeed, an additional inductor is used to obtain the DAB maximal series inductance needed in transfer Function (1):

$$P_{out} = \frac{V_{HVDC}\, V_{LVDC}}{L_{DAB}\, f_s\, \eta}\varphi(\frac{\pi - |\varphi|}{2\pi^2}) \tag{1}$$

where φ is the phase shift between primary and secondary bridges, V_{HVDC} and V_{LVDC} are the voltages of the DAB converter (Figure 2) and L_{DAB} is the maximum allowable DAB inductance value set to 100 µH.

3.3. Transformer Design and Description

The magnetic core is selected based on the product area method [33]. With this method, the product A_p (2) of the window area A_w and the core cross-section A_c is expressed as a function of the power and other electrical specifications:

$$A_p = A_w A_c = \frac{P_{max}}{K_f K_r B_m f_s J_w} \tag{2}$$

where P_{max} is the maximal power, K_f is the waveform coefficient, K_r is the window filling factor, B_m is the flux density, f_s is the switching frequency and J_w is the current density.

Based on design specifications given in [3], the needed product area is calculated: A_p = 54,000 mm^4. The magnetic core, association of an E-shape E64 with a plate PLT64 [34], is selected according to this value. Indeed, the combination of both have a product area value of 57,600 mm^4. 3C90 ferrite material is selected.

Regarding windings, limitation of leakage inductance and copper loss lead to a solution with a complete interleaving between primary and secondary layers. Then, the windings are divided on 9 conductive layers of copper foils (Figure 3a). Number of turns for the primary is set to N_p = 20: four layers with five turns connected in series. The secondary contains only one turn (N_s = 1): five layers of one turn connected in parallel. Thus, the transformer ratio is 0.05. Primary and secondary layer thicknesses are set differently: 200 µm for the primary and 350 µm for the secondary. As mentioned before, such thicknesses are difficult to realize with PCB technology while keeping constant insulation layer thickness. Primary and secondary track widths are set to fulfill the current density requirement.

Dimensions and positioning of windings are presented in Figure 3a. Figure 3b shows a complete 3-D FEA Model made with ANSYS Maxwell 3D [35]. On these figures, insulation Kapton sheets between layers are not represented.

Magnetizing inductance is adjusted considering a 225 µm gap between the three legs of the planar E core and the plate one. With this gap value, the magnetizing inductance (3) should be of 0.53 mH:

$$L_m = A_L N_p^2 \tag{3}$$

where N_p is the primary turn number and A_L is the inductance factor of the ferrite core.

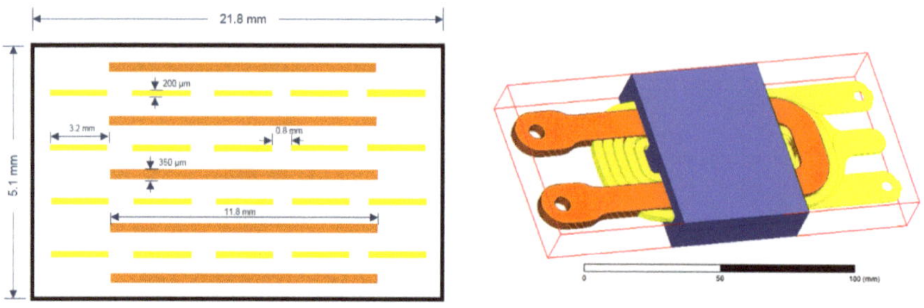

(**a**) (**b**)

Figure 3. Planar prototype: (**a**) transformer's window; (**b**) three-dimensional finite element analysis (3-D FEA) model.

Leakage inductance is calculated with the model detailed in [36]. The estimated value from the primary winding is L_{lk} = 4.45 μH.

In a HF transformer, the general formula for calculating the copper losses in a winding is:

$$P_{Cu} = R_{DC} \times I_{DC}^2 + \sum_{n=1}^{\infty} R_{AC}(nf_s) \times I_{rms}^2(nf_s) \tag{4}$$

where R_{DC} is the DC resistance, I_{DC} is the DC current, R_{AC} is the AC resistance depending on frequency, I_{rms} is the rms value of each current harmonic and n is the harmonic order.

In the design example, the DC part of the current is null. R_{AC} is estimated based on Dowell model [37] for the primary winding and Ferreira [38] for the secondary winding. Both applied models are different for the primary and the secondary due to the difference of winding porosity factors. For the primary, the latter is estimated at 0.73 while for the secondary it is 0.51. Considering the current fundamental and the first four harmonics (3rd, 5th, 7th and 9th), copper losses are estimated to 21.1 W. As a comparison, the copper losses due to the current fundamental only are 20.3 W.

Core losses are calculated with the Mulder model [39] at ambient temperature:

$$P_f = k \times f_s^{\alpha} \times \hat{B}^{\beta} \times \left(c_{t0} - c_{t1} \times T + c_{t2} \times T^2\right) \tag{5}$$

where k, α, β, c_{t0}, c_{t1} and c_{t2} are parameters for the magnetic material [39], $\hat{B}$ is the peak flux density and T is the temperature.

With this formula, for an ambient temperature of 25 °C and a peak flux density of 100 mT, the core losses are 4.84 W. This calculation corresponds to the worst case for core losses. Soft ferrite materials are usually optimized to have low losses between 80 °C and 100 °C.

In [40], McLyman gives an expression for the calculation of temperature rise in magnetics:

$$\Delta T = 450\left(\frac{P_{\Sigma}}{A_t}\right)^{0.826} \tag{6}$$

where ΔT is the temperature rise, A_t is the effective surface area in cm^2 and P_{Σ} is the total dissipated power.

Based on (6), the thermal resistance of the transformer becomes:

$$R_{th} = \frac{450}{P_{\Sigma}^{0.174}}\left(\frac{1}{A_t}\right)^{0.826} \tag{7}$$

Finally, copper and core loss values lead to an estimated increased temperature ΔT = 106.1 °C with a thermal resistance R_{th} = 4.09 °C/W for the E/PLT64 planar core.

4. Prototype Achievement in FabLab Environment

Based on the FabLab platform, the prototype described in the previous section is now developed. After the process overview, each step is detailed, highlighting difficulties and precautions that have to be taken to obtain good and functional transformer.

4.1. Manufacturing Process Overview

Figure 4 presents the process overview, introducing all the equipment that has to be used. Firstly, all the elements are modeled using Onshape Software [41]. Then, the numerical files are transferred to specific tools that allow the different transformer's parts to be made: copper windings for primary and secondary units, insulating Kapton and add-ons for the assembling parts. Finally, all these elements are assembled together on a planar magnetic core.

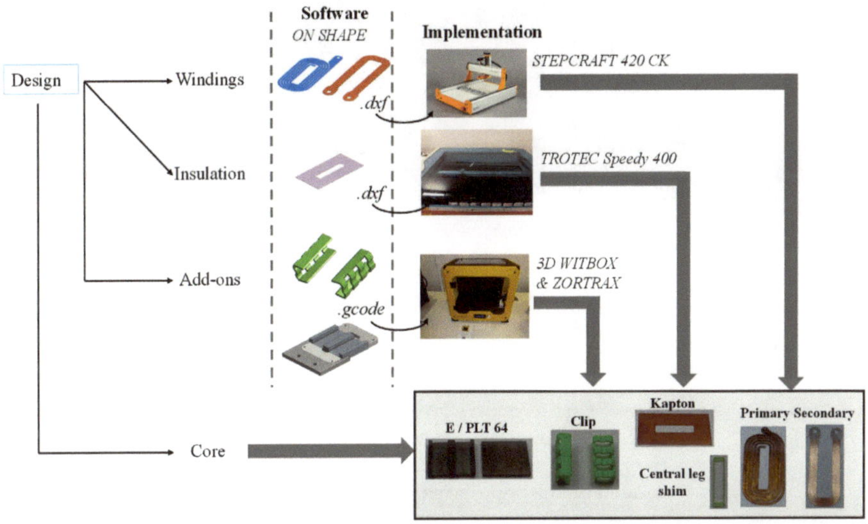

Figure 4. Process overview.

4.2. Parts Production

Three steps can be differentiated for the manufacturing of the different planar transformer's parts:

1. Copper windings: after their shape design, primary and secondary winding copper tracks are carved using a computer numerical control (CNC) machine Stepcraft 420 CK [42]. Figure 5 shows the milling process for two primary layers. During this process, lubricant is sprayed regularly. This provides two benefits: the drill run cooler and copper filings do not hang up. As a consequence, the drill's lifetime is extended. Then, endpoints of copper tracks are tined to ease connections between layers. Finally, winding layers are ready for assembling. One can note that all the secondary tracks are identical due to the parallel winding. On the opposite, primary layers are different in order to make possible series connections between the layers.
2. Kapton insulation: insulation between layers is made with Kapton film. Elementary 75 µm Kapton layers are then cut using the laser cutting machine Trotec Speedy 400 [43] to consider magnetic core and central leg size. Distances between winding layers will be set adding more or less of these elementary layers.
3. Add-ons: in order to assemble the transformer prototype, some supplementary elements have to be developed. Firstly, plexiglass clamps and secondary winding centering pins have to be cut with the laser cutting machine. All these elements are

made of polymethyl methacrylate (PMMA) material. Secondly, clips and central leg shim are printed in 3-D [44]. They are made of polylactic acid (PLA) material. Clips are used for maintaining both magnetic core parts together while the central leg shim is used for centering and spacing primary winding from the central core leg.

Figure 5. Milling windings.

4.3. Assembly Operations

Once all the parts have been made, everything is assembled layer by layer (Figure 6).

- Step 1: The plinth, including bottom clamps and centering pins is set around the E-part core.
- Step 2: Kapton insulated layers are added. Insulating thickness is adjusted adding more or less elementary layers.
- Step 3: First secondary layer is positioned using centering pins.
- Step 4: Kapton layers are added.
- Step 5: First primary winding is added. As can be seen in Figure 6, primary pins are located on the opposite side of the secondary winding. The biggest one corresponds to the winding connecting pin, while the smallest one, located close to the ferrite core, is used for connecting this layer to the next one. The primary winding's centering is ensured by the central leg shim that locks its positioning.
- Step 6: Stacking takes place, interleaving insulating layers, secondary winding layers and primary layers, respecting the layout introduced in Figure 3a. During this step, primary winding layers have to be soldered for setting the 20 turns of the winding.
- Step 7: When the last secondary layer is added, the stacking is almost over. Some insulated layers are inserted before the magnetic core to be closed with the ferrite PLT core part. The air gap is tuned adding some Kapton between E and PLT core parts. Then, the upper side of the plexiglass clamps is screwed to the bottom one to fix layers' stack. Clips are tightened around the magnetic core while the plinth with centering pins is removed.

The obtained final prototype is presented in Figure 7. This 2 kW planar transformer prototype weigh 295 g in a volume lower than 0.1 L.

It could be interesting to compare the obtained transformer to components that can be found in the manufacturer's catalogue. This benchmark comparison is quite difficult to realize because an HF transformer is usually designed for specific electrical constraints and cooling systems. Then, the comparison must be made with the same characteristics to be suitable.

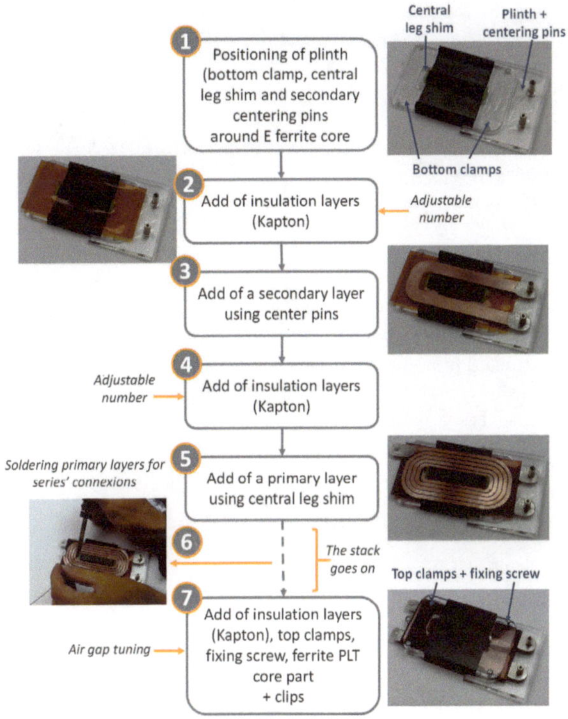

Figure 6. Assembling process layer by layer.

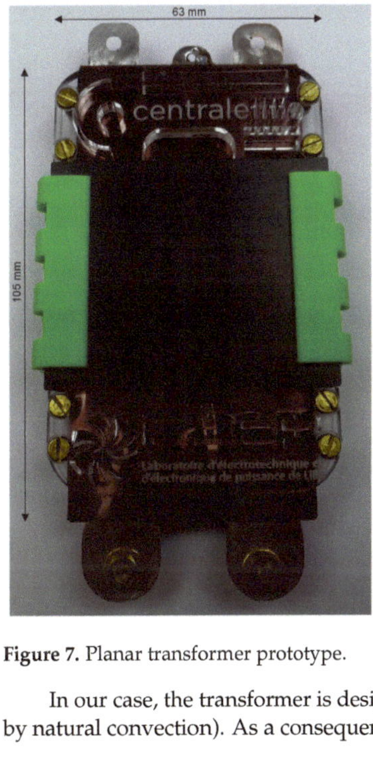

Figure 7. Planar transformer prototype.

In our case, the transformer is designed for use without a cooling system (heat transfer by natural convection). As a consequence, volume and weight of the prototype are higher than another transformer with the same power, voltage, frequency and cooling system. However, in general, the manufacturer Payton gives a 10 to 15 g per 100 W [45] for their planar transformer. In this case, the developed prototype is consistent with their components.

5. Prototype Validation: Characterization and Tests

In this section, the prototype is characterized and tested in order to validate its good functioning.

5.1. Small Signal Characterization Based on Impedance Measurements

The characterization is performed with impedance analyzer HP4294A [46]. The four measurements with open and short circuits are shown in Figure 8. The obtained impedances are typical for an HF transformer. This first characterization step enables us to conclude that the prototype is working as a transformer. The equivalent circuit parameters (Figure 9) are extracted from these measurements [47]. It can be noted that the parasitic capacitive effect is not shown in Figure 9. Table 2 compares some parameter measured values to the theoretical ones. For a better accuracy, the DC resistances are measured with a micro-ohmmeter CA 6250 [48].

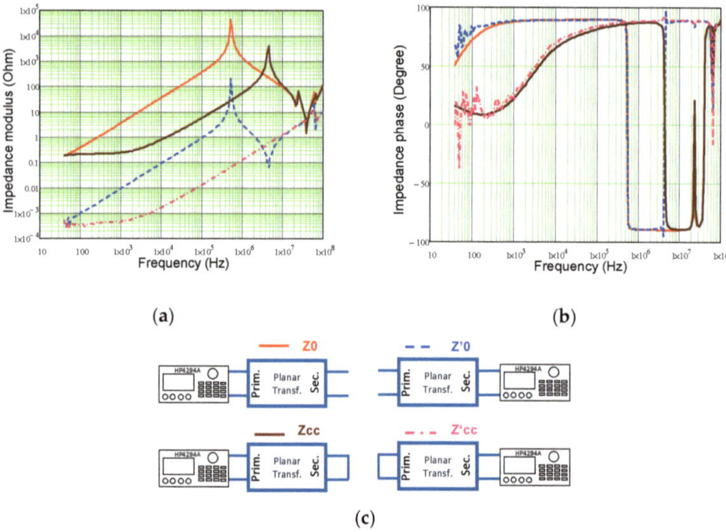

Figure 8. Planar transformer impedance measurements: (**a**) modulus; (**b**) phase; (**c**) measurement configurations.

These measurements are consistent with the theoretical ones, except for the low frequency leakage inductance seen from the primary winding. Indeed, the theoretical value, calculated with [36], is L_{lk} = 4.45 µH while the measured on is L_{lk_meas} = 16 µH. This difference can be explained by the short circuit (Figure 10a) used in the characterization process [49]. Regarding secondary impedance order of magnitude, the 3.2 cm short circuit copper conductor adds about 32 nH, i.e., 10 nH per centimeter, to the secondary leakage inductance. Subtracting this value (L_{sc}) from the measured primary value leads to consistent low frequency leakage value:

$$L_{lk\_LF} = L_{lk\_meas} - L_{sc} = 16{\cdot}10^{-6} - 32{\cdot}10^{-9}{\cdot}\frac{1}{\eta_{ps}^{2}} = 3.2\ \mu\text{H} \tag{8}$$

This lack of precision also could be attributed to the manual assembling process that prevent a precise positioning of the layers.

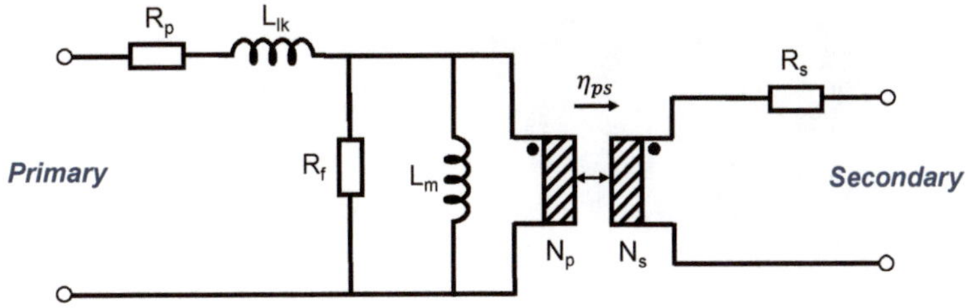

Figure 9. Planar transformer equivalent circuit.

Table 2. Planar transformer electrical characteristics.

Parameter	Symbol	Unit	Experimental Value	Theoretical Value
Transformer ratio	η_{ps}		0.05	0.05
Magnetizing inductance	L_m	mH	0.56	0.53
Leakage inductance (low frequency)	L_{lk}	µH	16	4.45
Primary DC resistance	R_p	mΩ	111.8	105.6
Secondary DC resistance	R_s	mΩ	0.198	0.22

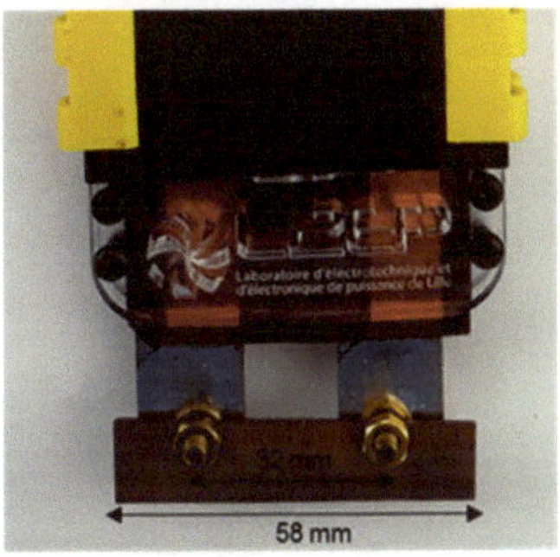

(**a**)

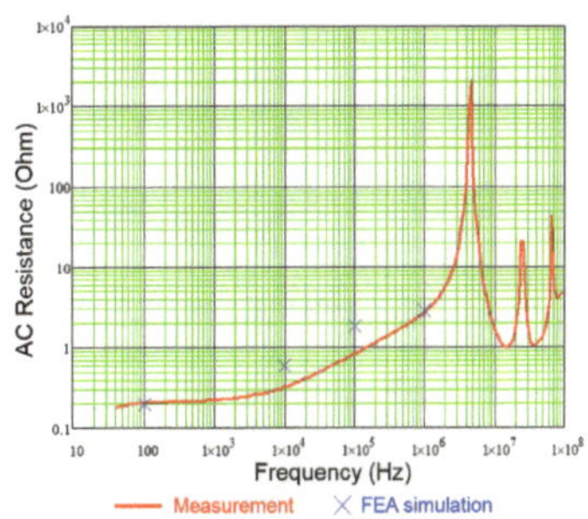

(**b**)

Figure 10. Planar transformer characterization: (**a**) short-circuit. (**b**) alternating current (AC) resistance vs. frequency.

The AC resistance is plotted versus frequency on Figure 10b. For frequency below 1 MHz, these measurements are consistent with FEA simulation performed with Ansys [35].

5.2. Thermal Characterization with Open and Short Circuit Power Measurements

Four power tests were performed for the thermal characterization of the transformer with a 100 kHz inverter. For each test, the temperature distribution is obtained with a thermal infra-red camera Fluke Ti95 [50]. Electrical waveforms are also presented.

5.2.1. Open Circuit Test

The primary planar transformer is supplied with a square voltage (±400 V) from full bridge inverter (Figure 11). Losses are mainly core ones and can be estimated to be 4.6 W. The temperature of the core is close to 43 °C for an ambient temperature of 23.3 °C.

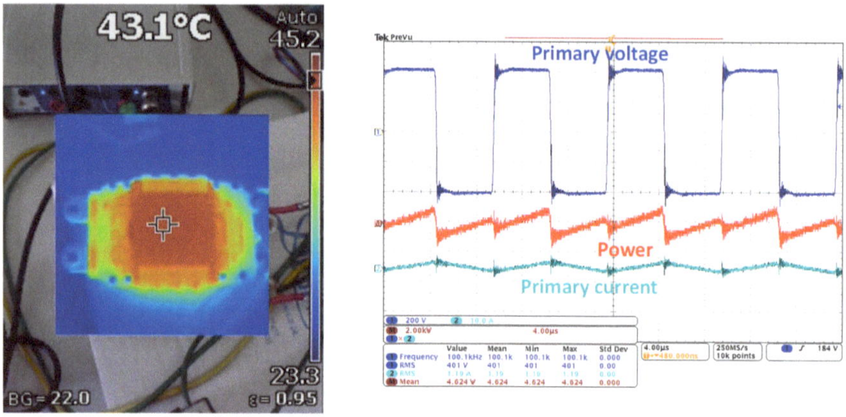

(**a**) (**b**)

Figure 11. Thermal characterization—open circuit test: (**a**) temperature distribution; (**b**) voltage, current and power waveforms.

5.2.2. Short-Circuit Test

For this test (Figure 12), the voltage is reduced to 40 V. At the secondary winding, the current reaches 100 Arms. The measured losses, mainly copper ones, are 23.7 W. The temperature rise is 76.9 °C with an ambient temperature of 23.4 °C. On Figure 12a, one can note hot spots close to secondary terminations. Due to the current value and the parallel secondary layers, specific attention must be paid to this winding. Termination soldering must be done with a highly conductive material in order to solve this issue.

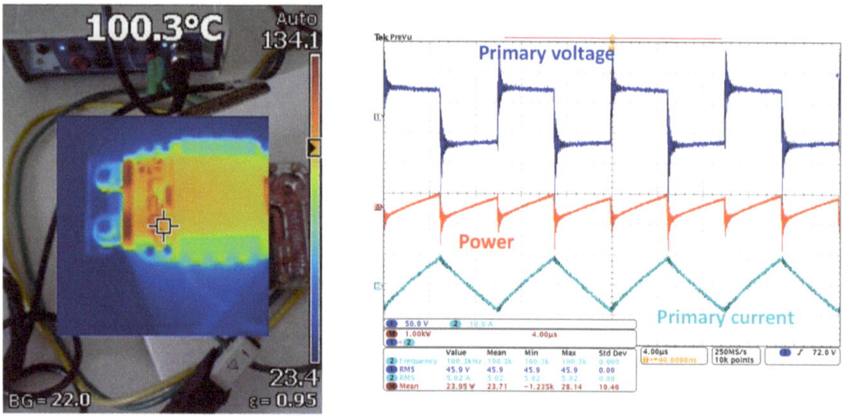

(**a**) (**b**)

Figure 12. Thermal characterization—short-circuit test: (**a**) temperature distribution; (**b**) voltage, current and power waveforms.

5.2.3. Load Tests

Two load tests were finally performed. For these tests, the output of the planar transformer is connected to a rectifier and a variable resistor.

The first load test (Figure 13) is performed with a 1 kW resistive load. The primary rms voltage is 350 V while the secondary current is 53.9 A. The obtained temperature is 59.6 °C for 21.4 °C ambient temperature. Despite the load being half the nominal power, the low temperature rise confirms the transformer capability to transfer power beyond 1 kW in natural convection.

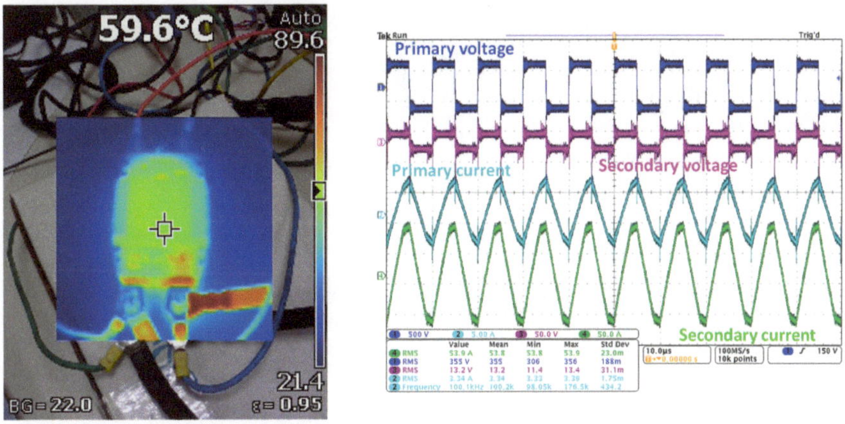

(a) **(b)**

Figure 13. Thermal characterization—350 V (primary)/53.9 A (secondary): (**a**) temperature distribution; (**b**) voltage and current waveforms.

The second load test (Figure 14) is performed with a resistive load closed to nominal power. The primary rms voltage is 400 V while the secondary current is 90.4 A. The temperature obtained is 121.7 °C for 23.9 °C ambient temperature. The temperature rise confirms the transformer capability to transfer this power. This result is consistent with the temperature rise computed in the Section 3.3. However, two comments can be made: firstly, the measured temperature with thermal infra-red camera is questionable and not so accurate, in particular for the winding temperatures. The use of thermo-couples could enable more accurate measurements to be made. Secondly, with such temperature, it would be necessary to use a fan to ensure normal operation for steady state uses. Such a cooling system will increase the lifetime of the transformer.

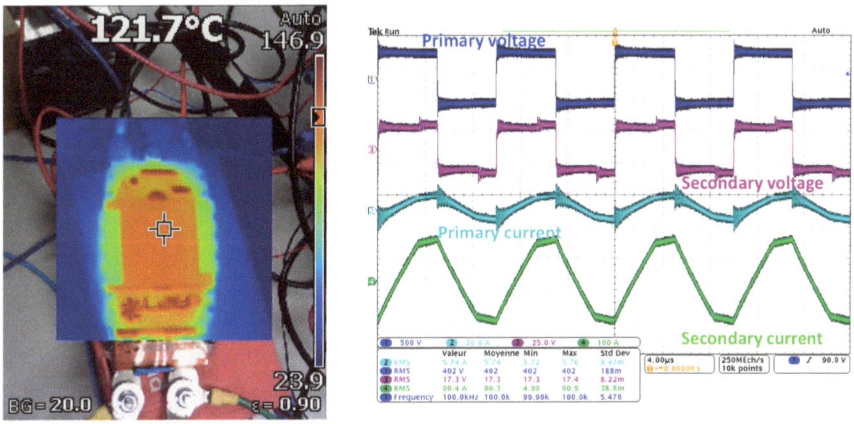

(a) **(b)**

Figure 14. Thermal characterization—400 V (primary)/90.4 A (secondary): (**a**) temperature distribution; (**b**) voltage and current waveforms.

5.3. Comments on Measurements and Process

The measurement results show that the design transformer is operational. Its parameters are closed to those estimated with analytical calculations and FEA modeling. Nevertheless, the leakage inductance theoretical value is too far from the measurement. This parameter must be investigated. This lack of precision could be attributed, in part, to the manual assembling process that prevent form a precise positioning of the layers. However, as mentioned in [49], this difference can also be attributed to the characterization process that must be revised. With such a step-down transformer with high current values at the output, precise measurements make it difficult to characterize secondary windings.

The described FabLab process has led to a functional planar transformer copper foil prototype. Compared to the multilayer PCB planar transformer technology, such a prototype presents benefits such as less complexity in a winding's connections, different and bigger copper layer thickness, and more tunable distance between windings as well as the potential add-on of FPC magnetic layer.

The prototype has been developed using only tools and machines that can be found in a traditional FabLab. With such an approach, making planar transformer becomes affordable for quite everybody, without recourse to complex, expensive and specific PCB multilayer machines. Of course, one can note that the process is only dedicated to power prototypes. Indeed, with lower power range, PCB solutions are acceptable and more realizable in a laboratory. Moreover, process time for each step of conception is not negligible and does not fill with a medium series production. For automated mass production, power planar transformers made of multilayer PCB remain the most attractive solution.

Regarding methodology improvements, insulated layers could be stuck to improve mechanical strength. This requires glue that tolerates functioning temperatures with a good coefficient of thermal expansion. Some good thermal conductor material could also be added in the windings to enhance thermal behavior of the power planar transformer.

At this stage, it might be interesting to investigate the aging of such planar transformer as in [51]. Indeed, copper windings and Kapton insulation can be subject to hot spot, as shown in Figures 13 and 14, that can affect the reliability and the lifetime of the planar transformer. The ecological impact of such a transformer could also be considered performing an eco-dimensioning study. Results could be compared to a planar PCB based technology ecological impact [52].

6. Conclusions

In this paper, the process for conceiving HF power planar transformer prototype with copper foil is described step by step based on machines and tools that can be found in a FabLab. Each elementary layer (copper windings, insulating Kapton) as well as add-ons are first realized independently before being stacked and assembled together. The focus is on difficulties and tricks that enable a functional planar transformer to be realized. A 2 kW planar transformer prototype is designed, developed and characterized with an impedance analyzer. Thermal measurements are also made with a HF power full bridge converter. The method and results are then discussed to highlight the limits of and improvements to the proposed process.

Author Contributions: Conceptualization, S.T., J.-S.N.T.M. and R.B.; methodology, S.T., J.-S.N.T.M. and R.B.; validation, X.M., J.-S.N.T.M. and R.B.; investigation, X.M., J.-S.N.T.M. and R.B.; writing—original draft preparation, X.M.; writing—review and editing, S.T., X.M., J.-S.N.T.M., R.B.; project administration, X.M. and P.L.M. All authors have read and agreed to the published version of the manuscript.

Funding: This research received no external funding.

Institutional Review Board Statement: Not applicable.

Informed Consent Statement: Not applicable.

Data Availability Statement: The data presented in the study are available in the article.

Conflicts of Interest: The authors declare no conflict of interest.

References

1. Quirke, M.T.; Barrett, J.J.; Hayes, M. Planar magnetic component technology-a review. *IEEE Trans.Compon. Hybrids Manuf. Technol.* **1992**, *15*, 884–892. [CrossRef]
2. Ouyang, Z.; Andersen, M.A.E. Overview of Planar Magnetic Technology–Fundamental Properties. *IEEE Trans. Power Electron.* **2014**, *29*, 4888–4900. [CrossRef]
3. Ngoua, T.; Magambo, J.S.; Bakri, R.; Margueron, X.; Le Moigne, P.; Mahe, A.; Guguen, S.; Bensalah, T. Planar Magnetic Components in More Electric Aircraft: Review of Technology and Key Parameters for DC-DC Power Electronic Converter. *IEEE Trans. Transp. Electrif.* **2017**, *3*, 831–842. [CrossRef]
4. Betts, B. Bringing the factory home [Manufacturing Personal]. *Eng. Technol.* **2010**, *5*, 56–58. [CrossRef]
5. Redlich, T.; Buxbaum-Conradi, S.; Basmer-Birkenfeld, S.-V.; Moritz, M.; Krenz, P.; Osunyomi, B.D.; Wulfsberg, J.P.; Heubischl, S. OpenLabs—Open Source Microfactories Enhancing the FabLab Idea. In Proceedings of the 2016 49th Hawaii International Conference on System Sciences (HICSS), Koloa, HI, USA, 5–8 January 2016; pp. 707–715. [CrossRef]
6. Fiaidhi, J.; Mohammed, S. Fab Labs: A Platform for Innovation and Extreme Automation. *IT Prof.* **2018**, *20*, 83–90. [CrossRef]
7. Stephenson, M.K.; Dow, D.E. The community FabLab platform: Applications and implications in biomedical engineering. In Proceedings of the 2014 36th Annual International Conference of the IEEE Engineering in Medicine and Biology Society, Chicago, IL, USA, 26–30 August 2014; pp. 1821–1825. [CrossRef]
8. Angrisani, L.; Arpaia, P.; Bonavolonta, F.; Lo Moriello, R.S. Academic FabLabs for industry 4.0: Experience at University of Naples Federico II. *IEEE Instrum. Meas. Mag.* **2018**, *21*, 6–13. [CrossRef]
9. Gendreau, D.; Mohand-Ousaid, A.; Rougeot, P.; Rakotondrabe, M. 3D-Printing: A promising technology to design three-dimensional microsystems. In Proceedings of the 2016 International Conference on Manipulation, Automation and Robotics at Small Scales (MARSS), Paris, France, 18–22 July 2016; pp. 1–5. [CrossRef]
10. Lim, C.W.J.; Le, K.Q.; Lu, Q.; Wong, C.H. An Overview of 3-D Printing in Manufacturing, Aerospace, and Automotive Industries. *IEEE Potentials* **2016**, *35*, 18–22. [CrossRef]
11. Dahle, R.; Rasel, R. 3-D Printing as an Effective Educational Tool for MEMS Design and Fabrication. *IEEE Trans. Educ.* **2016**, *59*, 210–215. [CrossRef]
12. Fiorillo, L.; Leanza, T. Worldwide 3D Printers against the New Coronavirus. *Prosthesis* **2020**, *2*, 87–90. [CrossRef]
13. Tarfaoui, M.; Nachtane, M.; Goda, I.; Qureshi, Y.; Benyahia, H. 3D Printing to Support the Shortage in Personal Protective Equipment Caused by COVID-19 Pandemic. *Materials* **2020**, *13*, 3339. [CrossRef]
14. Rendeki, S.; Nagy, B.; Bene, M.; Pentek, A.; Toth, L.; Szanto, Z.; Told, R.; Maroti, P. An Overview on Personal Protective Equipment (PPE) Fabricated with Additive Manufacturing Technologies in the Era of COVID-19 Pandemic. *Polymers* **2020**, *12*, 2703. [CrossRef]
15. Liang, W.; Raymond, L.; Rivas, J. 3-D-Printed Air-Core Inductors for High-Frequency Power Converters. *IEEE Trans. Power Electron.* **2016**, *31*, 52–64. [CrossRef]
16. Yan, Y.; Ding, C.; Ngo, K.D.T.; Mei, Y.; Lu, G. Additive manufacturing of planar inductor for Power Electronics applications. In Proceedings of the 2016 International Symposium on 3D Power Electronics Integration and Manufacturing (3D-PEIM), Raleigh, NC, USA, 13–15 June 2016; pp. 1–16. [CrossRef]
17. Simpson, N.; Tighe, C.; Mellor, P. Design of High Performance Shaped Profile Windings for Additive Manufacture. In Proceedings of the 2019 IEEE Energy Conversion Congress and Exposition (ECCE), Baltimore, MD, USA, 29 September–3 October 2019; pp. 761–768. [CrossRef]
18. Yu, Z.; Yang, X.; Wei, G.; Wang, L.; Wang, K.; Chen, W.; Wei, J. A Novel High-Current Planar Inductor With Cooling Fins Based on 3-D Printing. *IEEE Trans. Power Electron.* **2021**, *36*, 12189–12195. [CrossRef]
19. Hou, T.; Xu, J.; Elkhuizen, W.S.; Wang, C.C.L.; Jiang, J.; Geraedts, J.M.P.; Song, Y. Design of 3D Wireless Power Transfer System Based on 3D Printed Electronics. *IEEE Access* **2019**, *7*, 94793–94805. [CrossRef]
20. Gunewardena, T.R. Manufacturing design considerations for planar magnetics. In Proceedings of the Electrical Insulation Conference and Electrical Manufacturing and Coil Winding Conference, Rosemont, IL, USA, 25 September 1997; pp. 309–311. [CrossRef]
21. Zumel, P.; Prieto, R.; Cobos, J.A.; Uceda, J. Comparative study of flex-foil technology in HF planar transformer windings. In Proceedings of the 2002 IEEE 33rd Annual IEEE Power Electronics Specialists Conference, Cairns, Australia, 23–27 June 2002; pp. 1248–1253. [CrossRef]
22. de Jong, E.C.W.; Ferreira, B.J.A.; Bauer, P. Toward the Next Level of PCB Usage in Power Electronic Converters. *IEEE Trans. Power Electron.* **2008**, *23*, 3151–3163. [CrossRef]
23. Saket, M.A.; Shafiei, N.; Ordonez, M. LLC Converters with Planar Transformers: Issues and Mitigation. *IEEE Trans. Power Electron.* **2017**, *32*, 4524–4542. [CrossRef]
24. Chen, M.; Araghchini, M.; Afridi, K.K.; Lang, J.H.; Sullivan, C.R.; Perreault, D.J. A Systematic Approach to Modeling Impedances and Current Distribution in Planar Magnetics. *IEEE Trans. Power Electron.* **2016**, *31*, 560–580. [CrossRef]

25. Wang, Y.; Roodenburg, B.; Haan, d.S.W.H. Comparative study of three transformer concepts for high current dual active bridge converters. In Proceedings of the 5th International Conference on Integrated Power Electronics Systems, Nuremberg, Germany, 11–13 March 2008; pp. 1–4.
26. Carsten, B.W. The low leakage inductance of planar transformers; fact or myth? In Proceedings of the APEC 2001, Sixteenth Annual IEEE Applied Power Electronics Conference and Exposition (Cat. No.01CH37181), Anaheim, CA, USA, 4–8 March 2001; pp. 1184–1188. [CrossRef]
27. Planar E Cores. Available online: https://elnamagnetics.com/wp-content/uploads/library/Ferroxcube-Documents/Planar_E_Cores.pdf (accessed on 10 October 2021).
28. Chen, C.; Huang, Z.; Chen, L.; Tan, Y.; Kang, Y.; Luo, F. Flexible PCB-Based 3-D Integrated SiC Half-Bridge Power Module With Three-Sided Cooling Using Ultralow Inductive Hybrid Packaging Structure. *IEEE Trans. Power Electron.* **2019**, *34*, 5579–5593. [CrossRef]
29. Ho, G.K.Y.; Zhang, C.; Pong, B.M.H.; Hui, S.Y.R. Modeling and Analysis of the Bendable Transformer. *IEEE Trans. Power Electron.* **2016**, *31*, 6450–6460. [CrossRef]
30. Tan, W.; Margueron, X.; Taylor, L.; Idir, N. Leakage Inductance Analytical Calculation for Planar Components with Leakage Layers. *IEEE Trans. Power Electron.* **2016**, *31*, 4462–4473. [CrossRef]
31. TDK. Available online: https://www.tdk-electronics.tdk.com/en (accessed on 30 November 2020).
32. Kheraluwala, M.N.; Gascoigne, R.W.; Divan, D.M.; Baumann, E.D. Performance characterization of a high-power dual active bridge DC-to-DC converter. *IEEE Trans. Ind. Appl.* **1992**, *28*, 1294–1301. [CrossRef]
33. McLymnan, C.W.T. Chapter 5 Transformer Design Trade-Offs. In *Transformer and Inductor Design Handbook*, 4th ed.; CRC Press: Boca Raton, FL, USA, 2011.
34. Ferroxcube. Available online: http://www.ferroxcube.com (accessed on 10 October 2021).
35. Ansys. Available online: http://www.ansys.com (accessed on 10 October 2021).
36. Margueron, X.; Besri, A.; Jeannin, P.; Keradec, J.; Parent, G. Complete Analytical Calculation of Static Leakage Parameters: A Step toward HF Transformer Optimization. *IEEE Trans. Ind. Appl.* **2010**, *46*, 1055–1063. [CrossRef]
37. Dowell, P.L. Effects of eddy currents in transformer windings. *Proc. Inst. Electr. Eng.* **1966**, *113*, 1387–1394. [CrossRef]
38. Ferreira, J.A. Improved analytical modeling of conductive losses in magnetic components. *IEEE Trans. Power Electron.* **1994**, *9*, 127–131. [CrossRef]
39. Design of Planar Power Transformer. Available online: http://ferroxcube.home.pl/appl/info/plandesi.pdf (accessed on 10 October 2021).
40. McLyman, C.W.T. Chapter 6 Transformer-Inductor Efficiency, Regulation, and Temperature Rise. In *Transformer and Inductor Design Handbook*, 4th ed.; CRC Press: Boca Raton, FL, USA, 2011.
41. Onshape. Available online: https://www.onshape.com/en/ (accessed on 10 October 2021).
42. Stepcraft. Available online: https://www.stepcraft-systems.com/ (accessed on 10 October 2021).
43. Trotec. Available online: https://www.troteclaser.com/en/ (accessed on 10 October 2021).
44. Zortrax. Available online: https://zortrax.com/ (accessed on 10 October 2021).
45. Payton. Available online: https://www.paytongroup.com/ (accessed on 23 November 2021).
46. Agilent 4294A Precision Impedance Analyzer. Available online: http://literature.cdn.keysight.com/litweb/pdf/04294-90060.pdf (accessed on 10 October 2021).
47. Margueron, X.; Keradec, J.P. Identifying the Magnetic Part of the Equivalent Circuit of n-Winding Transformers. *IEEE Trans. Instrum. Meas.* **2007**, *56*, 146–152. [CrossRef]
48. Chauvin Arnoux. Available online: https://catalog.chauvin-arnoux.co.uk/uk_en/c-a-6255.html (accessed on 10 October 2021).
49. Skutt, G.; Lee, F.C.; Ridley, R.; Nicol, D. Leakage inductance and termination effects in a high-power planar magnetic structure. In Proceedings of the 1994 IEEE Applied Power Electronics Conference and Exposition-ASPEC'94, Orlando, FL, USA, 13–17 February 1994; pp. 295–301. [CrossRef]
50. Ti95. Available online: https://www.myflukestore.com/ (accessed on 10 October 2021).
51. Shen, Z.; Wang, Q.; Wang, H. Degradation Analysis of Planar Magnetics. In Proceedings of the 2020 IEEE Applied Power Electronics Conference and Exposition (APEC), New Orleans, LA, USA, 15–19 March 2020; pp. 2687–2693. [CrossRef]
52. de Freitas Lima, G.; Rahmani, B.; Rio, M.; Lembeye, Y.; Crebier, J.-C. Eco-Dimensioning Approach for Planar Transformer in a Dual Active Bridge (DAB) Application. *Eng* **2021**, *2*, 544–561. [CrossRef]

electricity

Article

Model Predictive Operation Control of Islanded Microgrids under Nonlinear Conversion Losses of Storage Units

Milad Gholami * and Alessandro Pisano

Department of Electrical and Electronic Engineering (DIEE), University of Cagliari, 09123 Cagliari, Italy; apisano@unica.it
* Correspondence: milad.gholami@unica.it; Tel.: +39-3898550651

Abstract: This paper proposes a certainty equivalence model predictive control (MPC) approach for the operation of islanded microgrids with a very high share of renewable energy sources. First, to make the MG model more realistic, the conversion losses of the storage units and the conversion losses of the power electronic devices are considered by the quadratic functions in the dynamic of units. Then, to mitigate the effect of errors in the storage units' state of charge prediction, the conversion loss functions are reformulated by the mixed-integer linear inequality functions and included in the proposed scheme. Finally, the effectiveness of the proposed certainty MPC is verified by a numerical case study.

Keywords: microgrid; operation control; model predictive control

Citation: Gholami, M.; Pisano, A. Model Predictive Operation Control of Islanded Microgrids under Nonlinear Conversion Losses of Storage Units. *Electricity* **2022**, 3, 33–50. https://doi.org/10.3390/electricity3010003

Academic Editors: Pavlos S. Georgilakis and Philippe Jacquod

Received: 2 October 2021
Accepted: 17 January 2022
Published: 26 January 2022

1. Introduction

Electric power systems have played a pivotal role in technological advances, infrastructure development, and economic growth since their invention [1]. However, conventional power systems typically use fossil fuels (for example, coal, natural gas, or oil) and nuclear and hydropower plants [2]. Unfortunately, the performance of most of them leads to a significant increase in greenhouse gas emissions. Hence, in recent years, researchers have been encouraged to reduce greenhouse gas emissions and fossil fuel consumption in power systems. One of the most effective ways to reduce greenhouse gas emissions is by replacing conventional generators with Renewable Energy Sources (RES) [3]—for example, Photo-Voltaic (PV) power plants or Wind Turbines (WT). Furthermore, to facilitate the integration of a sizeable number of renewable Distributed Generation (DG) units, the concept of microgrids (MGs) has become increasingly popular [4,5].

A microgrid is a small-scale power system, generally consisting of conventional generators, RESs, Energy Storage Sources (ESSs), and loads interconnected by transmission lines [6–8]. In general, MGs can be typically operated in grid-connected or island mode. Recently, the MG control system has been standardized into three layers [7]. The inner loop is called Primary Control (PC), and it provides the references for the DG's DC-AC power converters. The PC is decentralized to establish the desired sharing of power among DGs while preserving the MG stability by employing a droop control term. This layer typically operates on a fast timescale (in the range of tens of milliseconds to seconds). Then, because inverter-based DGs have no inertia, a Secondary Control (SC) layer is needed to compensate for the frequency and voltage deviations introduced by the PC's droop control terms and generally operates on a timescale from seconds to minutes. Finally, the operation control is designed to optimize the operation of the MGs by providing power setpoints to the lower control layers. The operation control typically operates on a timescale from minutes to fractions of hours. For this task, model predictive control (MPC) approaches are considered a good choice as they allow us to include constraints on the units explicitly and evaluate the system dynamics. This work mainly focuses on the optimal operation of microgrids with a high share of renewable sources.

1.1. Literature Review

Several control schemes have been reported for the operation of islanded MGs. According to how they handle uncertainties, these approaches can be categorized as: (i) certainty equivalence, wherein a forecast in the form of the mean value is fully reliable; (ii) worst-case, where no possibility information is presumed; (iii) risk-neutral stochastic, where a forecast probability distribution is thoroughly reliable, and (iv) risk-averse, where uncertainties in the forecast probability distribution are considered.

In this context, a two-stage operation control algorithm is recorded in [9], which includes a schedule and a dispatch layer. To dispatch generators of islanded MGs, an energy management approach is proposed in [10] by adapting the power setpoints and droop gains of the units. Furthermore, a method is provided in [11] to schedule islanded MGs. However, refs. [9–11] do not allow us to limit infeed from RES. By using genetic algorithms, a day-ahead schedule for MGs is also designed in [12]. Additionally, ref. [13] proposes an operation controller based on a rolling horizon strategy. Another operation control is introduced in [14] to address an energy management problem for deterministic forecasts of load and renewable generation. The authors in [15] formulated a multi-objective optimization problem based on model predictive control with the goals of minimizing fuel costs and changes in the power output of diesel generators, minimizing costs associated with the low battery life of energy storage. However, losses and line power flow limits are not considered. Using MPC, a centralized energy management system for isolated microgrids without considering power losses is designed in [16], allowing the proper dispatch of the energy storage units. However, refs. [13–16] disregard the power flow over the transmission lines. The model predictive operation control of microgrids in islanded operation mode is studied in [17]. Although there is a limit to the RES infeed, the conversion losses of storage units are ignored. A novel day-ahead EMS based on EV considering non-integer-hour energy flow is subsequently presented in [18] for pelagic island microgrid groups. The authors in [19] propose a real-time-driven primary regulation that mainly depends on the optimized P-f droop coefficient. This regulator reduces the power loss in all the operating conditions in an islanded microgrid. An MPC-based energy scheduling system is designed in [20] for an autonomous islanded microgrid with PV, a dispatchable generator, and hybrid ESS. The approach in [20] suffers from the fact that it is only applicable to grid-connected MGs where the transmission network acts as a slack node. In [21], a convex EMS is formulated for an islanded MG that optimizes its operating and emission costs. In addition, in [22,23], stochastic scenario-based MPC approaches for the operation of islanded MGs are reported. Furthermore, distributed conditional cooperation model predictive control approaches can be found in [24,25]. For the approaches pursued by [22–25], all their results rely on the assumption that there are no conversion losses in the storage units.

In brief, although all the mentioned strategies are promising, most models are hindered by one of the following limitations: (i) they do not consider a possible limitation of infeed from RES [9–11,14–16]; (ii) the dynamics of the storage units are not modeled [10]; (iii) the formulations do not include storage dynamics with power conversion losses [9–25]; (iv) it is not assumed that conventional units can be turned on and off [9,10]; (v) they do not explicitly model the power flow over a transmission network [9–11,13–16,21]. Additionally, the authors in [10,17,22,23] only take into account the power-sharing of grid-forming units.

1.2. Statement of Contributions

Thus motivated, we extend the works in [22,23] by including the conversion losses in the dynamics of storage units and in the proposed controller to reduce the prediction error in the storage units. We assume the nominal forecast of load and available renewable infeed in the proposed controller, while the uncertain RES and load are considered in the MG model. The closed-loop setup of the proposed certainty equivalence MPC scheme for the operation of an islanded MG is shown in Figure 1.

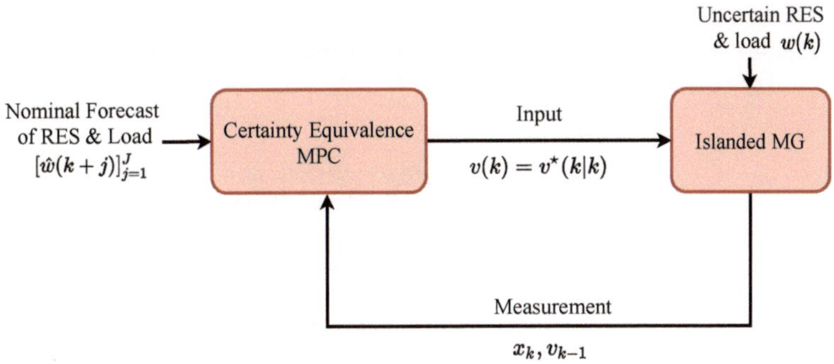

Figure 1. Certainty equivalence MPC scheme for operation of islanded MG at time instant k and $\forall j = 0, \ldots, J-1$.

In brief, the contributions of this work are as follows:

(i) We derive the model of an islanded MG with uncertain renewable generation and loads with a very high share of RES. This model, motivated by [22], considers a possible limitation of renewable infeed, while limitations on transmission lines are approximately accounted for using DC power flow approximations.
(ii) We model storage devices as grid-forming units, and, to make the MG model more realistic, we consider the conversion losses of the storage units, the losses of the power electronic devices when converting Alternating Current (AC) to Direct Current (DC) (and vice versa), as well as ohmic losses in the batteries as the quadratic functions in the dynamics of storage units.
(iii) We take into account power-sharing with the active conventional units. Therefore, the load fluctuations and renewable units influence all units' power and the storage units' state of charge. In this way, the model can also work where only RES and storage units are active, and no conventional unit is required.
(iv) We propose a novel MPC approach for the optimal operation of an islanded MG with a very high share of renewable energy sources. To solve the optimization problem and mitigate the effect of errors in the storage units' state of charge prediction, we reformulate the conversion loss functions as the mixed-integer linear inequality functions and include them in the proposed scheme.
(v) We confirm the properties of the proposed operation control scheme via realistic simulations with a high renewable share.

1.3. Paper Organization

This paper is organized as follows. In Section 2, we introduce the model of an islanded MG. Then, we formulate the certainty equivalence MPC problem in Section 3. In Section 4, we introduce the operating costs of the MG. In Section 5, we demonstrate the effectiveness of the resulting MPC in a numerical case study. Lastly, in Section 6, we provide the concluding remarks and possible directions for future research.

1.4. Mathematical Notation

The sets of complex, real, strictly positive, and negative real numbers are denoted by $\mathbb{C}$, $\mathbb{R}$, $\mathbb{R}_{<0}$, and $\mathbb{R}_{>0}$, respectively. Moreover, the set of non-positive real numbers is $\mathbb{R}_{\leq 0}$ and the set of positive real numbers including 0 is $\mathbb{R}_{\geq 0}$. The set of natural numbers is $\mathbb{N}$, and the set of non-negative integers is $\mathbb{N}_0$. Furthermore, the set of Boolean numbers is $\mathbb{B} = \{0, 1\}$. For $d \in \mathbb{N}$ and a column vector of $x \in \mathbb{R}^d$, let x' be its transpose. Given a matrix A, its transpose is denoted by A', while its hermitian (complex conjugate) transpose is A^H. The matrix A is also non-negative (psd), denoted by $A \succeq 0$, if A is Hermitian and $z^H A z \succeq 0$ for all $z \in \mathbb{C}^n$. Given a complex number z, its real part is denoted by $\mathrm{Re}(z)$, while the imaginary part of a complex number z is denoted by $\mathrm{Im}(z)$. Additionally, given a scalar a, its absolute value is defined by $|a|$. Finally, I_n denotes the n-dimensional identity matrix, and by 1_n and 0_n, respectively, all 1 and all 0 n-dimensional column vectors are represented.

2. Microgrid Model

We consider an MG that includes a grouping of conventional generators, renewable energy storage units, and loads connected by transmission lines. Figure 2 depicts a basic MG consisting of all these components. The electrical connection among units and loads due to power lines can be modeled by a di-graph $\mathcal{G}_N^e(\mathcal{V}, \mathcal{E}^e)$, where $\mathcal{V} = \{1, \ldots, N\}$ represents the set of agents and $\mathcal{E}^e \subseteq \{\mathcal{V} \times \mathcal{V}\}$ is the set of edges (transmission lines) between two distinct agents. In this manuscript, we consider that the set of agents includes four subsets $\mathcal{V}_T$, $\mathcal{V}_S$, $\mathcal{V}_R$, and $\mathcal{V}_L$, wherein these letters denote, respectively, the sets of conventional, energy storage, renewable units, and loads. We also denote by $\mathcal{Y}_{ij} = G_{ik} + \imath \cdot B_{ij} \in \mathbb{C}$ the admittance line between the i-th agent and j-th agent, wherein G_{ij} and B_{ij} show the line conductance and susceptance between the i-th agent and j-th agent. If no connection between the i-th agent and j-th agent exists, $G_{ij} = B_{ij} = 0$.

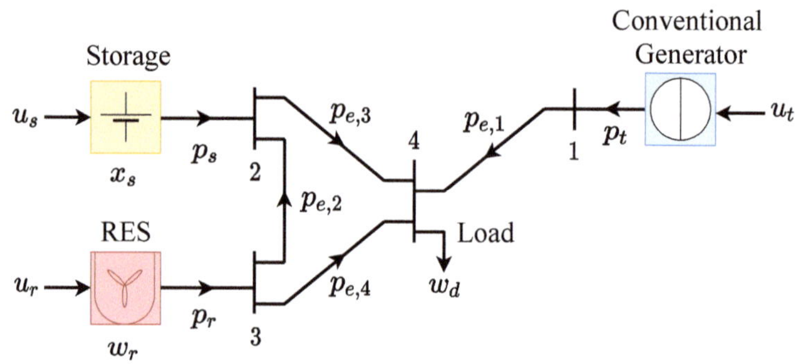

Figure 2. MG used for operation control as a running example.

With the variables summarized in Table 1, the behavior of the MG is modeled by:

$$x(k+1) = x(k) - T_s \cdot \Big\{ B \cdot \bar{q}(k) + \mathbb{F}(k) \Big\}, \tag{1a}$$

$$h_1 \geq H_1 \cdot x(k+1), \tag{1b}$$

$$h_2 \geq H_2 \cdot \left[v(k)' \quad \bar{q}(k)' \quad w(k)' \right]', \tag{1c}$$

$$g = G \cdot \left[v(k)' \quad \bar{q}(k)' \quad w(k)' \right]', \tag{1d}$$

where $k \in \mathbb{N}_0$ is the time index, $x(k) \in \mathbb{R}^S_{\geq 0}$ with $S \in \mathbb{N}$ is the state vector with initial value $x_0 \in \mathbb{R}^S_{\geq 0}$. In fact, this vector is inclusive of entries $x_i(k)$ that represent the stored energy of unit $i \in \mathbb{N}_{[1,S]}$. $\bar{q}(k) \in \mathbb{R}^Q$ is referred to as a vector of $Q \in \mathbb{N}$ auxiliary variables. $\mathbb{F}(k)$ is a vector of $S \in \mathbb{N}$ in which each of its elements represent the function of the conversion losses. $v(k) = [u(k)' \; \delta_t(k)']'$ is also the vector of control inputs, wherein $u(k) \in \mathbb{R}^U$ is denoted as the vector of the real-valued control inputs of all $U \in \mathbb{N}$ units and $\delta_t(k) \in \{0,1\}^T$ is defined to be the vector of $T \in \mathbb{N}$ Boolean inputs. The uncertain external input is also collected in the vector $w(k) = [w_r(k)' \; w_t(k)']'$, where $w_r(k)$ is the maximum infeed under weather conditions of all renewable units and $w_t(k)$ includes all loads, $W \in \mathbb{N}$. Let $B \in \mathbb{R}^{S \times Q}$ in (1a) and H_1, h_1 in (1b) be appropriate dimensions. Furthermore, we consider in (1c), H_2 and h_2, respectively, as a matrix and a vector of appropriate dimensions that reflect inequality constraints. Likewise, in (1d), G is a matrix and g a vector of appropriate dimensions that reflect equality constraints. In more detail, the real-valued control inputs are the power setpoints of the units $u(k) = [u_t(k)' \; u_s(k)' \; u_r(k)']'$, where $u_t(k)' \in \mathbb{R}^T_{\geq 0}$ is related to the conventional units, $u_s(k)' \in \mathbb{R}^S$ to the storage units, and $u_r(k) \in \mathbb{R}^R_{\geq 0}$ to the renewable units, such as wind turbines and PV power plants. Regarding the storage and conventional generators, $u_s(k)$ and $u_t(k)$ represent desired power values. For RES, $u_r(k)$ represents an upper limit on the weather-dependent power infeed. Hence, $u_r(k)$ is their maximum admissible value.

Table 1. Model-specific variables.

Symbol	Explanation	Unit	Size
x	Energy of storage units (state)	pu h	S
u_t	Setpoint inputs of conventional units	pu	T
u_s	Setpoint inputs of storage units	pu	S
u_r	Setpoint inputs of renewable units	pu	R
u	Setpoint inputs of all units	pu	U
δ_t	Boolean inputs of conventional units	-	T
v	Vector of all control inputs	-	Q
w_r	Uncertain available renewable power	pu	R
w_d	Uncertain load	pu	D
w	Vector of all uncertain inputs	pu	W
p_t	Active power of conventional units	pu	T
p_s	Active power of storage units	pu	S
p_r	Active power of renewable units	pu	R
p	Active power of all units	pu	U
p_e	Power over transmission lines	pu	E
δ_r	Boolean auxiliary variables	-	R
ρ	Real-valued auxiliary variable	-	1
$\bar{q}$	Vector of all auxiliary variables	-	Q

Furthermore, for each conventional generator, a Boolean control input $\delta_{t,i}(k) \in \{0,1\}$ is considered. This input shows whether generator $i \in \mathbb{N}_{[1,T]}$ is active ($\delta_{t,i}(k) = 1$) or inactive ($\delta_{t,i}(k) = 0$). The Boolean variables of all conventional generators are gathered in the vector δ_t.

In what follows, we will derive a control-oriented MG model of the form (1). We start by posing some assumptions.

Assumption 1 (Lower control layers). *The lower control layers, i.e., primary and secondary control, are considered to compensate for the frequency and voltage deviations by providing setpoints to the units, as well as establishing the power-sharing (see, e.g., [6,26,27]) among the grid-forming units. Notice that the MG can run autonomously in these layers for several minutes.*

Assumption 2 (Conventional units). *In terms of time, conventional units have shorter start-up and shutdown times than the sampling time of (MPC), meaning that switching actions are supposed to be instantaneous. Furthermore, changes in power are instantaneous, i.e., no climb rates need to be considered.*

Assumption 3 (RES units and loads). *It is assumed that the uncertain input follows the nominal forecast of load and available renewable infeed.*

Assumption 4 (Transmission lines). *It is assumed that the resistance of the electrical coupling among units and loads of MG as well as reactive power flow is negligible. Since the voltage amplitudes in the network are constant, and the phase angle differences small, the DC power flow approximations [28] can be employed.*

3. Certainty Equivalence Model Predictive Control

3.1. Plant Model Interface

The power setpoints of the units $u(k) = [u_t(k)'\ u_s(k)'\ u_r(k)']' \in \mathbb{R}^U$, called the real-valued manipulated variables containing $u_t(k)' \in \mathbb{R}^T_{\geq 0}$ as setpoints of the T conventional units, $u_s(k)' \in \mathbb{R}^S$ the setpoints of the S storage units and $u_r(k) \in \mathbb{R}^R_{\geq 0}$ the setpoints of the R (RES). To show that conventional units are enabled or disabled, we consider a Boolean input for each conventional unit and we collect all Boolean inputs in a vector $\delta_t(k) \in \{0,1\}^T$. Moreover, the stored energies of the storage units are gathered in the state vector $x(k) \in \mathbb{R}^S_{\geq 0}$. The uncertain external input of the model is expressed by $w(k) = [w_r(k)'\ w_d(k)']'$, where $w_r(k) \in \mathbb{R}^R_{\geq 0}$ shows the maximum available power of the renewable units under given weather conditions and $w_d(k) \in \mathbb{R}^D_{\geq 0}$ the load.

3.2. Power of Units

We consider $p(k) = [p_t(k)'\ p_s(k)'\ p_r(k)']'$ as the vector of power values, which consists of the power of conventional units, $p_t(k) \in \mathbb{R}^T_{\geq 0}$, storage units, $p_s \in \mathbb{R}^S$, and RES, $p_r(k) \in \mathbb{R}^R_{\geq 0}$. It is worth noting that, in islanded mode, since production, consumption, and storage power must be balanced in the presence of uncertain load and renewable infeed, the power of the units $p(k) \in \mathbb{R}^U$ is not necessarily equal to the setpoints $u(k)$.

3.2.1. Active Power at RES Units

Let the active power of renewable units, $p_{\mathrm{r}}(k)$, as well as the corresponding setpoints, $u_{\mathrm{r}}(k)$, be bounded by:

$$p_{\mathrm{r}}^{\min} \leq p_r(k) \leq p_{\mathrm{r}}^{\max}, \tag{2a}$$
$$p_{\mathrm{r}}^{\min} \leq u_r(k) \leq p_{\mathrm{r}}^{\max}. \tag{2b}$$

Furthermore, we consider that the power infeed $p_{\mathrm{r},i}(k) \in \mathbb{R}_{\geq 0}$ of any renewable unit $i \in \mathcal{N}_{[1,R]}$ can be constrained by the power setpoint $u_{\mathrm{r},i}(k) \in \mathbb{R}_{\geq 0}$. Notice that the power tracks the setpoint when the maximum possible infeed under current weather conditions $w_{\mathrm{r},i}(k) \in \mathbb{R}_{\geq 0}$ is greater than or equal to $u_{\mathrm{r},i}(k)$. This can be characterized by using the element-wise min operator as follows:

$$p_r(k) = \min(u_r(k), w_r(k)). \tag{3}$$

To solve the optimization problem, the authors in [22] reformulated (3) to a set of linear inequalities including integer variables as follows:

$$p_{\mathrm{r}}(k) \leq u_{\mathrm{r}}(k), \tag{4a}$$
$$p_{\mathrm{r}}(k) \geq u_{\mathrm{r}}(k) + (\mathrm{diag}(w_{\mathrm{r}}(k)) - \mathrm{M_r} I_R)\delta_{\mathrm{r}}(k), \tag{4b}$$
$$p_{\mathrm{r}}(k) \leq w_{\mathrm{r}}(k), \tag{4c}$$
$$p_{\mathrm{r}}(k) \geq w_{\mathrm{r}}(k) - (\mathrm{diag}(w_{\mathrm{r}}(k)) - \mathrm{m_r} I_R)(1_R - \delta_{\mathrm{r}}(k)). \tag{4d}$$

where $\delta_{\mathrm{r}}(k) \in \{0,1\}^R$ represents the free variable and $\mathrm{m_r} \in \mathbb{R}$, $\mathrm{m_r} < \min(p_{\mathrm{r}}^{\min})$ and $\mathrm{M_r} \in \mathbb{R}_{\geq 0}$, $\mathrm{M_r} > \max(p_{\mathrm{r}}^{\max})$ are the constant coefficients, which are computed offline.

3.2.2. Active Power at Conventional Units

We consider whether the conventional unit $i \in \mathcal{N}_{[1,T]}$ is enabled, i.e., if $\delta_{\mathrm{t},i}(k) = 1$, then its active power is bounded by $p_{\mathrm{t},i}^{\min} \in \mathbb{R}_{\geq 0}$ and $p_{\mathrm{t},i}^{\max} \in \mathbb{R}_{\geq 0}$. If the unit is disabled, i.e., $\delta_{\mathrm{t},i}(k) = 0$, then, naturally, $p_{\mathrm{t},i}(k) = 0$. The active power of conventional units with $p_{\mathrm{t}}^{\min} \in \mathbb{R}^T_{\geq 0}$, $p_{\mathrm{t}}^{\max} \in \mathbb{R}^T_{\geq 0}$ can be written in vector form as [22]:

$$\mathrm{diag}\left(p_{\mathrm{t}}^{\min}\right)\delta_{\mathrm{t}}(k) \leq p_{\mathrm{t}}(k) \leq \mathrm{diag}(p_{\mathrm{t}}^{\max})\delta_{\mathrm{t}}(k), \tag{5a}$$

The same holds for the active power setpoints, i.e.,

$$\mathrm{diag}\Big(p_t^{\min}\Big)\delta_t(k) \le u_t(k) \le \mathrm{diag}(p_t^{\max})\delta_t(k). \tag{5b}$$

3.2.3. Active Power at Storage Units

Since we assume that all storage units are always enabled, all their active power setpoints and active power values are bounded as:

$$p_s^{\min} \le p_s(k) \le p_s^{\max}, \tag{6a}$$

$$p_s^{\min} \le u_s(k) \le p_s^{\max}. \tag{6b}$$

where $p_s^{\min} \in \mathbb{R}^S_{\le 0}$ and $p_s^{\max} \in \mathbb{R}^S_{\ge 0}$ represent the known lower and upper power limits.

3.3. Power Sharing of Grid-Forming Units

Note that the power of all units does not necessarily equal the power setpoints that are assigned to the system due to variations in load and renewable infeed. Therefore, we assume that the secondary and primary control layers control all units to share the changes in load and renewable infeed in a desired proportional manner. The aforementioned so-called proportional power-sharing (see, e.g., [27,29]) depends on the design parameter $\chi_i \in \mathbb{R}_{>0}$ for all grid-forming units. A practical option for χ_i is, e.g., proportional to the nominal power of the corresponding units.

Power sharing can be described by (7), where two units $i \in \mathcal{N}_{[1,T+S]}$ and $j \in \mathcal{N}_{[1,T+S]}$, $i \neq j$ share their active power proportionally according to $\chi_i \in \mathbb{R}_{>0}$ and $\chi_j \in \mathbb{R}_{>0}$, if the next relation holds:

$$\frac{p_i(k) - u_i(k)}{\chi_i} = \frac{p_j(k) - u_j(k)}{\chi_j} \tag{7}$$

By using a new auxiliary free variable $\rho(k) \in \mathbb{R}$ and considering that only enabled units, namely units i with $\delta_{t,i}(k) = 1$, can participate in power-sharing, (7) can be recast for all grid-forming units with $K_t = \mathrm{diag}\Big([\frac{1}{\chi_1} \ \cdots \ \frac{1}{\chi_T}]'\Big)$ and $K_s = \mathrm{diag}\Big([\frac{1}{\chi_{(T+1)}} \ \cdots \ \frac{1}{\chi_{(T+S)}}]'\Big)$ as [22]:

$$K_t(p_t(k) - u_t(k)) = \rho(k)\delta_t(k) \text{ and} \tag{8a}$$

$$K_s(p_s(k) - u_s(k)) = \rho(k)1_S. \tag{8b}$$

For the formulation of the optimisation problem, ref. [22] transforms (8a) into the following set of linear inequalities with integer variables:

$$K_t(p_t(k) - u_t(k)) \le \mathrm{M}_t\delta_t(k), \tag{9a}$$

$$K_t(p_t(k) - u_t(k)) \ge \mathrm{m}_t\delta_t(k), \tag{9b}$$

$$K_t(p_t(k) - u_t(k)) \le 1_T\rho(k) - \mathrm{m}_t(1_T - \delta_t(k)), \tag{9c}$$

$$K_t(p_t(k) - u_t(k)) \ge 1_T\rho(k) - \mathrm{M}_t(1_T - \delta_t(k)). \tag{9d}$$

where $\mathrm{M}_t \in \mathbb{R}$ can be calculated offline and its value should be greater than the largest possible value of $\rho(k)$. Hence, with the largest possible value for the storage units, $\rho_s^{\max} = \max(K_s(p_s^{\max} - p_s^{\min}))$, and for the conventional units, $\rho_t^{\max} = \max(K_t(p_t^{\max} - p_t^{\min}))$, M_t has to be chosen such that $\max\left(\rho_s^{\max}, \rho_t^{\max}\right) < \mathrm{M}_t$. Moreover, $\mathrm{m}_t = -\mathrm{M}_t$.

3.4. Dynamics of Storage Units

The dynamics of all storage units can be formulated as:

$$x(k+1) = x(k) - \mathrm{T}_s p_s(k) - \mathrm{T}_s\mathbb{F}(p_s(k)), \tag{10}$$

where $T_s \in \mathbb{R}_{>0}$ is the sampling time and $x(k)$ represents the stored energy with initial state $x(0) = x_0$. The constraint of the stored energy is given by:

$$x^{\min} \leq x(k+1) \leq x^{\max}, \tag{11}$$

with $x^{\min} = 0_S$ and $x^{\max} \in \mathbb{R}^S_{\geq 0}$. In particular, $\mathbb{F}(p_s(k)) = [f_1(p_{s,1}(k)), \ldots, f_S(p_{s,S}(k))]'$ is a vector of $S \in \mathbb{N}$, where each of its elements represents the conversion loss terms, including the conversion losses of the storage units, the losses of the power electronic devices when converting AC to DC (and vice versa), as well as ohmic losses in the batteries, considered to be a convex quadratic function as follows:

$$f_i(p_{s,i}(k)) = ap_{s,i}(k)^2 + bp_{s,i}(k) + c, \quad a, b, c \in \mathbb{R} \tag{12}$$

where $p_{s,i}(k)$ is assumed to be limited as:

$$p_{s,i}(k) \in \mathcal{D}_i = \{p_{s,i}(k) | p^{\min}_{s,i} \leq p_{s,i}(k) \leq p^{\max}_{s,i}\} \tag{13}$$

Note that the polynomial function, $f_i(p_{s,i}(k))$, is an example and it could be replaced by some other complex functions for different storage technology. Moreover, the coefficients are found offline by looking at the storage units and analyzing measurements.

To solve the optimization problem, it is useful to reformulate the function $f(p_{s,i}(k))$ as piecewise affine functions (see, e.g., [30]), i.e.,

$$f_i(p_{s,i}(k)) = \begin{cases} A_{1,i}p_{s,i}(k) + B_{1,i}, & p_{s,i}(k) \in \mathcal{D}_{1,i}, \\ A_{2,i}ps, i(k) + B_{2,i}, & p_{s,i}(k) \in \mathcal{D}_{2,i}, \\ \vdots \\ A_{r,i}p_{s,i}(k) + B_{r,i}, & p_{s,i}(k) \in \mathcal{D}_{r,i}. \end{cases} \tag{14}$$

in which $A_{y,i}$, $B_{y,i} \in \mathbb{R}$ and the following holds:

$$\bigcup_{1 \leq y \leq r} \mathcal{D}_{y,i} = \mathcal{D}_i \tag{15a}$$

$$\bigcap_{1 \leq y \leq r} \mathcal{D}_{y,i} = 0 \tag{15b}$$

and, on the borders of sequential $\mathcal{D}_y$, the linear segments are connected, which means that $f_i(p_{s,i}(k))$ is continuous.

The condition $p_{s,i}(k)$ at each partition $\mathcal{D}_{y,i}$ can be associated with a binary variable $\delta_{y,i}(k) \in \{0,1\}$, $\forall y = 1, 2, \ldots, r$, satisfying the exclusive-or condition:

$$\bigoplus_{y=1}^{r} [\delta_{y,i}(k) = 1]. \tag{16}$$

such that:

$$[\delta_{y,i}(k) = 1] \longleftrightarrow p_{s,i}(k) \in \mathcal{D}_{y,i} \tag{17}$$

From (16), there exists some $\delta_{y,i}(k) = 1$, which implies $p_{s,i}(k) \in \mathcal{D}_{y,i}$, a contradiction by (15b). Therefore, (15)–(17) are equivalent to:

$$\mathbb{J}^T_{y,i}p_{s,i}(k) - \mathbb{H}^T_{y,i} \leq \mathbb{M}^\star_i[1 - \delta_{y,i}(k)] \tag{18a}$$

$$\sum_{y=1}^{r} \delta_{y,i}(k) = 1 \tag{18b}$$

with $\mathbb{J}_{y,i} = \begin{bmatrix} 1 & -1 \end{bmatrix}$ for $y = \{1, \ldots, r\}$ and $i = \{1, \ldots, S\}$, and $\mathbb{H}_{y,i}$ represents a vector of 2, where the first row of $\mathbb{H}_{y,i}$ is equal to the lower bound of the $D_{y,i}$ with a minus sign, while its second row is the upper bound. $\mathbb{M}_i^\star$ in (18a) can be computed as:

$$\mathbb{M}_i^\star \cong \max_{p_{s,i}(k) \in \mathcal{D}_i} \mathbb{J}_{y,i}^T p_{s,i}(k) - \mathbb{H}_{y,i}^T \tag{19}$$

By using this binary variable, we can recast (14) as follows (see, e.g., [30]):

$$f_i(p_{\mathrm{s,i}}(k)) = \begin{cases} A_{1,i} p_{\mathrm{s,i}}(k) + B_{1,i}, & \text{if } \delta_{1,i}(k) = 1, \\ A_{2,i} p_{\mathrm{s,i}}(k) + B_{2,i}, & \text{if } \delta_{2,i}(k) = 1, \\ \vdots & \\ A_{r,i} p_{\mathrm{s,i}}(k) + B_{r,i}, & \text{if } \delta_{r,i}(k) = 1. \end{cases} \tag{20}$$

Therefore, (20) can be rewritten as:

$$f_i(p_{\mathrm{s,i}}(k)) = \sum_{y=1}^{r} [A_{y,i} p_{\mathrm{s,i}}(k) + B_{y,i}] \delta_{y,i}(k). \tag{21}$$

Unfortunately, (21) is nonlinear since it involves products between logical variables and inputs. Therefore, we transform it into equivalent mixed-integer linear inequalities. This can be done using a similar strategy as proposed in [30]. To this aim, we set:

$$f_i(p_{\mathrm{s,i}}(k)) = \sum_{y=1}^{r} z_{y,i}(k) \tag{22a}$$

$$z_{y,i}(k) \cong [A_{y,i} p_{s,i}(k) + B_{y,i}] \delta_{y,i}(k). \tag{22b}$$

Then, (22b) is equivalent to:

$$\begin{aligned} & z_{y,i}(k) \le \widetilde{M}_i \delta_{y,i}(k), \\ & z_{y,i}(k) \ge \widetilde{m}_i \delta_{y,i}(k), \\ & z_{y,i}(k) \le A_{y,i} p_{s,i}(k) + B_{y,i} - \widetilde{m}_i (1 - \delta_{y,i}(k)), && (23) \\ & z_{y,i}(k) \ge A_{y,i} p_{s,i}(k) + B_{y,i} - \widetilde{M}_i (1 - \delta_{y,i}(k)). && (24) \end{aligned}$$

being

$$\widetilde{M}_i \cong \max_{y=1,\ldots,r} \Big\{ \max_{p_{s,i}(k) \in \mathcal{D}_i} A_{i,y} p_{s,i}(k) + B_{i,y} \Big\}. \tag{25a}$$

$$\widetilde{m}_i \cong \min_{y=1,\ldots,r} \Big\{ \max_{p_{s,i}(k) \in \mathcal{D}_i} A_{i,y} p_{s,i}(k) + B_{i,y} \Big\}. \tag{25b}$$

Remark 1. *Notice that, in [22], the dynamics of all storage units are considered without piecewise affine losses terms, namely,*

$$x(k+1) = x(k) - \mathrm{T_s} p_{\mathrm{s}}(k), \tag{26}$$

with

$$x^{\min} \le x(k+1) \le x^{\max}, \tag{27}$$

3.5. Transmission Network

Following [24], the DC power flow approximations can be employed to extract the power of transmission lines, $p_e(k) = [p_{e,1}(k) \quad \ldots \quad p_{e,E}(k)]'$. Hence, the power on lines can be formulated from the power of units and load using the following linear equation:

$$p_e(k) = F \cdot [p(k)' \quad w_d(k)']', \tag{28a}$$

where $F \in \mathbb{R}^{E\times(U+D)}$ represents a matrix that links the power flowing over the lines with the power produced or consumed by the units and loads. More details about the derivation of F are provided in [31]. It is assumed that $p_e(k)$ is bounded by:

$$p_e^{\min} \leq p_e(k) \leq p_e^{\max} \tag{28b}$$

with $p_e^{\min} \in \mathbb{R}^E_{\leq 0}$ and $p_e^{\max} \in \mathbb{R}^E_{\geq 0}$. This assumption is reasonable due to the physical limitation in the transmission capability of the lines. Moreover, the produced power must be equal to the consumed power at all times [24], e.g.,

$$1'_T p_t(k) + 1'_S p_s(k) + 1'_R p_r(k) = 1'_D w_d(k). \tag{28c}$$

3.6. Overall Model

In accordance with the equations considered for the different parts of an islanded MG, the overall model can be formulated as follows. The constraints on power and setpoint originate from (2), (5) and (6), namely,

$$\begin{bmatrix} \operatorname{diag}(p_{\mathrm{t}}^{\max})\delta_{\mathrm{t}}(k) \\ p_{\mathrm{s}}^{\max} \\ p_{\mathrm{r}}^{\max} \end{bmatrix} \leq u(k) \leq \begin{bmatrix} \operatorname{diag}(p_{\mathrm{t}}^{\min})\delta_{\mathrm{t}}(k) \\ p_{\mathrm{s}}^{\min} \\ p_{\mathrm{r}}^{\min} \end{bmatrix} \tag{29a}$$

and

$$\begin{bmatrix} \operatorname{diag}(p_{\mathrm{t}}^{\max})\delta_{\mathrm{t}}(k) \\ p_{\mathrm{s}}^{\max} \\ p_{\mathrm{r}}^{\max} \end{bmatrix} \leq p(k) \leq \begin{bmatrix} \operatorname{diag}(p_{\mathrm{t}}^{\min})\delta_{\mathrm{t}}(k) \\ p_{\mathrm{s}}^{\min} \\ p_{\mathrm{r}}^{\min} \end{bmatrix} \tag{29b}$$

By referring to (10), the dynamics of the storage unit are described by:

$$x(k+1) = x(k) - \mathrm{T_s} p_{\mathrm{s}}(k) - \mathrm{T_s}[f_1(p_{\mathrm{s},1}(k)), \dots, f_S(p_{\mathrm{s},S}(k))]', \tag{29c}$$

with constraint functions

$$x^{\min} \leq x(k+1) \leq x^{\max}, \tag{29d}$$

$$f_i(p_{\mathrm{s,i}}(k)) = \sum_{y=1}^{r} z_{y,i}(k) \tag{29e}$$

$$\begin{aligned} z_{y,i}(k) &\leq \widetilde{M}_i \delta_{y,i}(k), \\ z_{y,i}(k) &\geq \widetilde{m}_i \delta_{y,i}(k), \\ z_{y,i}(k) &\leq A_{y,i} p_{s,i}(k) + B_{y,i} - \widetilde{m}_i(1 - \delta_{y,i}(k)), \\ z_{y,i}(k) &\geq A_{y,i} p_{s,i}(k) + B_{y,i} - \widetilde{M}_i(1 - \delta_{y,i}(k)). \end{aligned} \tag{29f}$$

The renewable infeed, which is a function of the setpoint and the available power under weather conditions, is given by (4) as:

$$p_{\mathrm{r}}(k) \leq u_{\mathrm{r}}(k), \tag{29g}$$

$$p_{\mathrm{r}}(k) \geq u_{\mathrm{r}}(k) + (\operatorname{diag}(w_{\mathrm{r}}(k)) - \mathrm{M_r} I_R)\delta_{\mathrm{r}}(k), \tag{29h}$$

$$p_{\mathrm{r}}(k) \leq w_{\mathrm{r}}(k), \tag{29i}$$

$$p_{\mathrm{r}}(k) \geq w_{\mathrm{r}}(k) - (\operatorname{diag}(w_{\mathrm{r}}(k)) - \mathrm{m_r} I_R)(1_R - \delta_{\mathrm{r}}(k)). \tag{29j}$$

Power-sharing of the grid-forming units is described by (8), which, using (9), can be recast into:

$$K_t(p_t(k) - u_t(k)) \leq M_t\delta_t(k), \quad (29k)$$
$$K_t(p_t(k) - u_t(k)) \geq m_t\delta_t(k), \quad (29l)$$
$$K_t(p_t(k) - u_t(k)) \leq 1_T\rho(k) - m_t(1_T - \delta_t(k)), \quad (29m)$$
$$K_t(p_t(k) - u_t(k)) \geq 1_T\rho(k) - M_t(1_T - \delta_t(k)). \quad (29n)$$

Lastly, the power limit of the transmission lines is introduced by (28), i.e.,

$$p_e^{\min} \leq F \cdot [p_t(k)' \quad 'p_s(k)' \quad 'p_r(k)' \quad w_d(k)']' \leq p_e^{\max} \quad (29o)$$

$$1_T' p_t(k) + 1_S' p_s(k) + 1_R' p_r(k) = 1_D' w_d(k). \quad (29p)$$

Now, let us compute a compact form of (29). From (29c) and by denoting $\bar{q}(k) = [p(k)' \; \delta_r(k)' \; \rho(k)]'$ and $B = [0_{S\times T} \;\; I_S \;\; 0_{S\times 2R+1}]$, we can obtain Equation (1a), namely,

$$x(k+1) = x(k) - T_s \cdot \Big\{ B \cdot \bar{q}(k) + \mathbb{F}(k) \Big\}, \quad (30a)$$

By following (29d), we have

$$H_1 \cdot x(k+1) \leq h_1, \quad (30b)$$

with $H_1 = \text{diag}([1_S' \;\; -1_S']')$ and $h_1 = [(x^{\max})' \;\; (-x^{\min})']'$.

Finally, according to (29a) and (29b) and (29n)–(29p), the next equations can be yielded as

$$H_2 \cdot \left[v(k)' \quad \bar{q}(k)' \quad w(k)'\right]' \leq h_2, \quad (30c)$$
$$G \cdot \left[v(k)' \quad \bar{q}(k)' \quad w(k)'\right]' = g, \quad (30d)$$

where H_2 and h_2 in (1c) are formed such that they reflect (2), (4)–(6), (9) and (28b). Additionally, G and g in (1d) are formed such that they reflect (8b), (28a) and (28).This completes the introduction of the control-oriented MG model. The control scheme of the certainty equivalence MPC under the conversion loss functions is shown in Figure 3. Based on this section, we will now formulate a cost function for an islanded MG with a high renewable share.

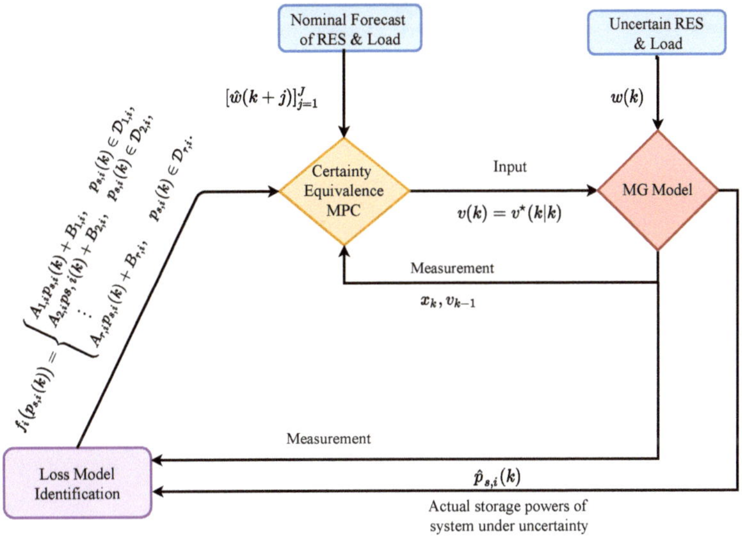

Figure 3. Control scheme of certainty equivalence MPC under the conversion loss functions.

4. Operating Costs

Motivated by [22], here, we extract an operating cost function for the MG. It consists of two parts. The first part is marked by $\ell_o(v(k-1), v(k), \bar{q}(k+1)) \in \mathbb{R}_{\geq 0}$ and is economically motivated. The second part is denoted by $\ell_s(\bar{q}(k)) \in \mathbb{R}_{\geq 0}$ and is related to the desired region of operation of the state of charge and the conversion losses. Thus, the cost function at time instant k, i.e., the stage cost, is:

$$\ell(v(k-1), v(k), \bar{q}(k+1)) = \ell_o(v(k-1), v(k), \bar{q}(k+1)) + \ell_s(\bar{q}(k)) \tag{31}$$

The economically motivated costs comprise (i) the fuel costs of conventional units, $\ell_t^{rt}(v(k), \bar{q}(k+1)) \in \mathbb{R}_{\geq 0}$; (ii) costs caused by switching conventional units on or off, $\ell_t^{sw}(v(k-1), v(k)) \in \mathbb{R}_{\geq 0}$, and (iii) costs incurred by low utilization of renewable sources, $\ell_r(\bar{q}(k+1)) \in \mathbb{R}_{\geq 0}$, i.e.,

$$\begin{aligned}\ell_o(v(k-1), v(k), \bar{q}(k+1)) = \ell_t^{rt}(v(k), \bar{q}(k+1)) + \ell_t^{sw}(v(k-1), v(k)) \\ + \ell_r(\bar{q}(k+1)).\end{aligned} \tag{32}$$

The cost of a conventional generator is generally described by four curves: fuel cost, heat rate, input/output (I/O), and incremental cost. Generator curves are typically represented as cubic or quadratic functions and piecewise linear functions. The conventional power plants use a quadratic fuel cost function such as the Fuel Cost Curve. Thus, by following [22,32], the fuel costs of conventional units are formulated as:

$$\ell_t^{rt}(v(k), \bar{q}(k+1) = c_t' \delta_t(k) + \tilde{c}_t' p_t(k) + p_t(k)' \operatorname{diag}(\hat{c}_t) p_t(k), \tag{33}$$

with weights $c_t \in \mathbb{R}^T_{>0}$, $\tilde{c}_t \in \mathbb{R}^T_{>0}$, and $\hat{c}_t \in \mathbb{R}^T_{>0}$.

In the islanded MGs with a high share of RES, grid-forming conventional units are usually regarded as backup generators in times of low renewable infeed and empty storage units. Moreover, conventional units are often turned off during periods of high renewable infeed or if the storage units can meet the load demand. Therefore, the decision to enable or disable them should be made by the operation control, taking into account their running costs and the cost associated with enabling or disabling them. Enabling or disabling a conventional generator entails costs. Hence, it is desirable to enable or disable conventional generators as little as possible. The switching cost of a conventional generator can be described by considering that it was disabled at the time instant $k-1$ and is enabled at time instant k, or was enabled at time instant $k-1$ and is disabled at time instant k, i.e,

$$\ell_t^{sw}(v(k), v(k-1)) = (\delta_t(k-1) - \delta_t(k))' \operatorname{diag}(c_t^{sw})(\delta_t(k-1) - \delta_t(k)) \tag{34}$$

with weight $c_t^{sw} \in \mathbb{R}^T_{>0}$.

Renewable sources, such as PV power plants or WTs, are usually associated with a very high initial investment and small operation costs after installation. Hence, RES heirs wish to keep the units' infeed as high as possible under given weather conditions. Restricting a renewable unit can be considered a financial loss as the available infeed is not sold to customers. Thus, the renewable unit costs can be adjusted by considering a penalty for using less than the maximal power p_r^{max}, i.e.,

$$\ell_r(v(k), \bar{q}(k+1)) = (p_r^{max} - p_r(k+1))' \operatorname{diag}(\hat{c}_r)(p_r^{max} - p_r(k+1)) + \tilde{c}_r' u_r(k) \tag{35}$$

with weight $\tilde{c}_r \in \mathbb{R}^R_{>0}$, $\hat{c}_r \in \mathbb{R}^R_{>0}$. Note that $u_r(k)$ is added to ensure that the setpoint does not exceed the maximum available power $w_r(k+1)$.

Very large or very low values of the state of charge, $x(k)$, can increase the ageing mechanism in storage units [33]. Hence, it is useful to keep the state of charge within a specific desired interval $[\tilde{x}^{\min}, \tilde{x}^{\max}]$. It should be pointed out that the desired interval, $[\tilde{x}^{\min}, \tilde{x}^{\max}]$, depends on the battery technology. In addition, storing energy usually causes conversion losses. Thus, the cost of storage units is modeled by:

$$\ell_s(\bar{q}(k)) = \bar{c}_s + p_s(k)' \operatorname{diag}(\tilde{c}_s) p_s(k) + c_x'(\max(\tilde{x}^{\min} - x(k), 0_s) + \max(x(k) - \tilde{x}^{\max}, 0_s)) \tag{36}$$

with $\tilde{c}_s \in \mathbb{R}^S_{>0}$, $\hat{c}_s \in \mathbb{R}^S_{>0}$, $c_x \in \mathbb{R}^S_{>0}$, and $\tilde{x}^{\min} \in \mathbb{R}^S_{>0}$, $\tilde{x}^{\max} \in \mathbb{R}^S_{>0}$.

Remark 2. *It should be noted that if Assumption 1 is not fulfilled, then we cannot have the optimal operation control. In general, we have to be sure that the lower control layers (primary and secondary layers) perform well; otherwise, it does not make sense to have optimal economic operation. Moreover, if Assumption 3 is not met, the charge's state may differ from what we expect. Thus, in this case, we need to use complex models of uncertain inputs, e.g., scenario trees (see [22]), to forecast load and renewable units. However, the main scope of this paper is to show a realistic model of the storage units considering the conversion losses.*

5. Case Study

In this section, we intend to verify the properties of the certainty equivalence model predictive control strategy proposed in Section 3. The microgrid structure depicted in Figure 2 is used for the simulations. It consists of a storage component and a conventional and renewable unit. The detailed parameters of the microgrid are summarized in Table 2.

Table 2. Parameters of the microgrid test system.

Parameter	Value	Weight	Value
$[p_t^{\min}, p_r^{\min}, p_s^{\min}]$	$[0.4, 0, -1]_{\text{pu}}$	c_t	0.1178
$[p_t^{\max}, p_r^{\max}, p_s^{\max}]$	$[1, 2, 1]_{\text{pu}}$	$\hat{c}_t$	0.0048 $_{1/\text{pu}}$
$[x^{\min}, x^{\max}]$	$[0, 7]_{\text{pu h}}$	$\tilde{c}_t$	0.751 $_{1/\text{pu}}$
$[\tilde{x}^{\min}, \tilde{x}^{\max}]$	$[0.5, 6.5]_{\text{pu h}}$	$\bar{c}_r$	0.0001
x^0	3 $_{\text{pu h}}$	$\hat{c}_r$	1 $_{1/\text{pu}}$
$[K_t, K_s]$	$[1, 1]$	$\tilde{c}_s$	0.09
$\widetilde{M}_i$	0.1	$\hat{c}_s$	0.01
$\widetilde{m}_i$	−0.17	c_t^{sw}	0.1

We considered the susceptance and conductance of the transmission lines between the units and the load equal to $b_{ij} = -20$ pu and $g_{ij} = 0$ pu, respectively. According to Equation (28a), the relation between the power of the units and the load and the power of the transmission is obtained as follows:

$$\begin{bmatrix} p_{e,1}(k) \\ p_{e,2}(k) \\ p_{e,3}(k) \\ p_{e,4}(k) \end{bmatrix} = \begin{bmatrix} 1 & 0 & 0 & 0 \\ 0 & -1/3 & 1/3 & 0 \\ 0 & 2/3 & 1/3 & 0 \\ 0 & 1/3 & 2/3 & 0 \end{bmatrix} \begin{bmatrix} p_{t,1}(k) \\ p_{s,1}(k) \\ p_{r,1}(k) \\ w_{d,1}(k) \end{bmatrix} \tag{37}$$

It is assumed that the transmission power of each line is between −1.3 pu and 1.3 pu. Simulations were performed using MATLAB 2018b with a sampling time of $T_s = 30$ min. It should be noted that this sampling time is chosen by the size of the storage unit, and the operation control typically operates on a timescale from minutes to fractions of hours. The results of the closed-loop simulations cover a simulation horizon of 7 d, i.e., 336 simulation steps. Note that the storage unit has a relatively high capacity compared to the rated power, which shows in $x^{\max}$ = 7 pu. The conversion losses are also considered by a quadratic function as $f_1(p_{s,1}(k)) = 0.09p_{s,1}(k)^2 + 0.01$. Moreover, we reformulate the function $f_1(p_{s,1}(k))$ as the following linear cases:

$$f_1(p_{s,1}(k)) = \begin{cases} -0.135p_{s,1}(k) - 0.035, & -1 \leq p_{s,1}(k) \leq -0.5, \\ -0.045p_{s,1}(k) + 0.01, & -0.5 \leq p_{s,1}(k) \leq 0, \\ 0.045p_{s,1}(k) + 0.01, & 0 \leq p_{s,1}(k) \leq 0.5, \\ 0.135p_{s,1}(k) - 0.035, & 0.5 \leq p_{s,1}(k) \leq 1. \end{cases} \quad (38)$$

In accordance with Equations (25a), (25b) and (38), we chose $\widetilde{M}_i = 0.1$ and $\widetilde{m}_i = -0.17$. We formulated the MPC problems in MATLAB using the YALMIP toolbox and solved them numerically with Gurobi. Here, we first compare the prediction error of the state of charge in the cases considering the dynamic storage with piecewise affine loss model (10)–(25) and the dynamic storage without piecewise affine loss model (26)–(29d) in the MPC problem. We formulated this error as $e(k) = x(k) - \tilde{x}(k)$, wherein $x(k)$ is the actual state of charge of the nonlinear loss model given the same power values, whereas $\tilde{x}(k)$ is the forecast of the state of charge. An analysis was carried out for both cases to compare the two cases. In the analysis, closed-loop simulations were performed over 366 simulation steps. For each simulation step, the state of charge prediction of the controllers (over 12 prediction steps) was compared to a prediction performed with the nonlinear plant model for the same storage power values. Then, at each prediction step, the probability distribution of the prediction errors was visualized in the form of box plots (see Figure 4). The circle of each box marks the median value of prediction errors of 336 data points in each prediction step. The box around the median values contains all data from the 25th to the 75th percentile. The downwards dash of each box represents the lowest occurring value of prediction error in each step, whereas the upwards dash marks the highest occurring value.

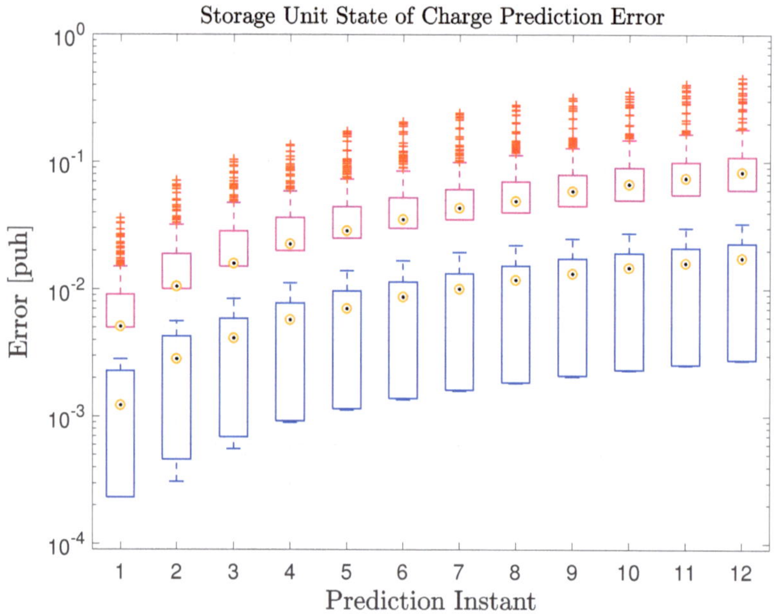

Figure 4. The prediction error of the state of charge (Up) with the dynamic storage without piecewise affine loss model (26)–(29d) in the controller; (Down) with the dynamic storage with piecewise affine loss model (10)–(25) in the controller.

It can be seen in Figure 4 that, including the conversion losses in the proposed model predictive controller, the prediction error is reduced. For example, at $N = 1$, when the conversion loss model is not employed in the controller, the median value is 5×10^{-3}. By adding the conversion loss model in the controller, this value is decreased 3 times to 1.5×10^{-3}. It is worth noting that this ratio increases as the prediction horizon rises. As can be observed from the last step of Figure 4, the median value of the error is equal to 9×10^{-2} when the conversion loss model is not included in the controller, whereas this value is reduced to 2×10^{-2} when the conversion loss model is added to MPC problem formulation. Therefore, the obtained results from Figure 4 illustrate that when we consider the dynamic storage with piecewise affine loss functions in the MPC problem, the actual state of charge $x(k)$ with much less error can be predicted by the controller.

The closed-loop simulation results of the power of units and load, as well as the stored energy and line power of the MG, are depicted in Figure 5. It can be noted that, at the beginning of the period, since the available power of the renewable unit is low, the storage unit is discharging. When the battery is empty, the conventional generator is enabled to provide power to the load. As soon as the available power of the renewable unit is sufficient to provide power to the load, the conventional unit is disabled, and the storage unit is charged. When the stored energy reaches the upper end of the desired state of charge, the setpoint of the wind turbine is set such that the wind power only covers the load. Thus, the stored energy approximately remains at $x^{\max} = 6.5$ pu h. At some point, the available renewable unit power cannot entirely cover the load, and the storage is discharged. When the renewable unit reaches the minimum value $x^{\min} = 0.5$ pu h. and the storage unit is totally discharged, the conventional unit is activated again to provide power to the load. At the end of the simulation, the available power of the renewable unit increases again such that the storage unit can be charged with the available renewable unit power. These results indicate that the conventional units can be enabled and disabled during the simulation. Thus, the model is also able to work when RES and storage units are enabled. However, if the storage unit is discharged and the available power of the renewable unit is low, the conventional unit is repeatedly enabled to provide power to the loads and charge the storage units. Then, after increasing the available power of the renewable unit, the conventional unit is turned off. Hence, it results that by choosing this proposed operation control strategy, the share of power from RES can be significantly increased. Finally, it is clear from Figure 5 that the line power in the lower plot was within the given bounds of ± 1.3 pu at all times.

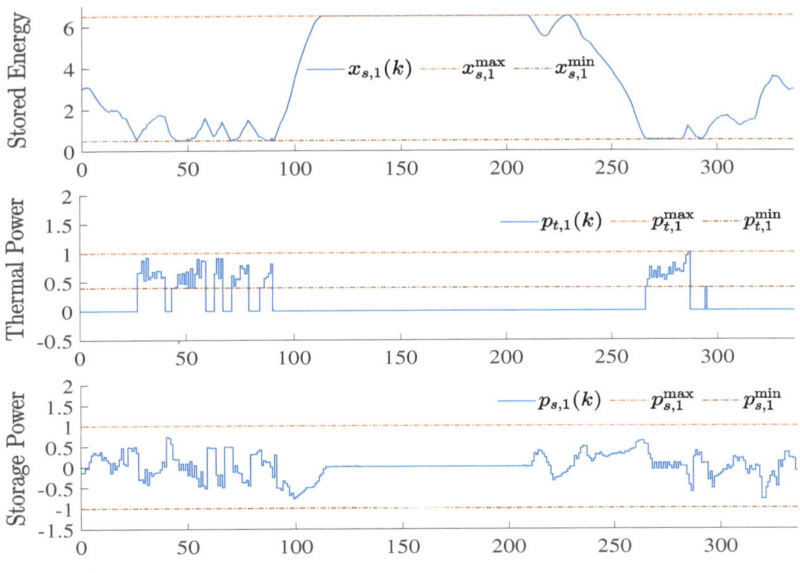

Figure 5. *Cont.*

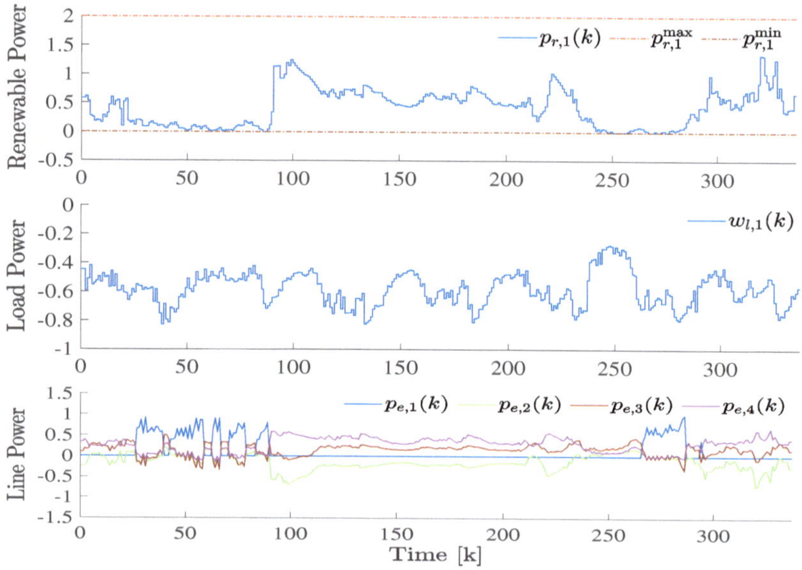

Figure 5. Power of units and load.

6. Conclusions

In this work, we designed a novel certainty model predictive control approach for the operation of an islanded MG with a very high share of renewable energy sources. To this aim, a mathematical model of an islanded MG was derived by considering the conversion losses as quadratic functions to realistically model the scenario. In the model, we included (i) grid-forming storage units; (ii) renewable energy sources, where the power infeed can be limited, e.g., if storage units are fully charged; and (iii) conventional generators that can be disabled, e.g., in times of high available renewable infeed. Moreover, the model allowed us to approximately include the power flow over the transmission lines and a limitation thereof, as well as power-sharing of grid-forming units. In order to reduce the state of charge prediction error, we reformulated the conversion losses of storage units as piecewise affine functions and included them in the proposed controller. The obtained results confirmed that when considering the piecewise affine conversion loss functions in the proposed MPC, the prediction error of the state of charge decreased. Moreover, it is shown that the proposed scheme leads to a very high share of renewable energy sources in closed-loop simulations. Future works will investigate AC optimal power flow (OPF) problems, which are more realistic than the widely used linearized DC power flow approximations. To solve the AC OPF problems, we intend to employ a convex relaxation of the original problem, which leads to a second-order cone program (SOCP) that can be solved by available commercial software.

Author Contributions: M.G.: Conceptualization, Methodology, Software, Investigation, Writing—Original Draft, Writing—Review and Editing. A.P.: Editing, Supervision, Funding Acquisition. All authors have read and agreed to the published version of the manuscript.

Funding: The research leading to these results received funding from Fondazione di Sardegna, under project "SISCO" (CUP:F74I19001060007), and the Sardinian Regional Government, under project "Estimation of unmeasurable variables and fault analysis in electrical smart grids".

Institutional Review Board Statement: Not applicable.

Informed Consent Statement: Not applicable.

Data Availability Statement: Not applicable.

Acknowledgments: Thanks to Christian A. Hans for his guidance and thanks to the editor and the referees for their exceptionally helpful comments during the review process.

Conflicts of Interest: The authors declare no conflicts of interest.

References

1. Anderson, P.A.; Fouad, A.A. *Power System Control and Stability*; John Wiley and Sons: Hoboken, NJ, USA, 2003.
2. Kundur, P.; Balu, N.J.; Lauby, M.G. *Power System Stability and Control*; McGraw-Hill: New York, NY, USA, 1994; Volume 7.
3. Pehnt, M. Dynamic life cycle assessment (LCA) of renewable energy technologies. *Renew. Energy* **2006**, *31*, 55–71. [CrossRef]
4. Peças, J.A.; Moreira, C.L.; Madureira, A.G. Defining control strategies for microgrids islanded operation. *IEEE Trans. Power Syst.* **2006**, *21*, 916–924.
5. Hatziargyriou, N.; Asano, H.; Iravani, R.; Marnay, C. Microgrids. *IEEE Power Energy Mag.* **2007**, *5*, 78–94. [CrossRef]
6. Gholami, M.; Pilloni, A.; Pisano, A.; Usai, E. Robust Distributed Secondary Voltage Restoration Control of AC Microgrids under Multiple Communication Delays. *Energies* **2021**, *14*, 1165. [CrossRef]
7. Guerrero, J.M.; Vasquez, J.C.; Matas, J.; De Vicuña, L.G.; Castilla, M. Hierarchical control of droop-controlled AC and DC microgrids—A general approach toward standardization. *IEEE Trans. Ind. Electron.* **2011**, *58*, 158–172. [CrossRef]
8. Guerrero, J.M.; Chandorkar, M.; Lee, T.L.; Chiang Loh, P. Advanced control architectures for intelligent microgrids—Part I: Decentralized and hierarchical control. *IEEE Trans. Ind. Electron.* **2012**, *4*, 1254–1262.
9. Kanagawa, M.; Nakata, T. Assessment of access to electricity and the socio-economic impacts in rural areas of developing countries. *Energy Policy* **2008**, *36*, 2016–2029. [CrossRef]
10. Barklund, E.; Pogaku, N.; Prodanovic, M.; Hernandez-Aramburo, C.; Green, T.C. Energy management in autonomous microgrid using stability-constrained droop control of inverters. *IEEE Trans. Power Electron.* **2008**, *23*, 2346–2352. [CrossRef]
11. Logenthiran, T.; Srinivasan, D. Short term generation scheduling of a microgrid. In Proceedings of the TENCON 2009—2009 IEEE Region 10 Conference, Singapore, 23–26 January 2009; pp. 1–6.
12. Chen, C.; Duan, S.; Cai, T.; Liu, B.; Hu, G. Smart energy management system for optimal microgrid economic operation. *IET Renew. Power Gener.* **2011**, *5*, 258–267. [CrossRef]
13. Palma-Behnke, R.; Benavides, C.; Lanas, F.; Sev, B.; Reyes, L.; Llanos, J.; Sáez, D. A microgrid energy management system based on the rolling horizon strategy. *IEEE Trans. Smart Grid* **2013**, *4*, 996–1006. [CrossRef]
14. Heymann, B.; Bonnans, J.F.; Martinon, P.; Silva, F.J.; Lanas, F.; Guillermo, J.E. Continuous optimal control approaches to microgrid energy management. *Energy Syst.* **2018**, *9*, 59–77. [CrossRef]
15. Mayhorn, E.; Kalsi, K.; Elizondo, M.; Zhang, W.; Lu, S.; Samaan, N.; Butler-Purry, K. Optimal control of distributed energy resources using model predictive control. In Proceedings of the 2012 IEEE Power and Energy Society General Meeting, San Diego, CA, USA, 22–26 July 2012; pp. 1–8.
16. Olivares, D.E.; Cañizares, C.A.; Kazerani, M. A centralized energy management system for isolated microgrids. *IEEE Trans. Smart Grid* **2014**, *5*, 1864–1875. [CrossRef]
17. Hans, C.A.; Nenchev, V.; Raisch, J.; ReinckeCollon, C. Minimax model predictive operation control of microgrids. *IFACProceedings* **2014**, *47*, 10287–10292. [CrossRef]
18. Sui, Q.; Wei, F.; Wu, C.; Lin, X.; Li, Z. Day-Ahead Energy Management for Pelagic Island Microgrid Groups Considering Non-Integer-Hour Energy Transmission. *IEEE Trans. Smart Grid* **2020**, *11*, 5249–5259. [CrossRef]
19. Tran, Q.T.T.; Riva Sanseverino, E.; Zizzo, G.; Di Silvestre, M.L.; Nguyen, T.L.; Tran, Q.-T. Real-Time Minimization Power Losses by Driven Primary Regulation in Islanded Microgrids. *Energies* **2020**, *13*, 451. [CrossRef]
20. Nair, U.R.; Costa-Castelló, R. A Model Predictive Control-Based Energy Management Scheme for Hybrid Storage System in Islanded Microgrids. *IEEE Access* **2020**, *8*, 97809–97822. [CrossRef]
21. Zia, M.F.; Elbouchikhi, E.; Benbouzid, M.; Guerrero, J.M. Energy Management System for an Islanded Microgrid with Convex Relaxation. *IEEE Trans. Ind. Appl.* **2019**, *55*, 7175–7185. [CrossRef]
22. Hans, C.A.; Sopasakis, P.; Raisch, J.; Reincke-Collon, C.; Patrinos, P. Risk-averse model predictive operation control of islanded microgrids. *IEEE Trans. Control Syst. Technol.* **2020**, *28*, 2136–2151. [CrossRef]
23. Lei, J.; Hans, C.A.; Sopasakis, P. Optimal operation of microgrids with risk-constrained state of charge. *arXiv* **2021**, arXiv:2104.08357. [CrossRef]
24. Hans, C.A.; Braun, P.; Raisch, J.; Grüne, L.; Reincke-Collon, C.R. Hierarchical distributed model predictive control of interconnected microgrids. *IEEE Trans. Sustain. Energy* **2019**, *10*, 407–416. [CrossRef]
25. Sampathirao, A.K.; Hofmann, S.; Raisch, J.; Hans, C.A. Distributed Conditional Cooperation Model Predictive Control of Interconnected Microgrids. *arXiv* **2021**, arXiv:1810.03361v2[cs.SY].
26. Gholami, M.; Pisano, A.; Usai, E. Robust Distributed Optimal Secondary Voltage Control in Islanded Microgrids with Time-Varying Multiple Delays. In Proceedings of the 2020 IEEE 21st Workshop on Control and Modeling for Power Electronics (COMPEL), Aalborg, Denmark, 9–12 November 2020; pp. 1–8.
27. Pilloni, A.; Gholami, M.; Pisano, A.; Usai, E. On the robust distributed secondary control of islanded inverter-based microgrids. In *Variable-Structure Systems and Sliding-Mode Control*; Springer: Cham, Switzerland, 2020; pp. 309–357.

28. Purchala, K.; Meeus, L.; Van Dommelen, D.; Belmans, R. Usefulness of DC power flow for active power flow analysis. In Proceedings of the IEEE Power Engineering Society General Meeting, San Francisco, CA, USA, 16 June 2005; pp. 454–459.
29. Krishna, A.; Hans, C.A.; Schiffer, J.; Raisch, J.; Kral, T. Steady state evaluation of distributed secondary frequency control strategies for microgrids in the presence of clock drifts. In Proceedings of the 2017 25th Mediterranean Conference on Control and Automation (MED), Valletta, Malta, 3–6 July 2017; pp. 508–515.
30. Bemporad, A.; Morari, M. Control of systems integrating logic, dynamics, and constraints. *Automatica* **1999**, *35*, 407–427. [CrossRef]
31. Petrollese, M.; Valverde, L.; Cocco, D.; Cau, G.; Guerr, J. Real-time integration of optimal generation scheduling with MPC for the energy management of a renewable hydrogen-based microgrid. *Appl. Energy* **2016**, *166*, 96–106. [CrossRef]
32. Zivic Djurovic, M.; Milacic, A.; Krsulja, M. A simplified model of quadratic cost function for thermal generators. In Proceedings of the 23rd International DAAAM Symposium, Zadar, Croatia, 24–27 October 2012; Volume 23, pp. 25–28.
33. Vetter, J.; Novak, P.; Wagner, M.R.; Veit, C.; Moller, K.-C.; Besenhard, J.; Winter, M.; Wohlfahrt-Mehrens, M.; Vogler, C.; Hammouche, A. Ageing mechanisms in lithium-ion batteries. *J. Power Sources* **2005**, *147*, 269–281. [CrossRef]

electricity

MDPI

Article

Determination of Dynamic Characteristics for Predicting Electrical Load Curves of Mining Enterprises †

Denis Anatolievich Ustinov and Konstantin Alekseevich Khomiakov *

Department of Electric Power and Electromechanics, St. Petersburg Mining University, 199106 St. Petersburg, Russia; ustinov_da@pers.spmi.ru

* Correspondence: s175094@stud.spmi.ru

† This paper is an extended version of our conference paper: Ustinov, D.A, Khomiakov, K.A. "Comprehensive assessment of factors determining electricity load curves concerning enterprises of mineral resource sector." In Proceedings of the Physical Studies and Digitalization in Mining Engineering (IPDME), St. Petersburg, Russia, 23–24 April 2020.

Abstract: The calculation of electrical loads is the first and most significant stage in the design of the power supply system. It is essential to make the right choice when choosing the power electrical equipment: transformers, power lines, and switching devices. Underestimation or overestimation of the calculated values can lead to large losses and an increase in capital costs. Therefore, the reliability of the results plays a key role. The use of energy-saving technologies and energy-efficient electrical equipment leads to a change in the nature and level of power consumption, which must be taken into account when determining the electrical loads. The existing methods leave out dynamic characteristics of electrical load curves, so the calculated values are overestimated by up to 40%. This study shows a load calculation method with the normalized correlation functions and its parameters at the level of the individual and group electricity consumers. As a result, the difference between the calculated and experimental values does not exceed 5%.

Keywords: calculation of electrical loads; utilization factor; normalized correlation function; probabilistic method; individual load curves; group load curves; energy-saving measures

Citation: Ustinov, D.A.; Khomiakov, K.A. Determination of Dynamic Characteristics for Predicting Electrical Load Curves of Mining Enterprises. *Electricity* **2022**, *3*, 162–181. https://doi.org/10.3390/electricity3020010

Academic Editor: Andreas Sumper

Received: 17 January 2022
Accepted: 29 March 2022
Published: 8 April 2022

1. Introduction

The existing high energy utilization rates in the energy sector mean that the required improvements must be made in terms of returns on capital investments and material costs. Currently, the consumption of electricity by industrial enterprises in certain regions of Russia can reach about 44% (in the Nenets Autonomous Area). For example, the Ural Federal District is one of the richest mineral regions in Russia, with a share of electricity consumption by industrial enterprises constituting about 30% and with the likelihood of such consumption increasing due to the active development of this region and the construction of new industrial enterprises [1]. The construction and further operation of any industrial enterprise is preceded by the design stage, involving the definition of the key characteristics of each system and the design of the power supply system (PSS). The efficiency of the PSS is determined by the reliable evaluation of the characteristics of electrical loads, with the calculated results then shown in graphs [2]. This is due to the fact that the calculated characteristics electrical loads provide information to inform the solutions for most of the technical and economic issues in the design stage.

It should also be noted that the issues regarding energy efficiency and energy savings are very acute, and the measures taken to achieve certain goals are widely implemented in all production cycles, meaning that it is important to take these changes into consideration when calculating electrical loads by using probabilistic calculation methods [3].

Reference [4] reviewed how the need for efficient electrical energy consumption has greatly expanded in the process industries. The article dealt with industrial enterprises for the recognition of electricity consumption in a two-echelon supply model with a stochastic lead-time demand and imperfect production, while considering the distribution-free approach. Reference [5] discussed the impact of electric energy on production cost recovery, transport discounts and process quality improvement in two algorithms are developed using an analytical approach to obtain the optimal solution. With the increase in the varieties products and the increasing uncertainty about product demand, the production preparation time is a significant factor in addressing these issues. The basic calculation methods applied today, such as the method of ordered diagrams (MOD) [6] and the modified statistical method (MSM) [7], lead to significant errors in estimating the design loads, as shown in References [7,8]. This is due to the fact that the applied methods are static modeling methods that involve uninformative mathematical model, while the indicated design characteristics can solve only a very limited range of technical and economic issues; as a consequence, the efficiency of the PSS is reduced. As such, increasing the energy efficiency of mineral resource enterprises by improving the methods for calculating electrical loads is an urgent task, as these initial datasets are used in design stage.

The methods for calculating electrical loads used currently in our country and abroad refer to probabilistic methods based on a mathematical model of the "random variable" [9,10]. It should be noted that we are talking about methods of raising mining enterprises and not about load forecasting based on an analysis of the current situation, and, for example, by using neural networks to predict both long-term and short-term load electrical curves [11]. The basis for all further research in the field of calculating electrical loads and, in fact, the first method based on the principles of the probability of a random process, is the statistical method; there is no need to dwell separately on the shortcomings, because they have already been repeatedly described in Reference [8].

The statistical method was replaced by the MOD [6], which made it possible to reduce errors in the estimation of electrical loads by creating a database of initial data of individual load schedules, but still remained insufficiently reliable; this is due to several shortcomings, especially the following: discrepancy between the design technological modes and the actual operating modes [12].

Today, all mining enterprises are designed by using an MSM [7]. Developed more than 25 years ago, these methods are the basis of approved methods for calculation of electrical loads [13]. Such a long break in research can be explained by the sufficient reliability of the MSM, combined with the simplicity of performing engineering calculations, but the last few years have been practically associated with a revolution in the mining and processing industry, increasing the automation of technological processes through the use of modern energy-efficient equipment [14,15]. Moreover, the effect of such changes must be taken into account, including in the calculation methods [16].

Separately, it is worth mentioning the alternative calculation methods considered in the works of Stepanova [17]; the main advantage of the method under consideration is the consideration of correlation dependencies both at the level of individual and group loads, as well as additional calculation of dynamic characteristics [18]. At the same time, the disadvantages include a small base of initial data on consumers, which, in turn, does not allow a reliable analysis of this method applied to consumers at mining enterprises, as well as the lack of accounting from the introduction of modern means to improve energy efficiency [19].

Table 1 presents the contributions to this paper.

Table 1. Contributions of authors.

Author(s)	Probabilistic Method	Large Accumulated Database Containing the Source Data	Simplicity of Performing Engineering Calculations	Correlation between Ordinates	Account the Modern Means to Improve Energy Efficiency
Sinchuk, O.N. Guzov, E.S. [6]	√	√	√		
Zhokhov, B.D. [7]	√	√	√		
Ali Hassan [9]	√	√	√		
Stepanov, V.P [17]	√		√	√	
This paper	√√	√√	√√	√√	√√

2. Problem Definition, Notation, and Assumptions

2.1. Problem Definition

An analysis of the MSM and MOD showed differences in the reference and actual values of the utilization factors in determining the calculated electrical loads of a conveyor unit with and without the use of a variable-frequency drive (VFD) (Tables 2 and 3).

Table 2. Average values of consumed active power (P_{av}), in kW, for the test period.

Consumer Type	Without a VFD	With a VFD
Conveyor	174.8	138.9
Hydrocyclone feed pump	44.3	30.3
Blower	217.5	159.5
Overhead crane	14.0	9.0

Table 3. Utilization factors.

Consumer Type	k_u. Reference Value	k_u Actual Value without VFD	k_u Actual Value with VFD
Conveyor	0.75	0.70	0.5
Hydrocyclone feed pump	0.80	0.78	0.55
Blower	0.75	0.72	0.56
Overhead crane	0.45	0.40	0.25

The discrepancy between the reference values and the actual value of the utilization factor (k_u) is about 35%, which goes beyond the range of permissible values. It should be noted that the existing database of coefficients does not take into account the usage of equipment with VFD [6,20].

Studies on the methods used for calculating electrical loads abroad have shown their similarities with the method of ordered diagrams and the statistical method. First of all, this is shown by the utilization factor, k_u; the demand factor, k_d; and the peak coincidence factor, k_M [8,21]. Unfortunately, this approach to determining the key parameters of a power supply system (PSS) has proven its inefficiency due to overestimating the calculated values [22]. When examining the indicators of power consumption of operating enterprises, the obtained calculated maximum loads, according to foreign methods, exceed the actual values by 40–60%. The same values were obtained by using the method of ordered diagrams. The situation is better with the use of the modified statistical method, the description of which is presented at the beginning of this section, whereby the obtained calculated values exceed the actual values by 20–40% [15].

The overestimation of the calculated values is associated with disadvantages that are inherent in all of the methods under study. The detailed analysis carried out revealed the main disadvantages, which are shown in Table 4.

Table 4. Advantages and disadvantages of existing load calculation methods used for calculating electrical loads.

Load Calculation Method	Disadvantages	Advantages
1. Method of ordered diagrams; 2. Statistical method 3. Modified statistical method.	- The use of approximate formulae according to which the thirty-minute maximum load is recalculated; - Correlation and inter-correlation between ordinates are not taken into account, both at the level of individual and group electrical load curves of an electric drive; - The average load is assessed based on the upper limits of the probabilistic values; - Overestimation of the average value of the load factors to assess the calculated heating load; - Discrepancy between the design technological modes and the actual operating modes.	- Large accumulated database containing the source data; - Simplicity of performing engineering calculations.

2.2. Notation

Therefore, just a revision of the estimated coefficients of existing methods is not enough to increase the reliability of the calculated values [6]. This is due to the fact that the applied methods are static modeling methods that involve uninformative mathematical model, while the indicated design characteristics can solve only a very limited range of technical issues. Moreover, an increase in the number and accuracy of the calculated characteristics is possible through the additional calculation of dynamic characteristics [16].

It should be noted that forecasting of electrical loads based on any type of probabilistic forecasting, for example, the use of neural networks, is best suited for the operation of the electrical complex of an enterprise; however, it is completely unsuitable for design [23].

2.3. Assumptions

The purpose of this paper is to increase the reliability of the calculated values through the applied method, which differs from classical methods. Such a calculation will help to make a right choice of power and switching electrical equipment for distribution and protection, which will reduce material inputs.

In summary, the significant contributions of this paper are as follows:

1. Determination of dynamic characteristics for predicting electrical load curves of mining enterprises and expanding the database containing the source data, as well as determining the difference in the characteristics of the same type of curves, using means to improve energy efficiency and without (Section 3).
2. Development of a methodology for the transition from individual load schedules to group ones and presentation of a calculation algorithm, taking into account the consideration of two laws of probability distribution (Section 4).
3. Carrying out a comparative analysis of existing classical calculation methods and the presenting a new method that makes it possible to take into account equipment, using means to improve energy efficiency (Section 5).

3. Dynamic Characteristics of Electrical Load Curves

As such, when determining the calculated loads, it is also necessary to take into account dynamic characteristics, such as peaks and valleys, surges and load dips, and load fluctuations, which will contribute to forecasting calculations of electrical loads with a higher confidence probability. It becomes possible by using a dynamic modeling method, such as the hierarchical–structural method (HSM) [24].

The method is realized for load curves of general-purpose industrial and special industrial consumers, and the base of the method is an extended database containing the source data with additional characteristics (Table 5), such as the normalized correlation function (NCF) and its parameters presented in the Table 5. The necessity of using additional characteristics is connected to a fundamental change of the initial information about the individual load curves of the consumers. Physically, the NCF describes the probabilistic causation for the sequence of ordinates of the random process of changing the load, and due to the fact that the technological processes in mining enterprises have a steady-state mode of operation, the load curves are stationary [25].

Table 5. Database with additional characteristics.

Effective number of power consumers	n
Rated power	p_{nom}
Utilization factor	k_u
Power factor	$tg\varphi$
Normalized correlation function	$K(\tau)$
Parameter characterizes the attenuation of correlations between the ordinates of the initial ELC	α
The natural frequency of oscillations of the CF, due to the repeatability of technological operations	ω_0

Our analysis of the literary sources [8,16,26] showed that the correlation functions (CF) for all general-purpose industrial and special industrial consumers can be approximated by using the following analytical expressions:

$$K(\tau) = Dp_{exp}(-\alpha|\tau|); \quad (1)$$

$$K(\tau) = Dp_{exp}(-\alpha|\tau|)Cos\omega_0\tau; \quad (2)$$

$$K(\tau) = Dp_{exp}(-\alpha|\tau|)(Cos\omega_0\tau - \frac{\alpha}{\omega_0} sin\omega_0|\tau|); \quad (3)$$

$$K(\tau) = Dp_{exp}(-\alpha|\tau|)(Cos\omega_0\tau + \frac{\alpha}{\omega_0} sin\omega_0|\tau|); \quad (4)$$

where Dp_{exp} is the variance of an individual electrical load curve (ELC), the parameter α characterizes the attenuation of correlations between the ordinates of the initial ELC, and ω_0 is the natural frequency of oscillations of the CF, due to the repeatability of technological operations. The main idea of this study is to identify differences in the NCF and its parameters for the same consumers, both with and without using means to improve energy efficiency.

Experimental studies aimed at determining the NCF and its parameters of the electrical load curves were carried out in operating mining enterprises, the main consumers of which are general-purpose industrial consumers. The load curves were compared with and without using energy-efficiency means—in this case, with and without the use of a variable-frequency drive. The initial curves and step-wise determination of the type and parameters of the NCF for a conveyor unit with power motor of p_{nom} = 250 kW are given by way of example. Earlier, in Reference [24], the authors described a detailed algorithm for determining the NCF, so this article presents only the main points.

1. Determination of the average load value (Table 6).
2. The given time values, namely the correlation time ($\mathbf{T_C}$), the sampling interval (Δt), and the recording time ($\mathbf{T_r}$) (Table 8), were determined according to the initial load curve (Figures 3 and 4). Then, using the MATLAB software package, the experimental NCF was calculated.
3. Using the least-squares method, we determined the parameters of a theoretical NCF, as shown in Table 7, and identified its type (Figures 1 and 2).

Table 6. Approximate average power values.

Without a VFD	With a VFD
$P_{av} = 175$ kW	$P_{av} = 139$ kW

Table 7. Theoretical NCFs.

Theoretical NCF without a VFD	Theoretical NCF with a VFD
$K(\tau) = exp(-1.723|\tau|)\cos(2.26\tau)$	$K(\tau) = exp(-3.37|\tau|)\cos(10.23\tau)$

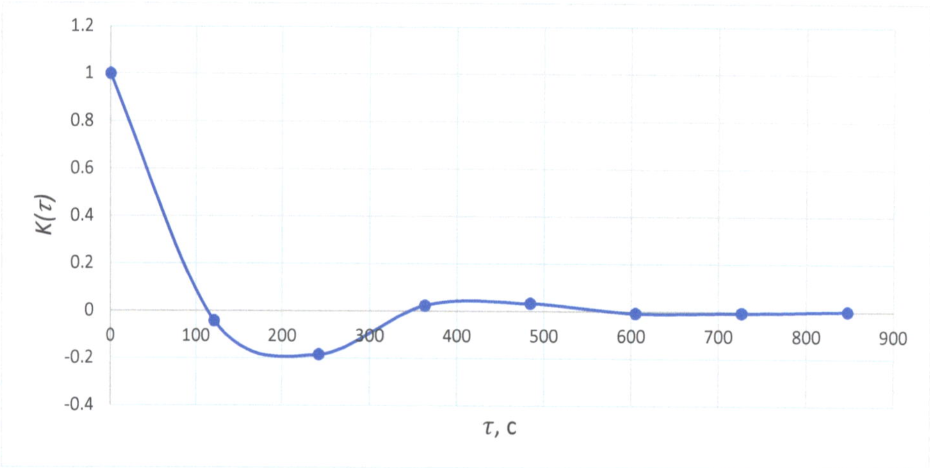

Figure 1. Theoretical NCF curve without a VFD.

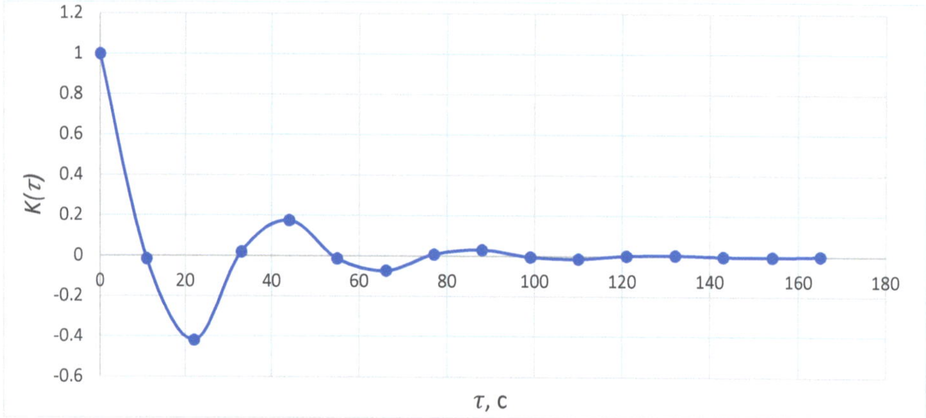

Figure 2. Theoretical NCF curve with a VFD.

Table 8. The time values.

Correlation Time, T_c		Sampling Interval, Δt		Recording Time, T_r	
Without a VFD	With a VFD	Without a VFD	With a VFD	Without a VFD	With a VFD
69.5 s	14.5 s	0.695 s	0.145 s	5560 s	1160 s

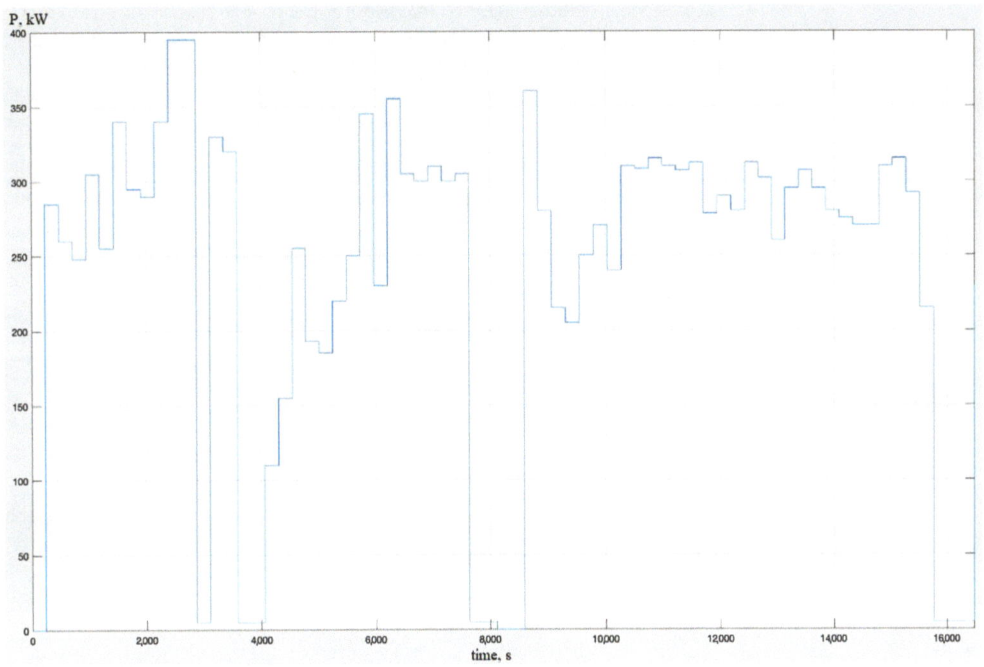

Figure 3. Fragment of the curve of active power consumption without a VFD.

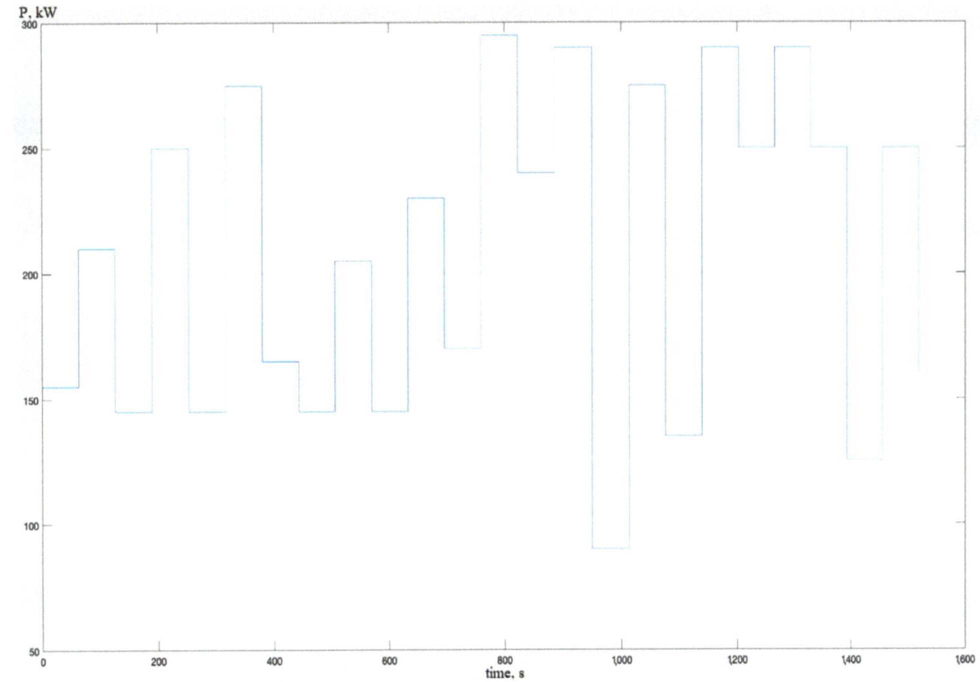

Figure 4. Fragment of the curve of active power consumption with a VFD.

The conducted studies showed that general-purpose industrial electric drives belong to the same type of NCF with and without the use of a variable-frequency drive, although the difference in the NCF parameters determines the calculated values of the entire power supply system. As a result, there is a need to supplement the new information database.

An additional series of loads were also studied, and the NCF and parameters were determined (Figures 5–8 and Table 9).

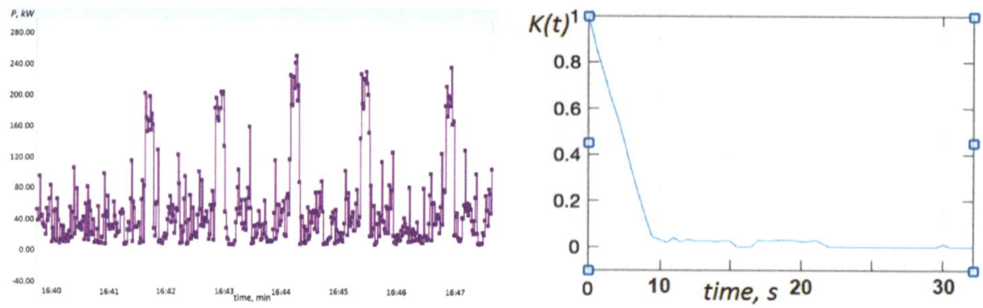

Figure 5. Fragment of the curve of active power consumption hydrocyclone feed pump with a VFD and his NCF.

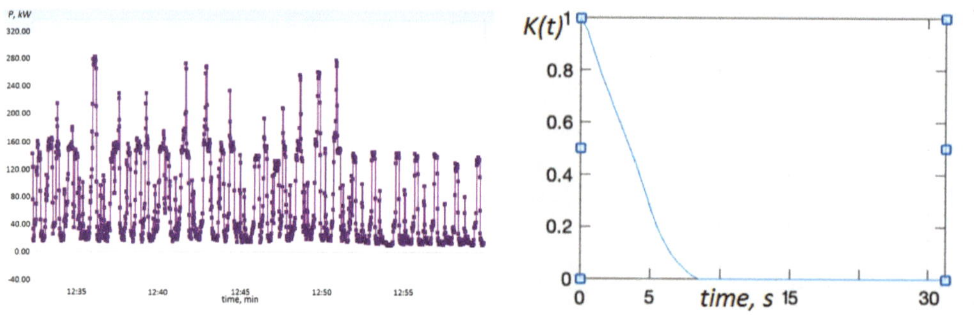

Figure 6. Fragment of the curve of active power consumption crushers with a VFD and his NCF.

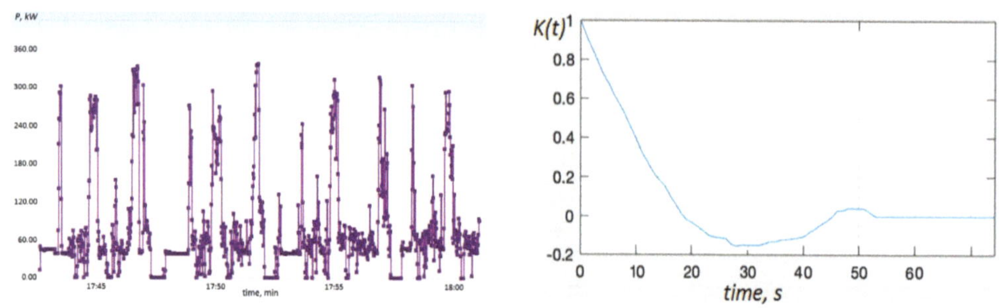

Figure 7. Fragment of the curve of active power consumption overhead crane with a VFD and his NCF.

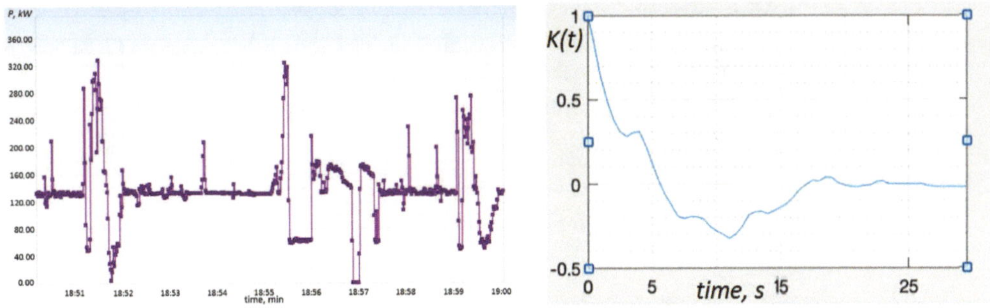

Figure 8. Fragment of the curve of active power consumption pneumatic–mechanical flotation tank cells without a VFD and his NCF.

Table 9. Parameters of the various types of NCFs, with and without the use of a frequency converter.

Electric Drive Name	Without a VFD			With a VFD		
	Type of the NCF	Parameters of the NCF		Type of the NCF	Parameters of the NCF	
		α, s^{-1}	ω_0, rad/s		α, s^{-1}	ω_0, rad/s
Conveyor	2	1.723	2.26	2	3.37	10.23
Hydrocyclone feed pump	1	0.70	1.18	1	1.95	3.04
Blower	1	0.29	0.73	1	0.87	1.08
Crushers	-	-	-	1	1.07	2.01
Pneumatic–mechanical flotation tank cells	2	0.98	1.34	-	-	-
Overhead crane	4	0.45	1.05	4	0.9	2.25
Fans	1	0.015	-	-	-	-
Cylindrical grinding machine	2	0.078	0.268	-	-	-

4. Group Load Curve Forecasting Using a Probabilistic Calculation Method

The above operations are also valid for group curves; however, due to the variations in PSS by structure and functionality, it is necessary to perform electrical load curve forecasting by summing individual curves in order to obtain equivalent parameters of the NCF [13]. To do this, at the level of the group curves, it is necessary to use the following analytical expression:

$$K(\tau) = DPexp(-\alpha_e|\tau|); \tag{5}$$

where DP is the variance of the active load curve, and $\alpha_{э}$ is the equivalent factor characterizing the attenuation of the correlation between the ordinates of the summed ELCs [27].

The group load curve *P(t)*, which is formed according to the individual curves, *p*(*t*), gives the following definitions for α_e:

$$\alpha_e = \frac{DP}{\sum_{i=1}^{m}\frac{DP_i}{\alpha_i} + \sum_{i=1}^{l}\frac{DP_i\alpha_i}{\alpha_i^2+\omega_{0_i}^2} + \sum_{i=1}^{r}\frac{2DP_i\alpha_i}{\alpha_i^2+\omega_{0_i}^2}}; \tag{6}$$

In the formula, the numbers of curves with NCF types 1, 2, 3, and 4, respectively, are taken for m, l, and r.

Taking into account the differences in the type and parameters of the NCF for individual consumers, in Formula (6), we determine the variance of DP_θ θ peaks and θ valleys of the ELCs in the averaging interval θ:

$$DP_\theta = \frac{2DP}{(\alpha_e\theta)^2}[exp(-\alpha_e\theta) + \alpha_e\theta - 1]; \tag{7}$$

Methodology for Calculating Load Estimation Parameters

The characteristics of electrical load curves, such as θ peaks and θ valleys, make it possible to determine the following main parameters:

- Determination of the conductor section according to the condition of heating and economic density;
- Determination of minimum and maximum power losses;
- Determination of voltage deviation.

The expression for determining the definition of θ peaks is as follows:

$$P_{p\theta} = P_{av} + \beta_1\sqrt{DP_\theta}\,; \tag{8}$$

and for θ valleys, it is as follows:

$$P_{v\theta} = P_{av} + \beta_2\sqrt{DP_\theta}\,; \tag{9}$$

Similar to other methods of calculation of electrical loads, the HS method has its own clear sequence of actions that must be performed to obtain a result, so the task is to simplify as much as possible and create an algorithm that would not require using other sources to search information. It makes the database for the individual consumers important. Such an algorithm was developed and is presented in Figure 9.

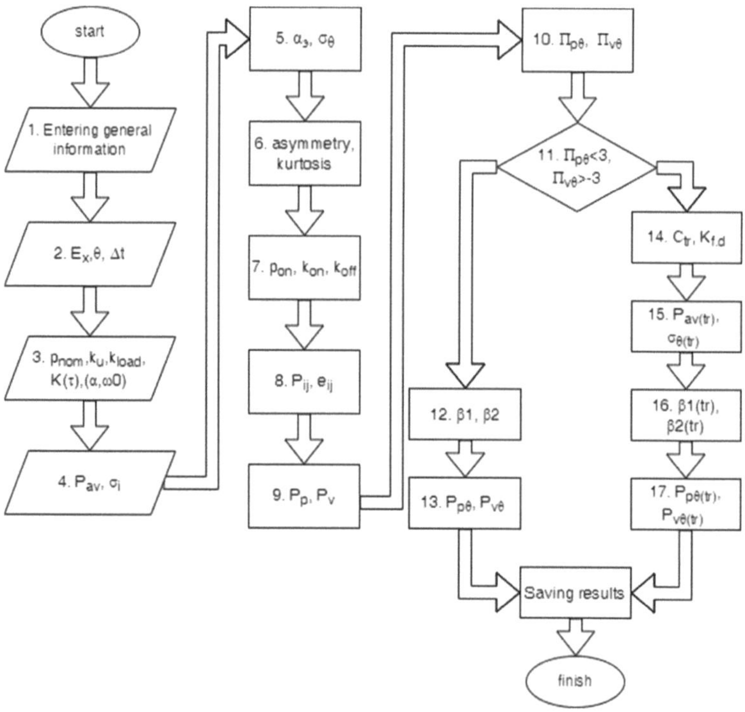

Figure 9. Algorithm for calculating the parameters of electrical load curves.

Following each block of the algorithm, it is necessary to consider what actions take place in the system:

1. It is necessary to indicate complete information on the design object, as well as information on the designer performing the calculation;

2. The value corresponding to the boundary probability (E_x), the required intervals of averaging (θ), and discretization (Δt);
3. Using the dataset, select the necessary equipment from the list, with the specified characteristics. Next, a group of consumers is formed, and the calculation is carried out according to the group schedule;
4. There is a calculation of the average active and reactive power, according to the following formulas:

$$P_{av} = \sum_{i=0}^{m} p_{nomi} k_{ui}; \tag{10}$$

$$Q_{av} = \sum_{i=0}^{m} p_{nomi} k_{ui} tg\varphi; \tag{11}$$

Moreover, the calculation of the standard deviation of the active load curve is as follows:

$$\sigma_i = \sqrt{\sum_{i=1}^{m} p^2{}_{nomi} k_{ui} (k_{loadi} - k_{ui})}\ ; \tag{12}$$

5. From this block, the transition to the group graph is implemented through an equivalent correlation function, and for this, as described above, it is necessary to determine the parameter $\alpha_э$ of the active load according to the following formula:

$$\alpha_э = \frac{DP}{\sum_{i=1}^{m} \frac{DP_i}{\alpha_i} + \sum_{i=1}^{l} \frac{DP_i \alpha_i}{\alpha_i^2 + \omega_{0_i}^2} + \sum_{i=1}^{r} \frac{2DP_i \alpha_i}{\alpha_i^2 + \omega_{0_i}^2}}; \tag{13}$$

Moreover, the variance of the load group graph is calculated as follows:

$$DP_\theta = \frac{2DP}{(\alpha_э \theta)^2} [exp(-\alpha_э \theta) + \alpha_э \theta - 1]; \tag{14}$$

6. In block 6, the coefficients of asymmetry (skewness) s_k and kurtosis (k) are calculated, and for cases when the values of these coefficients are not equal to zero, it is necessary to use a law other than normal, for example, the Gram–Charlier type A distribution law [16]:

$$s_k = \frac{\sum_{i=1}^{m} p^3{}_{nomi} k_{ui} (k_{loadi} - k_{ui})(k_{loadi} - 2k_{ui})}{\sigma^3}; \tag{15}$$

$$k = \frac{\sum_{i=1}^{m} p^4{}_{nomi} k_{ui} (k_{loadi} - k_{ui}) \left[(k_{loadi} - k_{ui})(k_{loadi} - 2k_{ui}) + k^2{}_{ui} \right] + 6 \sum_{i<j} \sigma_i^2 \sigma_j^2}{\sigma^4} - 3; \tag{16}$$

7. From block 7, we see that the algorithm for estimating the characteristics of group curve begins, taking into account the limitations of the limits of its change; thus, it is necessary to calculate additional characteristics for individual consumers, such as the following: the value of the average power during switch-on time t_{on},

$$p_{on} = p_{nom} k_{load}; \tag{17}$$

the value of the switch-on factor as the probability of the "on" mode,

$$k_{on} = \frac{k_u}{k_{load}}; \tag{18}$$

and the value of the switch-off factor as the probability of the "off" mode,

$$k_{off} = 1 - k_{on}. \tag{19}$$

8. Searching through the combinations of the ith modes "on" and the jth modes "off", we see that the possible ordinates for the ELCs are determined, and using the theorem of the product of probabilities, we obtain the probability of a single combination [28,29]:

$$e_{ij} = \prod_{i=0}^{m} k_{oni} \prod_{j=0}^{n-m} k_{offj}; \tag{20}$$

In Formula (11), the value m denotes the consumers operating in the "on" mode and with the corresponding k_{oni}, while the value n-m denotes the consumers operating in the "off" mode and with the corresponding k_{offj}.

The electrical load values correspond to the probabilities, e_{ij}:

$$P_{ij} = \sum_{i=0}^{m} p_{oni}; \tag{21}$$

If the same load arises as a result of different combinations of different options for the work of consumers, there is a general probability of the occurrence of a given load arising in a given complex situation of probabilities [30]. As a result, a statistical series is formed, according to which the step function of the probability distribution of the ordinates ELC *F(P)* is constructed. To determine the actual lower (valley) P_v and upper (peak) P_p values of the limits of variation of the ordinates of the ELC, the principle of practical confidence of the theory of probabilities is used [17]. According to this principle, practically impossible values of electrical loads are excluded from the obtained possible range of electrical load values from 0 to $\sum_{i=1}^{m} p_{oni}$, which have a lower probability than the boundary probability, $E_x = 0.05$.

9. Furthermore, according to Expressions (8) and (9) above, θ peaks and θ valleys are determined.
10. From the theory of electrical loads, it is known that the calculated values of the load curve are limited by the upper limit, which is obtained by summing all individual consumers, as well as the lower limit, which corresponds to the complete absence of load. In real industrial conditions, the upper design limits are always higher than the real maximum values, while for the lower limit, the load is usually greater than zero. Consequently, due to the disagreement between the real and theoretical limits, an error occurs in the calculations of the peaks, $P_{p\theta}$, and the valleys, $P_{\mathrm{pB}\theta}$. For this, in addition to the normal distribution law, it is also necessary to consider the "truncated" normal distribution law, and for this, it is necessary to determine the values of the normalized upper and lower limits:

$$\Pi_{p\theta} = \frac{(P_p - P_{av})}{\sigma_\theta}; \tag{22}$$

$$\Pi_{v\theta} = \frac{(P_v - P_{av})}{\sigma_\theta}; \tag{23}$$

11. Furthermore, according to the results obtained in block 10, the values are checked according to the normalized limits according to the following conditions:
    - For $\Pi_{p\theta} > 3$ and $\Pi_{v\theta} < -3$, the calculation is performed for blocks 12 and 13 by using the normal distribution law;
    - For $\Pi_{p\theta} < 3$ and $\Pi_{v\theta} > -3$, the calculation is performed in blocks 14–17, using the "truncated" normal distribution law.
12. Calculation of statistical coefficients and calculation of power peaks and valleys.

13. The use of the "truncated" normal distribution law is possible by introducing the truncation coefficient, C_{tr}, into probabilistic models and form factor, $K_{f.d\theta}$ [16,31]:

$$C_{tr} = \frac{\int_{-\infty}^{+\infty} f(P_\theta)dP_\theta}{\int_{P_{v\theta}}^{P_{p\theta}} f_{tr}(P_\theta)dP_\theta}; \tag{24}$$

After a series of mathematical transformations, the truncation coefficient, C_{tr}, expressed through the normalized function, Φ^*, of the normal distribution law, is defined by the following formula [32,33]:

$$C_{tr} = \left[\Phi^*(\Pi_{p\theta}) - \Phi^*(\Pi_{v\theta})\right]; \tag{25}$$

$$K_{f.d\theta} = \sqrt{1 + \frac{DP_\theta}{P_{av}^2}}; \tag{26}$$

14. Furthermore, taking into account the limits on the maximum and minimum values, the value of the average power and the standard deviation are determined as follows:

$$P_{av(tr)} = P_{av} \pm \sigma_\theta C_{tr} = \left[\varphi_*(\Pi_{p\theta}) - \varphi_*(\Pi_{v\theta})\right]; \tag{27}$$

where $\varphi_*(\Pi_{p,v\theta})$ is the density of the standard normal distribution [34]. The standard deviation of the load is as follows:

$$\sigma_{\theta(tr)} = \sigma_\theta \sqrt{\left\{1 + C_{tr}\left[\Pi_{v\theta}\varphi_*(\Pi_{v\theta}) - \Pi_{p\theta}\varphi_*(\Pi_{p\theta})\right] - C_{tr}^2\left[\varphi_*(\Pi_{v\theta}) - \varphi_*(\Pi_{p\theta})\right]^2\right\}}; \tag{28}$$

15. In block 16, the statistical coefficients $\beta_{1(tr)}$ and $\beta_{2(tr)}$ are as follows:

$$1 - E_x = \frac{C_{tr}}{\sqrt{2\pi}\sigma_\theta} \int_{P_{v\theta}}^{P_{p\theta(tr)}} \exp\left(-\frac{(P_\theta - P_{av})^2}{2\sigma_\theta^2}\right)dP_\theta; \tag{29}$$

$$E_x = \frac{C_{tr}}{\sqrt{2\pi}\sigma_\theta} \int_{P_{v\theta}}^{P_{p\theta(tr)}} \exp\left(-\frac{(P_\theta - P_{av})^2}{2\sigma_\theta^2}\right)dP_\theta; \tag{30}$$

16. Next θ—peaks $P_{p(tr)}$, and θ—valleys $P_{v(tr)}$ are calculated through the average power and statistical coefficients, taking into account truncation:

$$P_{p\theta(tr)} = P_{av(tr)} + \sigma_{\theta(tr)}\beta_{1(tr)}; \tag{31}$$

$$P_{v\theta(tr)} = P_{av(tr)} - \sigma_{\theta(tr)}\beta_{2(tr)}; \tag{32}$$

5. Discussion

Below are the results of the studies carried out according to the above-described methodology on the example of group curves of the load of conveyor units. A feature of this method is the determination of the actual consumption of active power, both with and without the use of variable-frequency drives. For this purpose, measurements of power consumption were taken at the facility during one of the most loaded shifts, with a duration of $T_r = 8$ h and with a sampling interval $\Delta t_a = 1$ min. [17,32] Such parameters allow for sufficient reliability when assessing the range of voltage deviations in electrical networks at the lower stages of the PSS (Table 10).

Table 10. Reference information for an ED.

Without Frequency Control								
Unit Name	Quantity, pcs.	P_{nom} kW	k_u	k_{load}	k_{on}	Parameters of the NCF		
						Type	α, s^{-1}	ω_0, rad/s
Main conveyer	5	250	0.7	0.83	0.84	(2)	1.723	2.26
With frequency control								
Main conveyer	5	250	0.5	0.61	0.82	(2)	3	11

The parameters of the NCF and the recalculated utilization factor (k_u) were defined by other authors earlier, with a detailed overview of the scientific research [4,26].

Furthermore, using the calculation formulae, we determined the probabilistic characteristics of the group ELC (Table 11).

Table 11. Calculation formulae for the probabilistic characteristics of individual and group ELCs.

Characteristics of Individual ELCs				Characteristics of Group ELCs		
Unit Name	p_{avi} kW	σ_i kW	σ	P_{av}	α_e	$\sigma_{\theta=1}$
Without frequency control						
Main conveyer 5IO3	175	171	171	875	4.69	165
With frequency control						
Main conveyer 5IO3	125	250	131	625	8.33	125

Furthermore, it is necessary to calculate the switch-on and -off factors according to (9) and (10), respectively, and also to calculate power consumption during the switch-on time according to (8). The data are summarized in Table 12.

Table 12. Probabilistic design characteristics for the group ELC.

Unit Name	p_{on} kW	k_{on}	k_{off}
Without frequency control			
Main conveyer	466	0.84	0.16
With frequency control			
Main conveyer 5	432	0.82	0.18

Using (11) and (12), we can determine the probabilities, e_{ij}, of the occurrence of an electric load (P_{ij}), according to which we can build two static series, which are summarized in Tables 13 and 14 and shown in graphical form in Figures 10 and 11 for a group curve, both without the use of and with the use of frequency control, respectively.

Table 13. Static series of the probability distribution of ordinates of the initial group ELC without using frequency control.

P_{ij}, kW	e_{ij}, rel.un.	$F(P_{ij})$, rel.un.
0	0.0001	0.0001
233	0.0028	0.0029
466	0.0289	0.0317
699	0.1517	0.1835
932	0.3983	0.5818
1165	0.4182	1.0000

Table 14. Static series of the probability distribution of ordinates of the initial group ELC with the use of frequency control.

P_{ij}, kW	e_{ij}, rel.un.	$F(P_{ij})$, rel.un.
0	0.0002	0.0002
216	0.0043	0.0045
432	0.0392	0.0437
648	0.1786	0.2224
864	0.4069	0.6293
1080	0.3707	1.0000

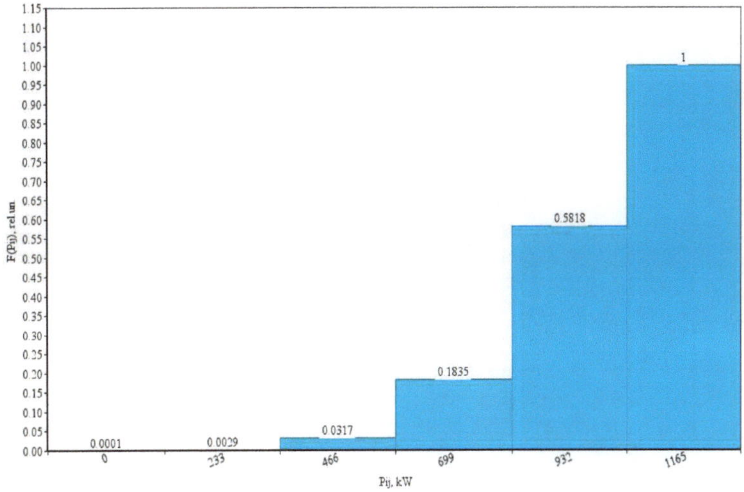

Figure 10. Step function of the probability distribution of the ordinates of the initial group ELC with the use of frequency control.

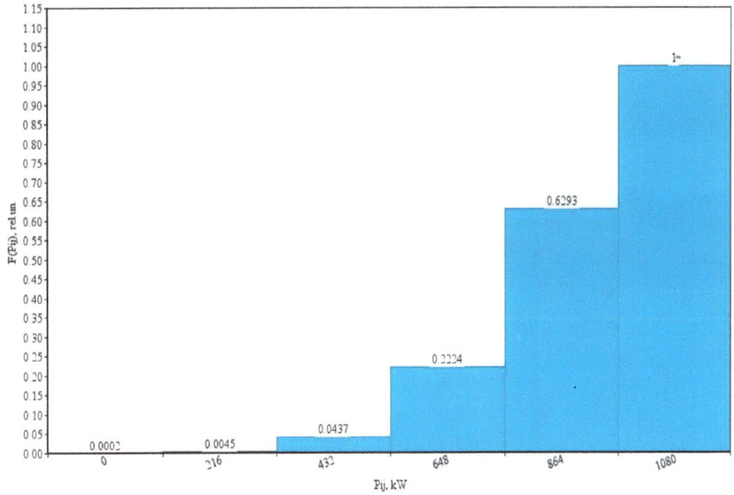

Figure 11. Step function of the probability distribution of the ordinates of the initial group ELC with the use of frequency control.

Having analyzed the obtained static series of the probability distribution of the group curve without using frequency control, we can now select the lowest $P_v = 466$ kW and the highest $P_p = 1165$ kW, and then with the use of frequency control, the lowest $P_v = 432$ kW and the highest $P_p = 1080$ kW.

Furthermore, using the specified averaging interval equal to 1 min, the upper and lower limits of change of θ-ordinates are calculated according to (13) and (14). The calculation results for the normalized limits are summarized in Table 15.

Table 15. Calculated values of the limits of change of the ordinates.

θ, min	P_v, kW	P_p, kW	P_{av}, kW	σ_θ, kW	$\Pi_{v\theta}$	$\Pi_{p\theta}$
			Without frequency control			
1	466	1165	875	165	−2.48	1.76
			With frequency control			
1	432	1080	625	125	−1.54	2.99

The results of intermediate studies proved the significant limitations in magnitude; therefore, further assessment of the calculated values of the peaks and valleys of conveyor units in order to compare the two distribution laws described above must be carried out by using these limitations, taking into account the truncation factor, C_{tr} [32,35]. Using Equations (24)–(32), we obtained the values which are summarized in Table 16.

Table 16. Calculated values of the limits of change of ordinates, taking into account the truncation by the upper and lower limits.

θ, min	$\Pi_{v\theta}$	$\Pi_{p\theta}$	C_{tr}	$k_{f.d\theta}$	$P_{av(tr)}$ kW	$\sigma_{\theta(tr)}$, kW	$\beta_{2(tr)}$	$\beta_{2(tr)}$	$P_{v\theta(tr)}$ kW	$P_{p\theta(tr)}$ kW
					Without frequency control					
1	−2.48	1.76	1.044	1.1	863.5	146.7	1.587	1.725	610	1096
					With frequency control					
1	−1.54	2.99	1.073	1.02	609	110.4	1.737	1.532	440	801

The resulting errors in the assessment of the calculated values of peaks and valleys relative to the two laws of distribution of the normal law and the truncated normal law are calculated by using the following expression [8]:

$$\delta_{p,v\theta} = \frac{\left|P_{p,v\theta(tr)} - P_{p,v\theta}\right|}{P_{p,v\theta(tr)}} * 100\%; \quad (33)$$

The error is no more than 6%, which does not go beyond the limits of permissible values of ±10% [36]; however, the use of a variable frequency drive is not considered, where the error is as follows:

$$\delta_{Pv\theta} = 44\%$$

$$\delta_{Pp\theta} = 38\%$$

This clearly goes beyond the limits of permissible values, causing us to consider a variable frequency drive as a separate load that has its own parameters.

Foreign methods, as well as the method of ordered diagrams and the modified statistical method (Table 17), do not allow the effects from using a variable-frequency drive to be assessed, and all attempts to do this by adjusting the utilization factor failed [37,38]. Figure 12 shows the values of the average calculated loads defined by using various methods.

Table 17. Errors in determining the average value of power, P_{av} , with various methods using a VFD relative to the hierarchical–structural method according to the truncated normal law.

Methods for Assessing Calculated Loads	P_{av}, kW	δ_{Pav}, %
Experimental values	615	-
The modified statistical method	966	37
The method of ordered diagrams	1181	48
Foreign method	1000	39
Hierarchical–structural method	625	1,4

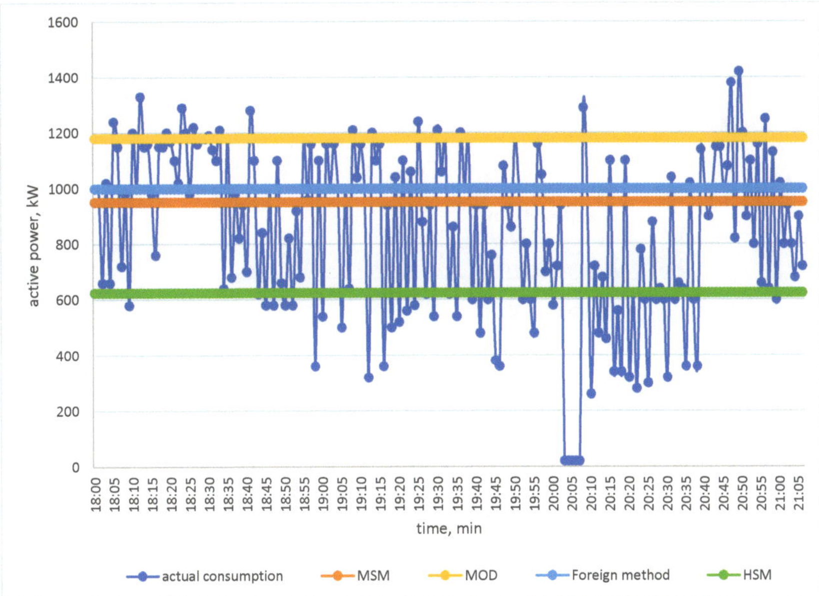

Figure 12. Actual values for the electrical load curves and the calculated values for the average load power.

6. Conclusions

In this study, a method was developed for determining the design calculated electrical loads of mining enterprises that also takes into account the dynamic characteristics of load changes, such as peaks and troughs, surges and load dips, and load fluctuations, which will solve the problems of performing electrical load calculations with a higher confidence probability. These load characteristics were not taken into account in previous studies. This has been shown to have a significant impact on the resulting design power. As a result of increasing the reliability of determining the calculated values of electrical loads, it became possible to make the right choice of power and switching electrical equipment for distribution and protection, which will lead to a reduction in material costs.

The experimental and theoretical studies were carried out personally at the mining facilities, while the gathering of initial data took about 2 years. The initial obtained data made it possible to expand the information base for various consumers, thereby increasing interest in using this calculation method, which ultimately will allow us to design high-quality power supply systems.

The technique proposed in this article can be extended in several aspects by including the technologies of neural networks and machine learning in the calculation.

Author Contributions: D.A.U. defined the general concept for the work and research and arranged for conducting experimental research at enterprises. K.A.K. collected initial data at the mining and processing enterprises, analyzed existing methods used for calculating electrical loads, determined the NCF for individual and group load schedules, and determined dynamic characteristics of the electrical loads graphs. All authors have read and agreed to the published version of the manuscript.

Funding: This research received no external funding.

Institutional Review Board Statement: Not applicable.

Informed Consent Statement: Inapplicable.

Data Availability Statement: Data available upon request, due to restrictions, e.g., privacy or ethical restrictions. The data presented in this study are available upon request from the corresponding author. The data are not publicly available (due to the fact that row data were obtained from industrial facilities).

Conflicts of Interest: The authors declare no conflict of interest.

Nomenclature

$p(t)$	Individual curves	(kW)
P(t)	Group load curve	(kW)
P_{av}	Average power	(kW)
P_h	Heating power	(kW)
T_c	Correlation time	(s)
Δt	Sampling interval	(s)
T_r	Recording time	(s)
P_p	Peak of the actual values of the ELC	(kW)
P_v	Valley of the actual values of the ELC	(kW)
p_{on}	Average power during switch-on time	(kW)
p_{nom}	Rated power	(kW)
P_{ij}	Electrical load values corresponding to the probabilities e_{ij}	(kW)
Dp	Variance of an individual electrical load curve	(kW^2)
DP	Variance of group electrical load curve	(kW^2)
Θ	Averaging interval	(min.)
DP_θ	Variance of group electrical load curve on the averaging interval θ	(kW^2)
σ_θ	Standard deviation of the load on the averaging interval θ	(kW)
α	Parameter characterizes the attenuation of correlations between the ordinates of the initial ELC	(1/s)
ω_0	The natural frequency of oscillations of the CF, due to the repeatability of technological operations	(1/s)
$\alpha_э$	Equivalent factor characterizing the attenuation of correlation between the ordinates of the summed ELCs	(1/s)
e_{ij}	Probability of a single combination	(rel.un.)
F(P)	Probability distribution function of the ELC	(rel.un.)
Ex	Boundary probability	(1)
$\Pi_{p\theta}$	Normalized upper limit	(rel.un.)
$\Pi_{v\theta}$	Normalized lower limit	(rel.un.)
$\beta_{1,2}$	Statistical coefficient	(rel.un.)
Φ^*	Normalized function of the normal distribution law	(rel.un.)
φ_*	Density of the normal distribution law	(rel.un.)
$\delta_{p,v\theta}$	Resulting errors in the assessment of the calculated values of peaks and valleys	(%)
k_u	Utilization factor	
k_d	Demand factor	
k_M	Peak coincidence factor	
$K(\tau)$	Normalized correlation function	
k_{load}	Load factor	
k_{on}	Switch-on factor	
k_{off}	Switch-off factor	
C_{tr}	Truncation coefficient	

Abbreviations

SM	Statistical method
MOD	Method of ordered diagrams
MSM	Modified statistical method
HSM	Hierarchical–structural method
VFD	Variable-frequency drive
PSS	Power supply system
CF	Correlation functions
NCF	Normalized correlation function
ELC	Electrical load curve

References

1. Uryadnikov, N.V. Electricity Consumption in the Russian Federation. Available online: https://www.fedstat.ru/indicator/43277 (accessed on 6 October 2021).
2. Kozyaruk, A.E.; Vasiliev, B.Y. Methods and means for increasing the energy efficiency of machines and technologies with asynchronous electric drives. *Bull. South Ural. State Univ.* **2014**, *1*, 47–53.
3. Khomiakov, K.A.; Belyakov, S.A. The use of RIVS-manufacture energy-saving electric motors in min-ing and processing facilities. *J. Min. Univ.* **2016**, *11*, 29–32.
4. Khan, I.; Jemai, J.; Lim, H.; Sarkar, B. Effect of Electrical Energy on the Manufacturing Setup Cost Reduction, Transportation Discounts, and Process Quality Improvement in a Two-Echelon Supply Chain Management under a Service-Level Constraint. *Energies* **2019**, *12*, 3733. [CrossRef]
5. Mahapatra, A.S.; Soni, H.N.; Mahapatra, M.S.; Sarkar, B.; Majumder, S. A Continuous Review Production-Inventory System with a Variable Preparation Time in a Fuzzy Random Environment. *Mathematics* **2021**, *9*, 747. [CrossRef]
6. Sinchuk, O.N.; Guzov, E.S. Improvement of methods for calculating electrical loads in the design of power supply systems at iron ore enterprises. *Bull. KrNU Mikhail Ostrogradsky* **2013**, *1*, 28–32.
7. Zhokhov, B.D. Analysis of the causes of the overestimation of calculated loads and their possible correction. *Ind. Power Eng.* **1989**, *7*, 7–9.
8. Khomiakov, K.A.; Ustinov, D.A. Improving the accuracy of calculation of electrical loads for industrial enterprises. *J. Phys. Conf. Ser.* **2019**, *1333*, 062009. [CrossRef]
9. Ali Hassan Electrical Load Estimation Course. Available online: http://www.electrical-knowhow.com/2012/12/electrical-pdf-courses.html (accessed on 1 November 2012).
10. Krotkov, E.A.; Stepanov, V.P. Assessment of errors of calculated electric power losses in industrial electrical networks. *Energy Sav.* **2000**, *3*, 47–79.
11. Kampelis, N.; Tsekeri, E.; Kolokotsa, D.; Kalaitzakis, K.; Isidori, D.; Cristalli, C. Development of Demand Response Energy Management Optimization at Building and District Levels Using Genetic Algorithm and Artificial Neural Network Modelling Power Predictions. *Energies* **2018**, *11*, 3012. [CrossRef]
12. Stepanov, V.P.; Krotkov, E.A. Theoretical substantiotion of the sampling interval for calculation the carrelation functions of electric load graphs Power engineering control quality and energy efficiency. In Proceedings of the Russian Scientific and Technical Conference "Energy: Management, Quality and Efficiency of Energy Resources Use", Blagoveshchensk, Russia, 14–16 October 1998; pp. 70–74.
13. Ustinov, D.A.; Khomiakov, K.A. Comprehensive assessment of factors determining electricity load curves concerning enterprises of mineral resource sector. *J. Phisics Conf. Ser.* **2020**, *1753*, 012061. [CrossRef]
14. Abramovich, B.N. Uninterruptible Power Supply System for Mining Industry Enterprises. *J. Min. Inst.* **2018**, *229*, 31–40.
15. Saidur, R. A review on electrical motors energy use and energy savings. In Renewable and Sustainable Energy Reviews. Elsivier: Amsterdam, The Netherlands, 2010; Volume 14, pp. 877–898. ISSN 1364-0321. [CrossRef]
16. Khomiakov, K.A.; Ustinov, D.A. Adjusting the method for calculating electrical loads of mining enterprises. *J. Eng. Sci. Technol. SPbPU Nat. Eng. Sci.* **2019**, *1*, 71–78.
17. Stepanov, V.P.; Krotkov, E.A.; Vedernikov, A.S.; Gudkov, A.V.; Idiatullin, R.F. Reasons for overestimating the calculated values of peaks and valleys of electrical load curves of machine-building plants. *Ind. Power Eng.* **2006**, *1*, 27–30.
18. Zhezhelenko, I.V.; Stepanov, V.P.; Gudkov, A.V.; Iduatulin, R.F. Influence of possible ordinates of electrical load curves on calculated values of peaks and valleys of load. *Energ. CIS High. Educ. Inst. Power Eng. Assoc.* **2006**, *2*, 5–12.
19. Nikolaev, N.I.; Shipulin, A.V.; Kupavykh, K.S. Improving the energy efficiency of well development. *J. Eng. Sci. Technol. SPbPU Nat. Eng. Sci.* **2015**, *2*, 48–57.
20. Ullah, K.R.; Thirugnanasambandam, M.; Saidur, R.; Rahman, K.A.; Kayser, M.R. Analysis of Energy Use and Energy Savings: A Case Study of a Condiment Industry in India. *Energies* **2021**, *14*, 4798. [CrossRef]
21. Seppälä, A. *Load Research and Load Estimation in Electricity Distribution*; VTT Publications 289; Technical Research Centre of Finland: Espoo, Finland, 1996; pp. 1–118.
22. Abramovich, B.N.; Sychev, Y.A. Problems of ensuring energy security for enterprises from the mineral resources sector. *J. Min. Inst.* **2016**, *217*, 132–139.

23. Evangelopoulos, V.; Karafotis, P.; Georgilakis, P. Probabilistic Spatial Load Forecasting Based on Hierarchical Trending Method. *Energies* **2020**, *13*, 4643. [CrossRef]
24. Strokov, A.Y.; Belyaev, A.N. Improving the dynamic stability of stand-alone energy grids based on the data from monitoring of transient processes. *J. Eng. Sci. Technol. SPbPU Nat. Eng. Sci.* **2015**, *1*, 48–58.
25. Anishchenko, V.A.; Pisaruk, T.V. Monitoring the reliability of measurements of load at industrial enterprises on the basis of the analysis of the dynamics of load alterations. *Energ. CIS High. Educ. Inst. Power Eng. Assoc.* **2020**, *63*, 393–410.
26. Shevyrev, Y.V.; Datskovsky, L.K.; Boguslavsky, I.Z. Selection of the type of electric drive for large-volume mills. *J. Min. Univ.* **2020**, *5*, 53–62.
27. Anishchenko, V.A.; Pisaruk, T.V. Monitoring the reliability of measurements in energy grids according to the first increment and on the basis of extrapolating filters. *Energ. CIS High. Educ. Inst. Power Eng. Assoc.* **2018**, *61*, 423–431.
28. Galkin, V.I. Extension of application range of tubular belt conveyors owing to constructive inno-vative solutions. *J. Min. Univ.* **2020**, *5*, 52–56.
29. Keka, I.; Betim, C.; Neki, F. Effect of parallelism in calculating the execution time during forecasting electrical load. *JNTS* **2017**, *44*, 169–180.
30. Khuntia, S.R.; Ruenda, J.L.; Van der Meijden, M.A. Long-term electricity load forecasting considering volatility using multiplicative error model. *Energies* **2018**, *11*, 3308. [CrossRef]
31. Chen, P. Stochastic Modeling and Analysis of Power System with Renewable Generation. Ph.D. Thesis, Aalborg University, Aalborg, Denmark, 2010; pp. 1–216.
32. Shklyarskiy, Y.; Hanzelka, Z.; Skamyin, A. Experimental Study of Harmonic Influence on Electrical Energy Metering. *Energies* **2020**, *13*, 5536. [CrossRef]
33. Shklyarskiy, Y.; Skamyin, A.; Vladimirov, I.; Gazizov, F. Distortion Load Identification Based on the Application of Compensating Devices. *Energies* **2020**, *13*, 1430. [CrossRef]
34. Sychev, Y.; Abramovich, B.; Prokhorova, V. The assessment of the shunt active filter efficiency under varied power supply source and load parameters. *Int. J. Electr. Comput. Eng.* **2020**, *10*, 5621–5630.
35. Zagrivniy, E.A.; Poddubniy, D.A.; Kovalchuk, M.S. Noncontact speed sensor for maintaining the autoresonant operating mode of the oscillating electric motor drive. In Proceedings of the 2019 IEEE Conference of Russian Young Researchers in Electrical and Electronic Engineering, Saint Petersburg, Russia, 28–31 January 2019; pp. 740–743.
36. Kovalchuk, M.S.; Zagrivniy, E.A.; Poddubniy, D.A. Simulation Model of a Dynamically Coun-terbalanced Drilling String with a Sensorless Control System for Controlling an Autoresonant AC Electric Drive. In Proceedings of the IEEE Conference of Russian Young Researchers in Electrical and Electronic Engineering, Saint Petersburg, Russia, 27–30 January 2020.
37. Baburin, S.V.; Zyrin, V.O.; Kovalchuk, M.S. Dependence of power supply systems reliability on the type of redundancy. In Proceedings of the IOP Conference Series: Materials Science and Engineering, Saint Petersburg, Russia, 23–24 May 2019; Volume 643.
38. Belsky, A.A.; Skamyin, A.N.; Vasilkov, O.S. The Use of Hybrid Energy Storage Devices for Balancing the Electricity Load Profile of Enterprises. *Energ. CIS High. Educ. Inst. Power Eng. Assoc.* **2020**, *63*, 212–222. [CrossRef]

electricity

MDPI

Article

Impact of Increased Penetration of Low-Carbon Technologies on Cable Lifetime Estimations

Xu Jiang [1], Edward Corr [1], Bruce Stephen [2,*] and Brian G. Stewart [2]

1 Power Network Demonstration Centre, University of Strathclyde, Cumbernauld G68 0EF, UK; x.jiang@strath.ac.uk (X.J.); edward.corr@strath.ac.uk (E.C.)

2 Institute of Energy and Environment, Royal College Building, University of Strathclyde, Glasgow G1 1XW, UK; brian.stewart.100@strath.ac.uk

* Correspondence: bruce.stephen@strath.ac.uk

Abstract: Cables are the largest assets by volume on power distribution networks and the assets with the least health information routinely gathered. Projections over the next 8 years suggest increased penetration of low-carbon technology (LCT) at the distribution level with higher loads resulting from electric vehicle (EV) and heat pump uptake. Over this period, increased cable loading will directly influence their lifetimes and may mean that existing asset management practices need to be revised to understand the specific impact on end-of-life assessment. Accordingly, this paper uses a physics-based thermal lifetime model based on projected uptake trends for LCTs to evaluate the impact on distribution cable lifetime. Two case studies are presented considering portions of network and the ultimate impact on asset life over the next decade. Two commonly used cables are considered to quantify the lifetime reduction caused by LCT for asset fleets. The paper shows that the projected uptake of EVs and heat pumps will have a detrimental effect on cable life with a 30% reduction in cable lifetime possible.

Keywords: electric vehicle; heat pump; cable; lifetime assessment; asset management

Citation: Jiang, X.; Corr, E.; Stephen, B.; Stewart, B.G. Impact of Increased Penetration of Low-Carbon Technologies on Cable Lifetime Estimations. *Electricity* **2022**, *3*, 220–235. https://doi.org/10.3390/electricity3020013

Academic Editor: Pavlos S. Georgilakis

Received: 13 April 2022
Accepted: 22 May 2022
Published: 27 May 2022

1. Introduction

As government and international bodies focus more on the impact of climate change, low-carbon technologies (LCTs) will become more prominent on power networks. However, as LCT penetration increases, the cable loading regimes will change significantly. For example, in the My Electric Avenue project [1], the British utility SSEN found that approximately one-third of LV networks in the UK need to be upgraded, when 40–70% customers have an EV, and this level of uptake is expected to occur by 2030. These upgrades would result in significant capital expenditure [2] and customer connection disruptions to enable reinforcement works and possible delays in the connection of LCTs. The UK government also endeavors to meet the target of net zero emissions across the economy by 2050 [1]. A key element of this will be switching over 20,000 homes per week from 2025 to 2050 to a low-carbon heat source [3]. Currently, 85% of UK households use natural gas to heat their homes [3]; this is one of the main contributors of the UK's carbon emissions. Heat pumps are one of the most promising solutions, and there is likely to be a targeted replacement strategy for households that are not on the gas network. In addition to the UK, other countries have also set up net-zero targets; for example, the European Union has agreed make climate neutral by 2050 [4]. To date, studies have focused on deferring reinforcement of the power network infrastructure [1,5–7] and the influence of LCT on the power flow [8,9]. No studies have discussed the influence of LCTs on the life of installed power cables, despite this being critical to both.

The lifetime estimation for power cables is important, as they are not visible and represent a large proportion of installed assets for network operators. A Remaining Useful

Life (RUL) predictor for assets can be classified into two potential approaches: data-driven and model-based. The data-driven approach normally applies machine learning algorithms on archived data, such as historical failure data, voltage, and current, to predict RUL. Common models are the Weibull model [10], inverse power law model [11], and Crow–AMSAA model [10]. The data-driven method is focused on the overall trends for a type of cable rather than an individual unit. The availability of run-to-failure and diagnostic data is sometimes limited in this case. Model-based prediction has also been widely investigated to date. Work in Reference [12] used the temperature, humidity, and voltage to calculate the material stress via a cable insulation model, and then the predicted lifetime was estimated on the stress value. Elsewhere, Reference [13] used load, ambient temperature, environmental parameters, and a thermal model of degradation to estimate cable lifetime. In addition to this, recent research has started to combine statistical analysis into model-based predictions. Work in Reference [14] develops a model-based case which used a state-space lifetime equation and a statistical uncertainty model to predict the RUL of cables by using thermal and load data.

The relevant literature indicates that the lifetime of power cables would be significantly affected by the presence of emerging LCTs. A lifetime prediction model is required to help asset managers target which cables need to be reinforced. This approach will mitigate failures, rather than invoke blanket replacement programs, which would represent a significant cost saving. To assess how the lifetime of power cables is affected by the increasing LCTs, the Arrhenius Inverse Power Model (IPM) [14] is employed to develop different scenarios of interest to estimate the average lifetime of the cable under investigation. The models and key parameters are based on the available literature/past studies [14–17]. The contribution of this paper is to combine modeling techniques with the projected loading to quantify the impact of LCTs on the lifetime of different cables. The model can be used in future studies for specific cable circuits or topologies of concern. The case studies in this paper consider an 11 kV distribution cable and 400 V mains' cable in a realistic network arrangement to assess the impact of prominent LCTs. The model results can be used to support utilities in determining the reinforcement and inspection/maintenance plans based on the projected lifetime of key cable types.

2. Low-Carbon Technology Adoption

There is increased focus in business and government internationally on climate change. More consumers and third parties are looking to make use of alternative technologies which are connected to the power system. This brings further challenges to the distribution utility providers around network operation, as power flows are more difficult to predict. These changes could affect cable life through increasing the loading in that portion of the network, and this may shorten life. Based on current trends, the immediate challenge is posed by electric vehicles (EVs) and heat pumps (HPs). To explore the impact on cable life, the relative uptake of these two technologies was investigated in the following subsections.

2.1. EV

With the cost of batteries falling rapidly and increased government incentives, it is anticipated that the stock of EVs would increase exponentially. Based on the prediction from the Global EV outlook 2021 [18], the global EV stock can grow by 13.8 times above 2020's level in 2028 and the global electricity demand from EVs would reach at least 525 TWh in 2030 (2% of global electricity total final consumption). Therefore, EVs would significantly impact power delivery and the life of assets on the distribution network. More consumers are making the switch to EVs as an environmentally friendly and cost-effective means of transport. EVs that need to be recharged can be classified as two types: plug-in hybrid electric vehicles (PHEVs) and battery electric vehicles (BEVs). A PHEV can be powered by both petrol and electricity and BEV is a pure electric vehicle. The battery of PHEVs has a lower capacity than BEVs and a reduced electric only range. A BEV normally has a DC charging port (used for rapid charging) and an AC charging port (used for destination charging). In contrast, PHEVs normally have an AC charging port for destination charging. In the UK, in 2020, 108k BEVs were sold and 66k PHEV were sold [19]; this was 10% of the 2020 UK market share. The battery capacity of current BEVs range from 17.6 kWh in the Smart EQ ForTwo with a range of 93.3 km up to 100 kWh in the Tesla Model S and Model X, which have over 482.8 km range [20]. EV charging activity can be classified into four different typical arrangements: home, work, destination, and fast/rapid charging station. The chargers at home are typically slow (supplies 3 to 11 kW AC to the vehicle [21]), and 87% of charging activity is performed at consumers' homes [21]. Workplace charging is typically 3–11 kW AC, but charging is more likely to happen during daytime hours (approximately 9 a.m.–5 p.m.). The destination type denotes charging at shops and car parks; the chargers are generally higher power than household level but lower than rapid chargers (normally capable of supplying 7 to 22 kW) [21]. The final type is rapid chargers (can supply 22 to 120 kW—a 50 kW rapid charger could have the same impact on the network as 25 homes [21]). This enables the EV to be topped up in minutes to hours; however, such a large load interacting with the distribution network might be a challenge to manage at a large scale. DC charging speed is seen as a critical factor in the future, and charging speeds of 350 kW [22,23] will become more common as new vehicles accept this speed of charging. AC charging speed is not likely to increase much beyond the current capabilities, as there is no clear trend for increasing the rating of the on-board chargers on EVs (AC to DC converters) beyond the 7–22 kW range. This paper considers the impact of AC EV charging infrastructure at household and industrial customers.

In general, an increase in the number of EVs charging on the network would bring a higher burden for cables in distribution networks. To avoid heavy reinforcement work, an interesting option is to use a smart charging control system or incentivized energy tariffs to shift the peak of demand to off-peak periods. However, there is still a long way to go before smart charging solutions are fully implemented, and the EV stock is increasing exponentially. It can therefore be predicted that some cable runs may experience a sharp increase in the number of EVs based on local hotspots of EV uptake. This may occur before smart charging schemes are fully implemented, which may result in a significant reduction in cable life in this interim period. Even if smart charging solutions were to be fully implemented, this would not completely eliminate the impact of increased load on cables caused by EV charging. Therefore, cable life could be reduced in this period, as an increasing number of EVs are connected to the power network. As cables are the least monitored asset, modeling tools can support network operators over this interim period and beyond as further data are made available [14].

2.2. Heat Pump

Currently, 85% of UK households use fossil-fuel-based natural gas to heat their homes [3], making it one of the main contributors of the UK's carbon emissions. Utility providers, government, and policymakers have already conducted a range of projects [3,24,25]. The analysis [24] shows that the yearly installation rate of HP would sharply increase over the next 10 years. As an example, the yearly installation rate in 2030 would be more than 28 times the installation rate in 2020.

If conventional fossil-fuel-based natural gas boilers are replaced by HP, the electricity consumption would increase significantly. HPs come as air-source and ground-source units. The two technology types suit different installation conditions. An air-source heat pump (ASHP) is better suited to an urban environment. The ASHP absorbs heat from the air via a fluid, the fluid passes through a compressor, and this, in turn, increases the temperature of the fluid. A heat exchanger is used to transfer the heat from the fluid to the hot-water circuits of the home. A ground-source heat pump (GSHP) is more suited to a rural environment, as significant civil works are required during installation. A GSHP is more efficient, as it extracts heat from the round via buried pipework; this is termed the ground loop and has a water/antifreeze fluid inside. The system uses the same operating principal to elevate the temperature of the fluid and transfer the heat to the hot-water circuits of the home [26]. According to References [27,28], the average power consumption ranges from 1.2 to 1.7 kW per unit, and a peak value of 4 kW was noted when one dwelling was considered. Although the demand of a single ASHP is less than a single EV, the operating time is longer, and customers are unlikely to wait until a certain time to heat their home. The flexibility options that are apparent for EV charging do not exist for ASHP, given the lack of energy storage in the power network.

An ASHP needs electricity to run, but for every 1 kWh of electricity, an ASHP can produce 3 kWh of heat [29]. The average annual thermal demand for most homes in the UK is at 12,000 kWh [29]. Therefore, this paper assumes that the average annual power consumption of the ASHP is 4000 kWh. This represents a significant increase in the average household usage (3781 kWh [30]).

3. Cable Lifetime Assessment Method

The previous section introduced a range of potential challenges for the most prominent LCTs. The impact will ultimately be reflected in the cables' operating temperature based on the applied load regime. This study employs a thermal aging model to estimate the cable lifetime based on the cable loading. The end-to-end workflow is given in Figure 1.

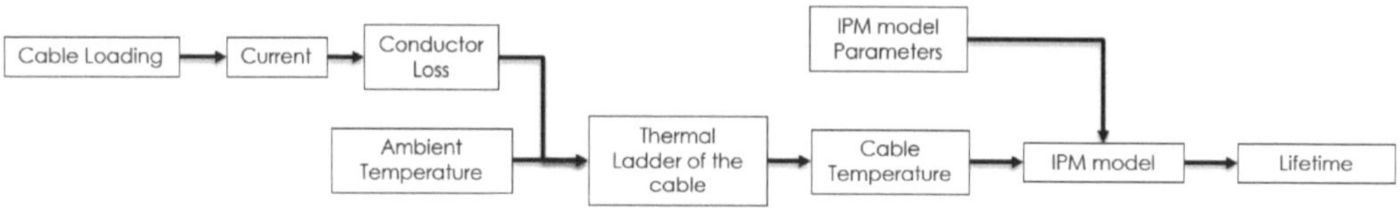

Figure 1. Thermal aging model for cable lifetime estimation.

As Figure 1 shows, the cable loading will increase the temperature of the conductor via the conductor losses. The conductor losses were calculated by using the conductor resistance at the rated operating temperature of the cable and the applied current due to the specified loading regime. The conductor temperature can be estimated based on the ambient temperature and the predicted temperature difference from the thermal ladder calculation. This was used as a sense check to confirm that the cable under investigation was not operating beyond the thermal capabilities of the insulation system. Finally, the Arrhenius Inverse Power Model (IPM) is used to project the lifetime with the cable temperature. To estimate the aging of a cable, the cable temperature is required. In this paper, all the heating of the cable system is assumed to come from the ambient temperature and power loss of the cable. All models were implemented in MATLAB for this study.

3.1. Thermal Ladder Evaluation

In this study, the transient temperature was ignored. Therefore, the cable temperature can be given as follows:

$$\theta_{cable} = \theta_{am} + \theta_{loss} \tag{1}$$

where θ_{am} was assumed to be 30 °C within this study. The temperature rise due to power losses can be simplified to a thermal ladder, which is given in Figure 2.

As Figure 2 shows, the θ_{loss} is related to R_a, R_b, and W_c; and R_a and R_b can be further decomposed into T_A, T_B, Q_A, and Q_B. Among them, T_A and T_B are the thermal resistance, Q_A and Q_B are thermal capacitances, and W_c is the conductor loss. According to Reference [14], the θ_{loss} of the XLPE cable can also be written as follows:

$$\theta_{loss} = W_c\left(R_a\left(1 - e^{-at}\right) + R_b\left(1 - e^{-bt}\right)\right) \tag{2}$$

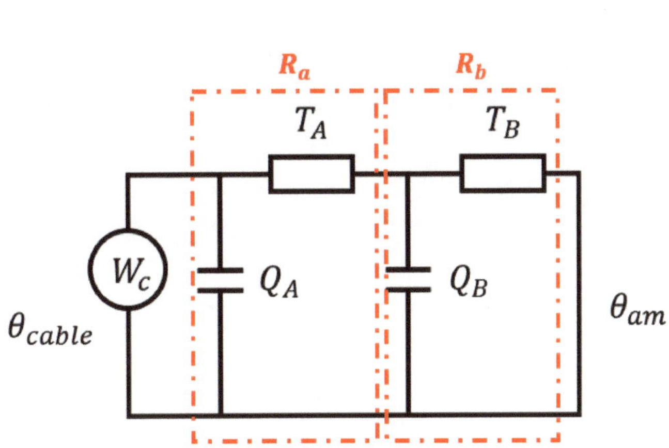

Figure 2. The two-loop equivalent thermal ladder representation [31].

This investigation aims to consider the average impact; therefore, θ_{loss} is assumed to be a constant value, and t is assumed to be infinite. Thus, the representation of θ_{loss} can be simplified as follows:

$$\theta_{loss} = W_c(R_a + R_b) \tag{3}$$

where the variables are represented by the following [14]:

$$R_a = \frac{1}{a-b}\left(\frac{1}{Q_A} - b(T_A + T_B)\right) \tag{4}$$

$$R_b = T_A + T_B - R_a \tag{5}$$

where a and b are the scale parameters derived from the two-loop equivalent diagram; T_A, T_B, Q_A, and Q_B can be further decomposed into T_1, T_2, T_3, Q_1, Q_2, and Q_3; and the calculation details can be found in Reference [14]. The thermal resistance (T_1, T_2, and T_3) and capacitance (Q_1, Q_2, and Q_3) can be calculated by using IEC 60287-2-1 [32]. Additionally, W_c is the conductor loss, which can be calculated by using the following equation:

$$W_c = I^2 R * pf \tag{6}$$

where I is current on the cable, R is conductor resistance, and pf is the power factor, which is assumed to be 1 in the study. The power factor is the ratio of active power to apparent power; hence, a power factor of 1 could indicate the worst-case conductor loss.

The thermal ladder approximation for the two cable topologies is derived in the following sections; the two topologies require specific equations from the standard [32] in the derivation of T_1, T_2, T_3, T_A, T_B, Q_A, Q_B, and Wc.

3.1.1. The 11 kV Distribution Cable

This section details the derivation of the thermal ladder model for the 400 mm^2 3 core 11 kV XLPE cable. The key dimensions and previous modeling data were available in a past publication [33]. The specification [15] of the 400 mm^2 XLPE cable is given in Table 1, and the dimension is given in Figure 3.

Table 1. Specification of the 400 mm^2 11 kV XLPE cable.

Parameter	Value	Parameter	Value	Parameter	Value	Parameter	Value
Voltage Level (kV)	11	Conductor Screen Thickness (mm)	0.7	Sheath Diameter (mm)	-	Armor Diameter (mm)	87.5
Rated Current (Amp)	522	Insulation Diameter (mm)	31.8	Sheath Thickness (mm)	-	Armor Thickness (mm)	3.15
Conductor Area (mm^2)	400	Insulation Thickness (mm)	3.4	Concentric Neutral Diameter (mm)	35.3	Jacket Diameter (mm)	95.5
Conductor Diameter (mm)	23.6	Insulation Screen Diameter (mm)	33.6	Concentric Neutral Thickness (mm)	0.85	Jacket Thickness (mm)	4
Conductor Screen Diameter (mm)	25	Insulation Screen Thickness (mm)	0.9	Armor Bedding Diameter (mm)	81.2		

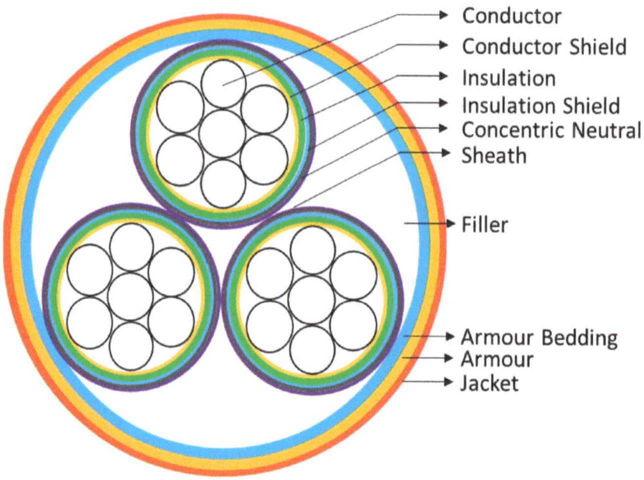

Figure 3. Layout of the 11 kV distribution cable.

According to IEC 60287-2-1 [32], thermal resistance, T_1, is the resistivity between the conductor and metallic sheath/screen; T_2 is the resistivity between metallic sheath and armor; T_3 is between armour and surroundings; and T_4 represents the surroundings. T_1 is defined below.

$$T_1 = \frac{\rho_{ins}}{2\pi} G + 0.031\left(\rho_f - \rho_{ins}\right) e^{0.67\frac{t_1}{d_c}} \tag{7}$$

where ρ_{ins} and ρ_f are the thermal resistivity of the insulation and the filler material respectively; t_1 is the thickness of the material between the conductors and outer covering; d_c is the diameter of conductor; and G is geometric factor, which also depends on the ratios $\frac{t_1}{d_c}$ and can be obtained from the geometric factor curve in Reference [32]. Looking up the curve, $G \approx 0.8$ can be used for the 11 kV XLPE cable in this study.

T_2 and T_3 are given by the following relationships:

$$T_2 = \frac{1}{2\pi} \rho_{ins} In\left(1 + \frac{2t_2}{D_s}\right) c \tag{8}$$

$$T_3 = \frac{1}{2\pi} \rho_{ins} In\left(1 + \frac{2t_3}{D'_a}\right) \tag{9}$$

where t_2 is the thickness of the bedding (in mm); D_s is the external diameter of the sheath (in mm); t_3 is thickness of outer covering; and D'_a is the external diameter of sheath, screen, and bedding.

The thermal capacitance ($Q_1, Q_2, Q_3,$ and Q_4) can be calculated by using the following:

$$Q = \frac{\pi}{4}\left(D_{ext}{}^2 - D_{int}{}^2\right)C \tag{10}$$

where C is volumetric specific heat [33]. Based on the dimensional properties of the 11 kV XLPE cable [15], the steady-state thermal resistances and capacitors are as follows: $T_A = 0.3366$ K/W, 0.3366 K/W, $T_B = 0.5197$ K/W, $Q_A = 1935$ J/K, $Q_B = 573$ J/K, and $W_c = 18.25$ W/m.

3.1.2. Mains' Cable

This section details the derivation of the thermal ladder model for the 3 core 300 mm^2 XLPE waveform cable. The waveform cable has sectorial shaped solid aluminium conductors. The specification and layout of the 300 mm^2 waveform cable are shown in Table 2 and Figure 4.

Table 2. Specification of the 300 mm^2 waveform cable [16].

Parameter	Value	Parameter	Value
Voltage Rating (V)	1000	Insulation Thickness (mm)	1.8
Rated Current (Amp)	435	Earth Wires (mm)	1.85
Conductor Area (mm^2)	300	Sheath Thickness (mm)	2.8
Conductor Diameter (mm)	36.8	Overall Diameter (mm)	55.1
Rubber Bedding Layer (mm)	0.9	Rated Operating Temperature (°C)	80

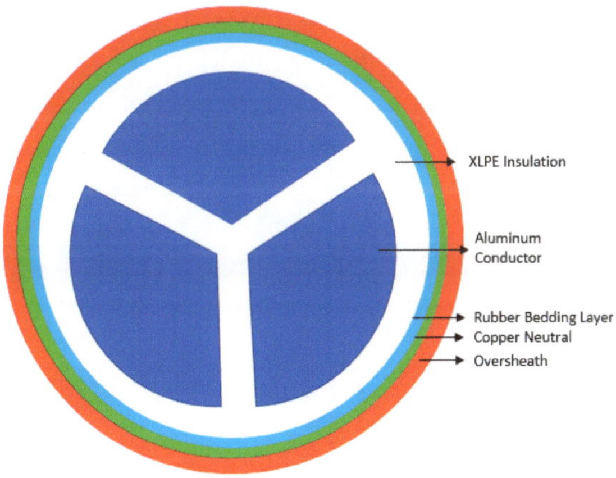

Figure 4. Layout of the 300 mm^2 waveform cable.

According to IEC 60287-2-1 [32], the waveform cable model also employed Equation (7) to calculate T_1. However, the geometric factor, G, is calculated in a different way, as outlined in Equation (12), and F_2 is a coefficient for the belted cable, given by the following relationships:

$$G = 3F_2 In\left(\frac{d_a}{2r_1}\right) \tag{11}$$

$$F_2 = 1 + \frac{3t}{2\pi(d_x + t) - t} \tag{12}$$

where d_a is the external diameter of the belt insulation (in mm), r_1 is the radius of the circle circumscribing the conductor, d_x is the diameter of a circular conductor, and t is the insulation thickness between conductors. T_2 and T_3 are given by the following relationships:

$$T_2 = 0 \tag{13}$$

$$T_3 = \frac{1}{2\pi}\rho_{ins} In\left(1 + \frac{2t_3}{D'_a}\right) \tag{14}$$

where t_3 is thickness of outer covering, and D'_a is the external diameter of sheath, screen, and bedding.

The thermal capacitance ($Q_1, Q_2, Q_3,$ and Q_4) can be calculated by using Equation (10). Using the equations above, the steady-state thermal resistances and capacitances are as follows: $T_A = 0.1798$ K/W, $T_B = 0.4840$ K/W, $Q_A = 3744.1$ J/K, $Q_B = 357.9$ J/K, and $W_c = 23.84$ W/m. The AC resistance of the 11 kV XLPE cable is 0.0645 Ω/km [34]. The AC resistance of the waveform cable is 0.126 Ω/km [16].

3.2. Lifetime Assessment with IPM

The hypothesis of cable aging is based around the IPM. The model [14] assessed how the two LCTs influenced the cable life under different scenarios. The IPM model is given below:

$$TTF = TTF_0 \frac{E}{E_0}^{-(\eta_0 - b(\frac{1}{\theta_0} - \frac{1}{\theta_{cable}}))} \exp\left(-B\left(\frac{1}{\theta_0} - \frac{1}{\theta_{cable}}\right)\right) \tag{15}$$

$$\theta_{cable} = (\theta_{am} + \varepsilon_{am}) + (\theta_{loss} + \varepsilon_{loss}) \tag{16}$$

where E is electric field (kV/mm); θ_{cable} is the cable temperature in Kelvin, and it can be further decomposed into the ambient temperature $(\theta_{am} + \varepsilon_{am})$ and the joule heating in the cable conductor $(\theta_{loss} + \varepsilon_{loss})$; and ε_{am} and ε_{loss} represent the temperature errors. The temperature error could occur as a measurement or calculation error rather than a dynamic demand variation. This paper therefore explores the impact of ±5% temperature error on the cable lifetime [14]. Moreover, θ_0 is a reference temperature, η_0 is the voltage endurance coefficient at $\theta_{cable} = \theta_0$, E_0 is a value of electric field below which electrical aging is deemed as negligible (kV/mm), TTF_0 is time-to-failure at $\theta_{cable} = \theta_0$, and $E = E_0$. B is the ratio of $\Delta W/k$ (ΔW is the activation energy of the main thermal degradation reaction and k is the Boltzman constant), and b is a parameter that models the synergism between electrical and thermal stresses (K.mm/kV). The thermal rating of XLPE and PILC are at 90 °C (363.15 Kelvin) and 70 °C (343.15 Kelvin), respectively. The lifetime estimation parameters are given in Table 3.

Table 3. The general parameters for IPM lifetime estimation [14].

Parameter	Value	Parameter	Value
TTF_0 (h)	1×10^6	E_0 (kV/mm)	5
b (K mm/kV)	4420	E (kV/mm)	7.2
η_0 (non-dimensional)	15	B (K)	12,430

4. Case Study

This section will provide two case studies to analyze the demand impact on typical distribution network cables. Two representative cable types (11 kV XLPE cable and 400 V waveform cable) are considered in this study and are introduced in the following sections.

4.1. 11 kV Distribution Network

For the 11 kV distribution cable, the example network has 24 distribution transformers which step down to 400 V, and there are a total of 5032 customers [17]. Among these customers, approximately 5.3% are commercial and the remainder are domestic. According to a 2018 report from the Department for Business, Energy, and Industry Strategy [30], the mean annual domestic electricity consumption per meter in GB in 2016 was 3781 kWh. The mean annual industrial/commercial electricity consumption in 2016 was 68,460 kWh. The example network has an average loading of 4.14 MW. By employing the workflow presented in Figure 1, the lifetime projections for different loading regimes can be derived, and the impact of temperature error is also considered. The variation of cable lifetime with applied current is outlined in Figure 5.

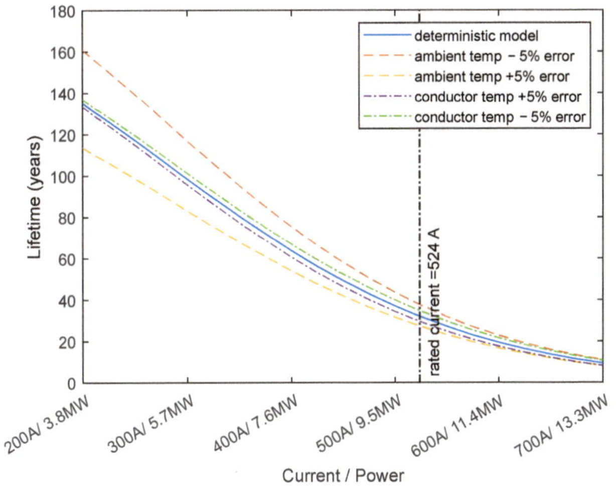

Figure 5. The lifetime curve for applied current and loading for the 11 kV XLPE cable.

As Figure 5 shows, if the rated current (for 10 MW rated power) is applied throughout the cable operation, a lifetime of 31.7 years is projected with the deterministic model. A 5% ambient temperature error could result in a ~15% variation lifetime, and a 5% conductor temperature error could incur an 8% variation on lifetime estimate. Ambient temperature error had a more significant impact on the cable lifetime estimation when the cable loading was low. Based on the analysis above, the 5% error in ambient temperature has a larger impact on the lifetime prediction. Generally, heavier cable loading would exceed the operating temperature of the cable and, thus, would significantly reduce the cable life or cause premature failure. The lifetime of the cable with a constant load of 4.14 MW is projected as 129 years $\pm$ 22 years. The use of mean cable loading does not account for the peak-times in the load profile. Therefore, this could lead to an underestimate of lifetime. Based on this example, the 11 kV cable is operating at 50% of its rated capacity in the normal baseline scenario. This will not be the case for all portions of the network, and the ability to pinpoint these issues before they occur will be paramount to future network operation. Some variations are presented based on the impact of EVs and ASHPs in the following sections.

4.2. *Mains' Cable Network*

The mains' cable is assumed to supply 200 domestic customers from the local distribution transformer. The example network has an average loading of 86.32 kW. This level of loading, on average, would equate to a lifetime of 65 years. The following sections discuss the impact of EVs and ASHPs on cable lifetime. The following calculations assume that the operating voltage for the cable is 400 V. The impacts of applied current and loading on the lifetime are given in Figure 6.

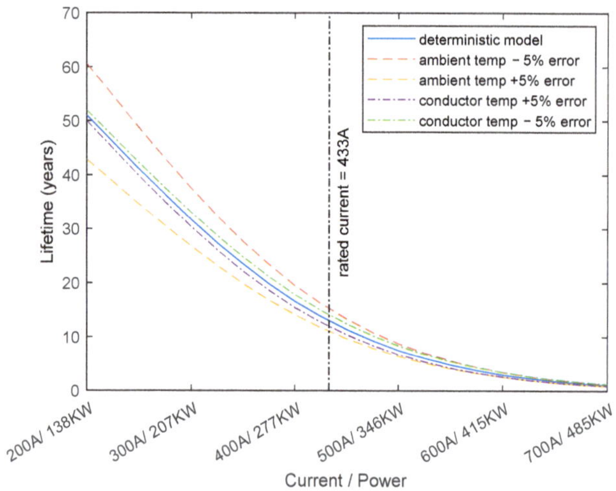

Figure 6. The lifetime curve for applied current and loading for the 300 mm^2 waveform cable.

As Figure 6 shows, the loading will determine the lifetime of the mains' cable, and when operating at the rated current, the lifetime drops significantly to 12.8 years (deterministic model value). The conductor temperature and ambient temperature error result in ±8% and ±15% lifetime variation, respectively. To investigate how the technologies influence the cable life, investigations were based on an example network.

4.3. EV Impact

This section investigates the impact of EV charging on the example networks introduced above. According to the background review in Section 2.1, each domestic charger consumes around 3 to 7 kW, and an industrial/commercial charger could be from 7 to 22 kW. This investigation assumes that each household has 6 h for charging every day; this may be an over estimation in itself. The EV loading is assumed to be an additional load at each domestic/non-domestic property.

The new average cable loading, L_{new}, can be calculated with the following equation:

$$L_{new} = L + (p_d \times r_d \times n_d + p_i \times r_i \times n_i) \times w \tag{17}$$

where L is the conventional cable loading; p_d and p_i are the average power consumption of domestic and industrial customers, respectively; r_d and r_i are the usage rate of the LCT technologies for domestic and industrial customers, respectively; n_d and n_i are the number of domestic and industrial customers on the operational cable, respectively; and w is the adoption rate of EVs.

4.3.1. The 11 kV Distribution Cable

The EV loading on the 11 kV distribution cable was considered first. The best (lowest possible average consumption increase), median, and worst scenarios (largest increase) are given in Table 4.

In all scenarios, the introduction of EVs would increase the loading of the cable. This would significantly reduce the lifetime of the cable. With an adoption rate of 100% in the best-case scenario, the cable lifetime would drop down to 54 years. In the worst-case scenario, the cable would fail in 7.3 years if the rated conditions of the cable were ignored. In reality, the adoption rate would not be 100% in the near future. Figure 7 outlines the effect on penetration rate for the three scenarios.

Table 4. Three different scenarios for EV consumption at 11 kV.

Scenario	Average Consumption Increase		Demand at 100% Adoption Rate
	Household Customer	**Industrial Customer**	
Best-Case Scenario	3 kW	7 kW	8.18 MW
Median Scenario	7 kW	11 kW	13.21 MW
Worst-Case Scenario	7 kW	22 kW	13.94 MW

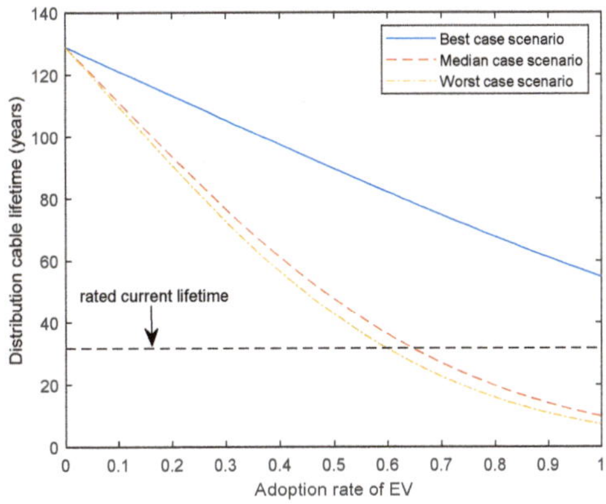

Figure 7. The lifetime of the distribution cable with EV adoption rate.

4.3.2. Mains' Cable

Two scenarios are used to analyze the EV impact on the LV mains' cable; only domestic customers are considered in this case. If each household has an EV, the cable loading would be given by two different assumed load levels (3 and 7 kW). In this case, the new load equation is expressed as follows:

$$L_{new} = L + (p_d \times r_d \times n_d) \times w \tag{18}$$

As Figure 8 shows, the presence of EVs can significantly reduce the lifetime of a mains' cable. For the worst-case scenario, the lifetime can be reduced to approximately 2.5 years if the rated conditions are exceeded. An EV adoption rate of 60% in the worst-case scenario represents the threshold at which the rated conditions of the cable have been exceeded.

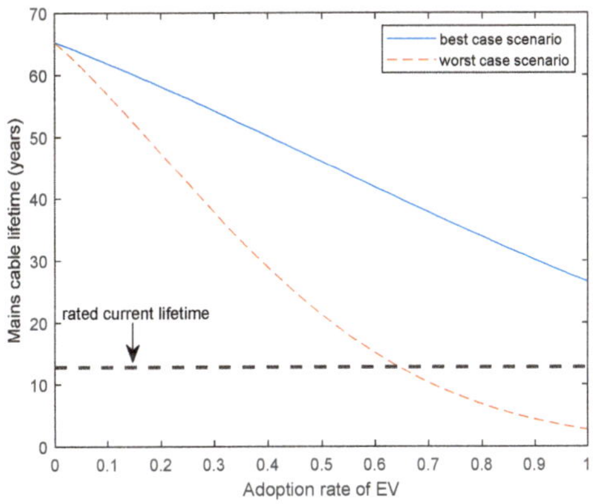

Figure 8. The lifetime of the mains' cable with EV adoption rate.

4.4. Heat-Pump Impact

As discussed in Section 2.2, the average power consumption of an ASHP is assumed as 4000 kWh per year. Based on government incentives/goals, the adoption rate of ASHP is expected to rise, and the load on distribution cables will rise sharply. This work assumes that the domestic consumption for a single household is the same as that of an industrial customer. The impact on cable life with ASHP adoption is shown in Figure 9.

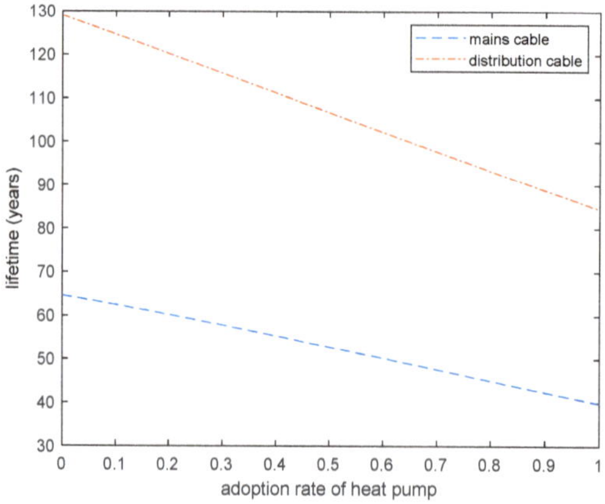

Figure 9. Impact of adoption rate of ASHP on the life of the mains' cable.

If every household on the example distribution network installed an ASHP (adoption rate of 1), the total demand would be 6.44 MW, and the cable life would be reduced to approximately 84 years from a base case of 127 years with no ASHP penetration. The lifetime of the distribution cable generally drops faster than the mains' cable.

The mains' cable is assumed to have only domestic customers. In the example network, when conventional boilers are replaced with ASHPs, the lifetime estimation for the mains' cable decreases sharply. Based on the 200-customer example network, when the adoption rate reaches 100% (cable loading is 177.6 kW), life would decrease to around 40 years. In the example network, this level of penetration could be permitted without exceeding (on average and with ASHPs the only contributor to a customer load increase) the rating of the mains' cable.

5. Impact of Mixed LCT

As discussed in the previous section, the growth of EVs and ASHPs could have a significant impact on the life of cables in the near future. This section uses the projections for EV and ASHP uptake combined with the IPM model to estimate how *LCT* loading could impact the end-life of cables. This analysis will employ the network examples in Section 4. According to Reference [33], the number of households in the UK is around 27.8 million. The adoption rate ($Rate_{LCT}$) of EVs and ASHPs are considered separately and employs the following relation:

$$Rate_{LCT} = \frac{S_{LCT}}{N_H \times (1 + g_r)^n} \tag{19}$$

where S_{LCT} is the stock of the *LCT*, N_H is the number of households under consideration, g_r is the growth rate, and n is the number of years under consideration. Additionally, this analysis assumes that the number of UK households will grow by 0.9% every year [35] and that the EV stock growth rate follows the global estimate. Based on Reference [36], in the UK, there were approximately 450,000 EVs in 2020. Therefore, the EV adoption rate in 2020 was calculated as 1.6%. The global EV stock was 10.2 million in 2020 and is projected to be

141.1 million in 2028 [18]. Based on this projection, the UK EV adoption rate in 2028 would be 20.66%

According to Reference [24], the UK HP installation rate in 2020 was projected as 36,000, and the UK HP installation rate in 2028 is projected at 714,000. According to Reference [37], the installed capacity of HPs was 238,823 by 2019. This analysis also assumes that the household size in 2019 was 27.8 million and the growth rate is constant at 0.9%. Therefore, the HP adoption rate in 2020 was calculated as 0.98%, and the adoption rate in 2028 was calculated as 9.52%.

This study assumes that the mean cable lifetime is based on the loading regime projected by the best/worst-case scenarios. The combined effect from EVs and HPs is based on the anticipated adoption rates discussed above. The projected values are given in Tables 5–8. Based on Figure 5, the largest change in the lifetime prediction was observed for variations in ambient temperature. Only the effect of ambient temperature error is considered in Table 8.

Table 5. Adoption rate of EVs and HPs in 2020 and 2028.

2020		2028	
$Rate_{EV}$	$Rate_{HP}$	$Rate_{EV}$	$Rate_{HP}$
1.6%	0.98%	20.66%	9.52%

Table 6. Distribution cable loading for 2020 and 2028.

Property		2020		2028	
		EV	HP	EV	HP
Rate		1.6%	0.98%	20.66%	9.52%
Consumption	Best case	64.5 kW	21.1 kW	0.8333 MW	0.2187 MW
	Worst case	156.6 kW		2.022 MW	
Total Consumption	Best case	85.6 kW		1.052 MW	
	Worst case	177.7 kW		2.241 MW	

Table 7. Mains' cable loading for 2020 and 2028.

Property		2020		2028	
		EV	HP	EV	HP
Rate		1.6%	0.98%	20.66%	9.52%
Consumption	Best case	2.4 kW	0.895 kW	30.99 kW	8.69 kW
	Worst case	5.6 kW		72.31 kW	
Total Consumption	Best case	3.295 kW		39.68 kW	
	Worst case	6.495 kW		81 kW	

Table 8. Lifetime prediction changes (years) for 2020 and 2028, considering ±5% ambient temperature error.

Property	Projected Lifetime (Years)			
	Distribution Cable		Mains' Cable	
	Best Case	Worst Case	Best Case	Worst Case
2020	127 ± 20	125 ± 20	64 ± 10	63 ± 10
2028	108 ± 18	85 ± 13	54 ± 9	42 ± 7
Difference	−19	−40	−10	−21

The scenarios in 2020 have a limited impact on the cable lifetime for both distribution and mains' cables. As the adoption rate increases, as projected in 2028, the impact becomes more significant. When compared to the 2020 best-case scenario, the average lifetime of both distribution and mains' cable could reduce by up to 30% (it could be up to 67% out if the worst-case error is considered) for the projected worst-case scenario in 2028. This estimation is based on the average load; the actual lifetime may reduce more significantly if the peak loading is considered. Generally, the rapid uptake of both electric vehicle and heat pump adoption in the next eight years will have a significant impact on cable life, and a large number of cables would need to be upgraded. The proposed model with more specific network data can be used to strategically target replacement and reinforcement, which will significantly save costs for utilities, whilst still maintaining standards of service.

6. Conclusions

The targets for net zero around the world have placed further emphasis on the adoption of LCTs in domestic and industrial electricity customers. This increased adoption of LCT can bring further loading for the power network as a whole. The demand of both customer profiles is projected to increase, and this presents a challenge for network operators to ensure that the network will cope with this changing need. The widespread deployment of sensors and measurement systems is not practical or cost effective on the LV distribution network. This paper has demonstrated a temperature-based end-of-life estimation model to derive a relation between cable loading and lifetime. Two representative cable topologies and associated example networks were employed to explore the challenge. The study found that EVs and HPs are most likely to be the prominent technologies adopted in the short-to-longer term. Both technologies increase the cable loading and could reduce the cable life by up to 30% by 2028, based on the projected uptake rate of EVs and HPs.

In general, EVs would have a significant impact on the customer demand and cable lifetime. The scenarios mentioned above are very simplified; however, this can be developed further to be more realistic for specific network scenarios. An additional complication may be the development of vehicle-to-grid chargers; trials are ongoing in this area, so EVs may become a source, as well as a load. A network of EVs working as sources could be a potential future mitigation method to reduce the overall power imported to portions of network if constraints exist.

HPs alone might not cause cable overload problems, but the combination of this load increase along with other LCTs may pose challenges to the power system. It is anticipated that the initial uptake of HPs will be highly dependent on the prior source of heating to the customer. It is likely that government incentives will target the most polluting heating systems first. Furthermore, some customers will not be on the gas mains network and will be reliant on oil heating systems. This initial uptake may lead to localized pressures on the distribution network in areas where no gas mains are laid/available for customers to be connected.

The calculations within this paper are based on the average load increase at the two customer types, and this may present an underestimate of the challenge facing network operators. This work represents an initial step in the use of modeling tools to support the end-of-life assessment of cables, whilst further monitoring tools are deployed on the power network. Advantages of this approach center on the ability to perform long-term assessments for installed cable assets. This can enable the study of a range of possible future scenarios and varied installation conditions. The disadvantages of the approach may involve the quality of models or input data, with any errors at these stages impacting the accuracy of the lifetime prediction. Future work will consider the temperature impact on the AC conductor resistance, additional cable topologies, time-based behavior of the identified loads/sources, and potentially the uptake of DC EV chargers. The ultimate aim is to provide a tool for asset managers to make informed investment decisions and reduce the reliance of reactive maintenance/replacement schemes.

Author Contributions: X.J., methodology, software, validation, formal analysis, investigation, data curation, writing—original draft preparation, visualization, and writing—review and editing; E.C., conceptualization, resources, writing—review and editing, visualization, validation, supervision, project administration, and funding acquisition; B.G.S., conceptualization, methodology, writing—review and editing, validation, supervision, and funding acquisition; B.S., conceptualization, methodology, writing—review and editing, validation, supervision, and funding acquisition. All authors have read and agreed to the published version of the manuscript.

Funding: This work was funded by the industrial members of the Power Network Demonstration Centre (PNDC) through the Core Research Programme, and this support is gratefully acknowledged.

Institutional Review Board Statement: Not applicable.

Informed Consent Statement: Not applicable.

Data Availability Statement: Not applicable.

Acknowledgments: The technical support and input from industrial partners through the lifecycle of the project is gratefully acknowledged.

Conflicts of Interest: The authors declare no conflict of interest.

References

1. EA Technology & Southern Electric Power Distribution. SSET205—My Electric Avenue Project Close-Down Report. 2016. Available online: https://www.ofgem.gov.uk/sites/default/files/docs/2016/04/my_electric_avenue_i2ev_close-down_report_v2_3_clean.pdf (accessed on 11 September 2021).
2. Ma, K.; Li, R.; Li, F. Quantification of Additional Asset Reinforcement Cost From 3-Phase Imbalance. *IEEE Trans. Power Syst.* **2015**, *31*, 2885–2891. [CrossRef]
3. Gahan, L.; East, J. Heating our homes in a Net Zero Future: Understanding What Matters to Consumers. 2020. Available online: https://www.nationalgrid.com/document/134296/download (accessed on 11 September 2021).
4. European Union. 2050 Long-Term Strategy. Available online: https://ec.europa.eu/clima/policies/strategies/2050_en#tab-0-0 (accessed on 11 September 2021).
5. UKPN. Smart Charging Architecture Roadmap (SmartCAR). 2018. Available online: https://innovation.ukpowernetworks.co.uk/wp-content/uploads/2019/05/UKPN-Smart-Charging-Architecture-Roadmap-Final-Report.pdf (accessed on 11 September 2021).
6. Duke, R. Electric Nation (Car Connect) Closedown Report. 2019. Available online: https://www.westernpower.co.uk/downloads-view-reciteme/101713 (accessed on 11 September 2021).
7. SPEN. Charge Refuelling Tomorrow'S Electrified Transport. 2019. Available online: https://www.spenergynetworks.co.uk/pages/charge.aspx#:~{}:text=Charge%3A%20Refuelling%20tomorrow\T1\textquoterights%20electrified%20transport&text=The%20project%20brings%20together%20transport,Group%20and%20Smarter%20Grid%20Solutions (accessed on 11 September 2021).
8. Shafiee, S.; Fotuhi-Firuzabad, M.; Rastegar, M. Investigating the Impacts of Plug-in Hybrid Electric Vehicles on Power Distribution Systems. *IEEE Trans. Smart Grid* **2013**, *4*, 1351–1360. [CrossRef]
9. Taylor, M.J.; Alexander, A. Evaluation of the impact of plug-in electric vehicle loading on distribution system operations. In Proceedings of the IEEE Power & Energy Society General Meeting, Calgary, AB, Canada, 26–30 July 2009; pp. 1–6.
10. Tang, Z.; Zhou, W.; Zhao, J.; Wang, D.; Zhang, L.; Liu, H.; Yang, Y.; Zhou, C. Comparison of the Weibull and the Crow-AMSAA Model in Prediction of Early Cable Joint Failures. *IEEE Trans. Power Deliv.* **2015**, *30*, 2410–2418. [CrossRef]
11. Zhang, X.; Gockenbach, E.; Wasserberg, V.; Borsi, H. Estimation of the Lifetime of the Electrical Components in Distribution Networks. *IEEE Trans. Power Deliv.* **2006**, *22*, 515–522. [CrossRef]
12. Santhosh, T.; Ghosh, A.; Fernandes, B. Remaining life prediction of I&C cables for reliability assessment of NPP systems. *Nucl. Eng. Des.* **2012**, *245*, 197–201. [CrossRef]
13. Biçen, Y. Trend adjusted lifetime monitoring of underground power cable. *Electr. Power Syst. Res.* **2017**, *143*, 189–196. [CrossRef]
14. Aizpurua, J.I.; Stewart, B.G.; McArthur, S.D.J.; Jajware, N.; Kearns, M.; Garro, U.; Muxika, E.; Mendicute, M. A Diagnostics Framework for Underground Power Cables Lifetime Estimation Under Uncertainty. *IEEE Trans. Power Deliv.* **2021**, *36*, 2014–2024. [CrossRef]
15. Fernandez, E.; Patrick, J. Emergency and Cyclic Ratings of HV Cables. 2019. Available online: https://elek.com.au/wp-content/uploads/2019/02/Emergency-and-cyclic-ratings-of-HV-cables.pdf (accessed on 11 September 2021).
16. Prysmian Waveform Cable Low Voltage 600 1000V BS7870. Available online: https://www.powerandcables.com/wp-content/uploads/2017/07/Prysmian-Waveform-Cable-Low-Voltage-600-1000V-BS7870.pdf (accessed on 11 September 2021).
17. Csányi, G.M.; Tamus, Z.Á.; Varga, Á. Impact of Distributed Generation on the Thermal Ageing of Low Voltage Distribution Cables Impact of Distributed Generation on the Thermal Ageing of Low Voltage Distribution Cables. In *Doctoral Conference on Computing, Electrical and Industrial Systems*; Springer: Cham, Switzerland, 2017; pp. 251–258.

18. IEA. *Global EV Outlook 2021;* IEA: Paris, France, 2021.
19. SMMT. SMMT VEHICLE DATA. 2021. Available online: https://www.smmt.co.uk/vehicle-data/car-registrations/ (accessed on 11 September 2021).
20. Kane, M. Compare Electric Cars: EV Range, Specs, Pricing & More. *Eric Loveday*, 2020. Available online: https://insideevs.com/reviews/344001/compare-evs/(accessed on 11 September 2021).
21. Ofgem. Implications of the transition to Electric Vehicles. 2018. Available online: https://www.ofgem.gov.uk/sites/default/files/docs/2018/07/ofg1086_future_insights_series_5_document_master_v5.pdf (accessed on 11 September 2021).
22. ABB. Terra HP Charger—Up to 350 kW. Available online: https://new.abb.com/ev-charging/produkte/car-charging/high-power-charging (accessed on 10 May 2021).
23. Brown, A.; Schayowitz, A.; Klotz, E. *Electric Vehicle Charging Infrastructure Trends from the Alternative Fueling Station Locator: Third Quarter 2020 Electric Vehicle Charging Infrastructure Trends from the Alternative Fueling Station Locator: Third Quarter 2020;* National Renewable Energy Lab.(NREL): Golden, CO, USA, 2021.
24. Heat Pump Association. Delivering Net Zero: A Roadmap for the Role of Heat Pumps. 2019. Available online: https://www.heatpumps.org.uk/wp-content/uploads/2019/11/A-Roadmap-for-the-Role-of-Heat-Pumps.pdf (accessed on 10 May 2021).
25. Department for Business Energy & Industrial Strategy. Heat Pump Manufacturing Supply Chain Research Project. 2020. Available online: https://assets.publishing.service.gov.uk/government/uploads/system/uploads/attachment_data/file/943712/heat-pump-manufacturing-supply-chain-research-project-report.pdf (accessed on 10 May 2021).
26. MCS. Domestic Heat Pumps A Best Practice Guide. Available online: https://mcscertified.com/wp-content/uploads/2020/07/Heat-Pump-Guide.pdf (accessed on 10 May 2021).
27. UKPN. *Impact of Electric Vehicle and Heat Pump Loads on Network Demand Profiles;* UK Power Networks: London, UK, 2014; pp. 20–22.
28. Love, J.; Smith, A.Z.; Watson, S.; Oikonomou, E.; Summerfield, A.; Gleeson, C.; Biddulph, P.; Chiu, L.F.; Wingfield, J.; Martin, C.; et al. The addition of heat pump electricity load profiles to GB electricity demand: Evidence from a heat pump field trial. *Appl. Energy* **2017**, *204*, 332–342. [CrossRef]
29. IMS Heat Pumps. Air Source Heat Pump Running Costs. Available online: https://www.imsheatpumps.co.uk/blog/air-source-heat-pump-running-costs/ (accessed on 11 September 2021).
30. Department for Business Energy & Industrial Strategy. Subnational Electricity and Gas Consumption Statistics. 2021. Available online: https://assets.publishing.service.gov.uk/government/uploads/system/uploads/attachment_data/file/1043464/subnational_electricity_and_gas_consumption_summary_report_2020.pdf (accessed on 10 May 2021).
31. Anders, G.J. *Rating of Electric Power Cables in Unfavorable Thermal Environment;* IEEE Press Series on Power Engineering; Wiley: Hoboken, NJ, USA, 2005.
32. *BS IEC 60287-2-1:2015;* Electric Cables. Calculation of the Current Rating—Thermal Resistance. Calculation of Thermal Resistance. International Electrotechnical Commission: Geneva, Switzerland, 2015.
33. Yang, L.; Qiu, W.; Huang, J.; Hao, Y.; Fu, M.; Hou, S.; Li, L. Comparison of Conductor-Temperature Calculations Based on Different Radial-Position-Temperature Detection for High-Voltage Power Cable. *Energies* **2018**, *11*, 117. [CrossRef]
34. Nexans. 6-36kV Medium Voltage Underground Power Cables. Available online: https://www.powerandcables.com/wp-content/uploads/2016/12/Nexans-6-33kV-Medium-High-Voltage-Underground-Power-Cables.pdf (accessed on 12 May 2022).
35. Office for National Statistics. *Families and Households in the UK: 2019;* Office for National Statistics (ONS): London, UK, 2019.
36. Next Greencar. Electric car MARKET Statistics. 2021. Available online: https://www.nextgreencar.com/electric-cars/statistics/ (accessed on 11 September 2021).
37. Statista Research Department. Annual Amount of Heat Pumps in Operation in the United Kingdom (UK) from 2013 to 2019. Available online: https://www.statista.com/statistics/740491/heat-pumps-in-operation-uk/ (accessed on 11 September 2021).

electricity

Article

Optimal Coordinated Control of DC Microgrid Based on Hybrid PSO–GWO Algorithm †

Zaid Hamid Abdulabbas Al-Tameemi [1,*], Tek Tjing Lie [1], Gilbert Foo [1] and Frede Blaabjerg [2]

1 Department of Electrical and Electronic Engineering, Auckland University of Technology, Auckland 1010, New Zealand
2 Department of Energy Technology, Aalborg University, 9220 Aalborg, Denmark
* Correspondence: zaid.altameemi@autuni.ac.nz

† This paper is an extended version of our paper published in Al-Tameemi, Z.H.A.; Lie, T.T.; Foo, G.; Blaabjerg, F. Optimal Power Sharing in DC Microgrid under Load and Generation Uncertainties Based on GWO Algorithm. In Proceedings of the 2021 IEEE PES Innovative Smart Grid Technologies—Asia (ISGT Asia), Brisbane, Australia, 5–8 December 2021.

Abstract: Microgrids (MGs) are capable of playing an important role in the future of intelligent energy systems. This can be achieved by allowing the effective and seamless integration of distributed energy resources (DERs) loads, besides energy-storage systems (ESS) in the local area, so they are gaining attraction worldwide. In this regard, a DC MG is an economical, flexible, and dependable solution requiring a trustworthy control structure such as a hierarchical control strategy to be appropriately coordinated and used to electrify remote areas. Two control layers are involved in the hierarchy control strategy, including local- and global-control levels. However, this research focuses mainly on the issues of DC MG's local control layer under various load interruptions and power-production fluctuations, including inaccurate power-sharing among sources and unregulated DC-bus voltage of the microgrid, along with a high ripple of battery current. Therefore, this work suggests developing local control levels for the DC MG based on the hybrid particle swarm optimization/grey wolf optimizer (HPSO–GWO) algorithm to address these problems. The key results of the simulation studies reveal that the proposed control scheme has achieved significant improvement in terms of voltage adjustment and power distribution between photovoltaic (PV) and battery technologies accompanied by a supercapacitor, in comparison to the existing control scheme. Moreover, the settling time and overshoot/undershoot are minimized despite the tremendous load and generation variations, which proves the proposed method's efficiency.

Keywords: DC microgrid; voltage regulation; power sharing (PS); local control layers; GWO; hybrid PSO–GWO

Citation: Al-Tameemi, Z.H.A.; Lie, T.T.; Foo, G.; Blaabjerg, F. Optimal Coordinated Control of DC Microgrid Based on Hybrid PSO–GWO Algorithm. *Electricity* **2022**, 3, 346–364. https://doi.org/10.3390/electricity3030019

Academic Editors: Pavlos S. Geor-gilakis and Oriol Gomis-Bellmunt

Received: 29 June 2022
Accepted: 4 August 2022
Published: 8 August 2022

1. Introduction

Electrical energy is deemed as the fuel for today's computerized world, and in the not-too-distant future, electricity scarcity will be a serious issue for many countries, especially in remote areas [1]. The distance from the central electrical grid system or the high cost of installing the grid line to such a distance are the primary reasons for the lack of electrification for the population in these areas [2]. Moreover, load demand is much lower in rural locations than in more populated ones. Therefore, using grid electricity is not the ideal option because of the transmission and regular maintenance expenditures. Furthermore, power outages, grid shutdowns, CO_2 emissions, degradation of the ozone layer, biomass fuel reliance, and global warming, as summarized in Figure 1, have all contributed to a greater awareness of the need for safe, renewable, environmentally friendly, and clean energies. The direct integration of available renewable energy sources, involving photovoltaic (PV) and wind turbine generation (WTG), into the utility grid, is not easy due to their sporadic nature [3]. In this regard, the standalone MGs power system can combine such resources in one place as an excellent choice for isolated locations where grid power is unavailable [4]. In the literature, several categories of MGs, comprising AC, DC, and hybrid AC/DC MGs, have been employed to coordinate both renewable energy resources and energy storage systems to cater to required demand [5]. However, DC MGs are expanding more rapidly as compared to conventional AC MGs [6]. There are no harmonics or frequency conflicts, no synchronization is needed in the islanded mode, and no concerns about reactive power regulation exist [7,8]. In order to gain the full utilization of this type of microgrid, there are a few obstacles that need to be considered, including a seamless transition from islanded mode to on-grid mode operation, along with compatibility with AC loads [9]. Furthermore, microgrid protection is a problematic issue owing to the unavailability of zero-crossing current in addition to grounding [10,11]. The stability of the DC microgrid is a significant problem during fault circumstances, because of the resistive impedance characteristic of DC microgrid schemes and the absence of physical inertia [12]. Standardization seems to be another obstacle to the adoption of DC MGs [13]. Regardless of the abovementioned issues, DC MGs have a bright future owing to their improved compatibility with distributed renewable energy sources (DRES), better efficiency, and increased reliability [4]. It is essential to mention that power electronics converters are commonly employed as interfaces in each MG, to link each source to the shared bus [14].

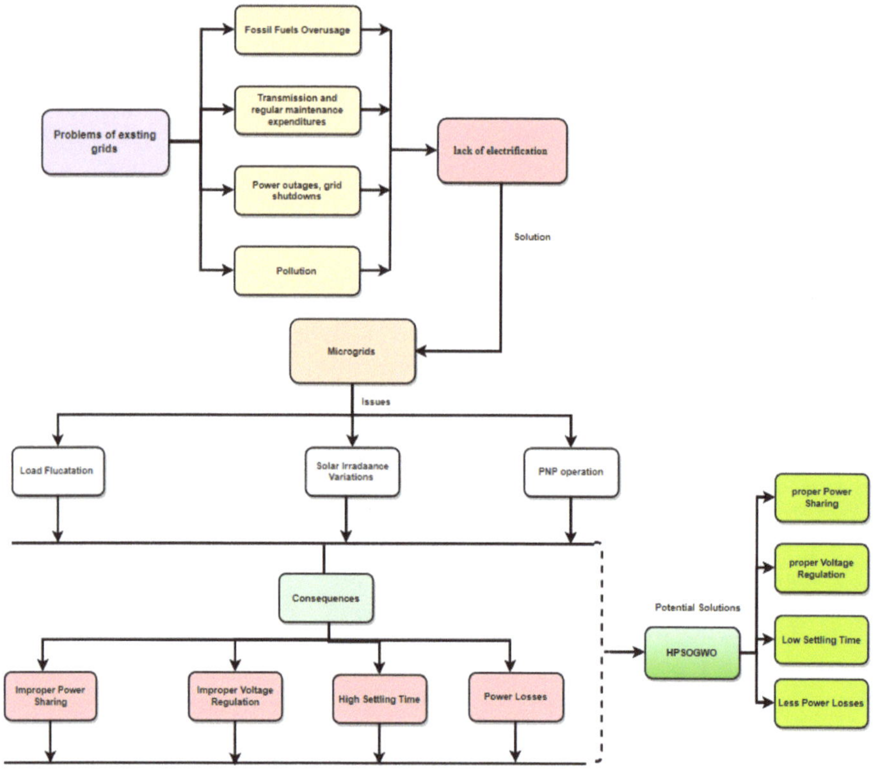

Figure 1. The main issues of the existing grids and microgrids.

Power electronics technology improvements have allowed DC electric grid schemes to achieve the criteria cost-effectively and smoothly [15]. With the advent of the DC MG approach, DC loads were widely used throughout different industries, including telecom facilities and data centers [16]. DC voltage rates are required for the operation of most modern electronic devices, including LED bulbs, phones/laptops chargers, and televisions. Therefore, DC MGs may address the rising demand in remote and tiny towns worldwide because of these qualities [17]. It is worth mentioning that the control of DC-bus voltage, besides power-sharing in the DC MGs, is a crucial aspect that scholars always need to be focused on [6]. Providing voltage support and efficient power distribution are the main challenges in a multi-source DC MG. In the literature, the conventional droop control approach of DC MGs has been used because it has simplicity in implementation due to the non-existence of a communication link [18]. This control strategy, nevertheless, results in inappropriate current distribution, voltage fluctuations, and circulating current regulation because of droop and line resistance amongst converters in an MG [4]. These converters must be properly managed to achieve an MG's desired performance. In this regard, the DC MG has employed a hierarchical control method consisting of local and global control layers. In the first control layer of the MG, the primary control goals are determined to achieve adequate power and bus voltage stabilization within allowable constraints. Numerous control strategies have been put forth at the local control levels [14–25] to achieve these objectives. For realizing such objectives, there are various extensive studies on solving control issues in the local layers, especially droop control methods, including voltage/current, current/voltage, or voltage/power, which are often used to coordinate autonomously distributed units and manage the flow of power between load demand and generation [19–21]. The droop control approach is a popular decentralized approach for distributing power based on using virtual resistance. It is important to mention that power-sharing accuracy and voltage stability are intimately related to the droop coefficient of the system [22]. More specifically, as the droop coefficient increases, current sharing accuracy improves with an increase in voltage variations and vice versa [23]. This confirms that the selection of the droop coefficient value is the controller's intrinsic trade-off. Although the key benefit of the droop control approach is that it eradicates the need for crucial communication links between parallel-connected converters since there are no communication links between parallel-connected converters, there is a lot of flexibility and reliability [24]. In contrast, there are several drawbacks to the conventional droop strategy: sluggish transient response, the intrinsic trade-off between voltage control and load sharing, line-impedance misalignment between parallel converters that impact active power-sharing, and unsatisfactory performance renewable resource [22,25]. Therefore, droop control is the subject of many related studies in the literature to enhance its performance. In [12], an adaptive droop method has been suggested to adjust the virtual resistance to follow the load current's fluctuation across parallel-linked DC–DC converters. Owing to the sporadic nature of renewable resources (PV and WTG), energy storage systems (ESSs) consisting of batteries, supercapacitors (SCs), and ultracapacitors need to be included to overcome any load-generation mismatch that requires reliable control approaches to work effectively. Many control techniques to regulate power-sharing between supercapacitors and batteries have been discussed in the literature. In [26], state-of-charge feedback control is employed to ensure the battery's charging level is within its allowable threshold, whereas the scheme of battery-energy storage (BES) is utilized to smooth out a wind farm's production fluctuations. Combining the two into a hybrid electrical system is important because batteries have a higher power density and SCs have a high-power density. The batteries supply the main source of energy, and the SCs handle momentary load interruptions and power peaks during unforeseen circumstances [27–29]. Various control strategies have been proposed to enhance the performance, to keep the new hybridization system operating more efficiently. In [30], a novel control method has been presented based on the disconnection of low- and high-frequency power components of the hybrid battery/SCr energy supplier. The proposed technique leverages the battery current error component to control the SC. This method offers the advantages of faster voltage control and less battery-current stress. In [31], a new control approach is given to regulate power-sharing between batteries and SC and, therefore, keep MG voltage stability even with large energy-generation fluctuations. The authors of [32] concentrated on MG control schemes that involved PV, WTG, batteries, and SCs with hybrid AC/DC buses. The approach proposed a basic control technique for managing power between renewable resources and storage systems. It is noticed that the control methods used in local layers suffer from some issues under load disruptions and fluctuations of power generation because of dependency on classical proportional–integral (PI) controllers [33,34]. Under such conditions, these controllers do not provide satisfactory performance. Converters, on the other hand, are non-linear and time-variable systems [35]. Therefore, some attempts have been made to overcome such issues by replacing classical PI controllers with non-linear and advanced controllers. For instance, the authors in [27] adopted a proper control approach to reject the transient disturbance that results from the load current and limit the oscillations of the battery current. A new disturbance-rejection control approach based on a sole DC/DC converter has been devised for a hybrid battery/SCs system. The suggested control approach uses a super-twisting (STC) and PI controller in the internal and external voltage control loops. Furthermore, a Pulse Wave Modulator (PWM)-based sliding-mode control method is presented in [35] to adjust the voltage of the DC–DC boost converter. Although this method confirms its effectiveness in controlling the DC voltage, power-sharing has not been considered. In order to coordinate the power-sharing among renewable resources, battery packs, and load, the authors [36] developed a multilayered feed-forward artificial neural network (ANN)-based power management system (PMS). To realize the farthest utilization of the energy resources, while also controlling the voltage and power-sharing amongst hybrid renewable resources, a hybrid-modified Bat search algorithm/ANN control approach has been presented in [37]. The main issues with these methods are that hybrid energy storage systems have not been taken into account, so they may not be applied to experience high energy variations. Furthermore, ANN-based DC MG power management with a hybrid battery–SC energy system was proposed in [38,39] to accomplish a better voltage regulation besides well-organized power-sharing under various operating conditions, and the results were compared to the traditional control technique to confirm the superiority of the proposed approach. However, the proposed control strategy in [38,39] has not been examined simultaneously under extreme load changes and PV power fluctuations, to assess its effectiveness in such critical conditions in a real-life scenario. Moreover, the improvement of the droop control of the PV system and hybrid battery/SC control sides has not been explored. Furthermore, the authors of the aforementioned articles have not addressed the additional battery power provided to the simulated MG, which might possibly reduce its efficiency, especially in [38,39]. It has been discovered that there is a lack of focus on using meta-heuristic optimization approaches, which are featured because of their capability in addressing complex problems and optimizing controller parameters [34] to tackle the major challenges in the local control layer of DC MGs that have been discovered in earlier studies. These problems can be addressed using a variety of metaheuristic algorithms. A few examples of these algorithms are the ant lion algorithm, the PSO method, and the gravitational search algorithm. It is important to note that no single optimization algorithm can guarantee optimal system performance due to inherent strengths and drawbacks. As an illustration, PSO is characterized by its uncomplicated concept, relative resilience to control parameters, and computing efficiency; on the other hand, PSO's restricted local/global search functionality causes it to get stuck in a local minimum when dealing with situations that are both highly constricted and highly robust [40]. Another example is the GWO algorithm, which has been demonstrated to outperform all the above mentioned algorithms [41]. Although GWO is featured by its speedy convergence, simplicity of implementation, and higher performance in uncertain and challenging search spaces, particularly in engineering applications [42], its delayed late-stage convergence and proneness to local optimality are major disadvantages [43,44]. There is no doubt that each of the two algorithms has a special search strategy. Their updating strategies guided by two or three agents leads to a reduction in variety, early convergence, or collapsing into the local optimum [45]. In order to address such issues, hybridization among algorithms has been proposed in previous studies to achieve a proper performance of the system. Hybrid PSO–GWO can be adopted to utilize the strengths of GWO and PSO. The primary goal of this hybridization is to enhance the capability of exploitation in PSO with the power of exploration in GWO [46]. As a result of this, this paper is focused on solving all of the problems depicted in Figure 1 using the hybrid PSO–GWO algorithm, which is characterized by its simplicity and ease of use as a tool to be used in successful execution, as well as completive performance, in comparison to those other optimization algorithms [47]. Using this algorithm to locate the global-best solution to an optimization problem has been shown to be effective [48], and this can boost the simulated MG's local control layer's performance. It is important to mention that improved power-sharing across paralleled DC–DC converters, improved MG voltage regulation within the specified level of 5%, reduced battery power losses, minimized settling time, and less overshoot/undershoot under diverse operating situations are significant contributions of the suggested technique. It is worth noting that this study is an extension of [49], with the new proposed approach achieving more consistent performance than [49]. For this study, a standard DC MG comprises several distributed energy resources (DERs), including a PV system and hybrid battery/supercapacitor, which are connected to a DC-bus by using boost and buck-boost converters, respectively, to support DC load in the system, as illustrated in Figure 2. It is important to mention that this microgrid is explained in detail in Section 2.

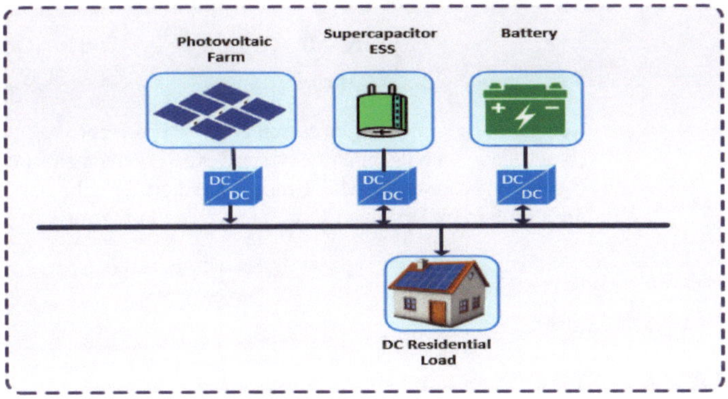

Figure 2. DC MG configuration.

The structure of this article is as follows. DC MG control techniques, including traditional and new ones, are discussed in Section 2. In Section 3, HPSO–GWO is reviewed and formulated in order to address the problem under consideration. The graphic comparison of the significant outcomes of the traditional technique and the new control approach is shown in Section 4. A summary of what has been accomplished is provided in Section 5.

2. Modeling of the Local Control Layer in DCMG

2.1. DC MG Control

A typical MG incorporates several forms of distributed energy resources such as PV, WTG, and fuel cells accompanied by ESSs to deal with the intermittent nature of such sources [3,50]. Most of these resources provide inherently DC electric power to the system, and it is possible that WTGs that provide AC power may be connected to the DC MG if they have been rectified [33]. To optimally coordinate distributed renewable energies, a DC MG necessitates an optimum control approach to accomplish appropriate power-sharing amongst all coupled sources in the system, while also managing the voltage within the IEEE standard limit of 5%. In this respect, hierarchical control mechanisms involving local and global control stages have been proposed to fulfill the above control objectives [23,51]. The three control layers of a hierarchical control technique are the primary or local control layer, which is followed by a secondary level, and the third level of control, classified as a global control layer. In detail, the goal of the local layer is to facilitate the transfer of power among these DGs. The secondary level is intended to compensate for voltage variations.

The objective of the tertiary stage is to govern power transfer between MGs and the electric grid [33].

2.2. PV Side Control Strategy

PV system is the dominant source in DC MG due to availability and inherently producing DC power, so it is not vital for it to be converted to another form of power. However, it cannot be directly coupled to a common DC-link without adopting DC–DC converters because both voltage and power of PV systems are uncontrollable and are not always compatible with the system specifications [10]. These converters need to be efficiently controlled to stabilize a DC-bus under different operating conditions. In this regard, two operating modes, as illustrated in Equation (1), droop and maximum power point tracking (MPPT) modes are adopted to control the DC–DC converter [52]. The first mode is activated in case a low-load demand is required or a fully charged battery reaches, so this mode matches the altering load to retain the voltage of the DC-bus [53]. Otherwise, MPPT mode needs to be initiated for compensating the power and the DC-link voltage destabilization, if not addressed. PV systems commonly work in this mode to wholly exploit renewable energy sources [53,54].

The output voltage of the droop-controlled converter is presented as Equation (1).

$$D\begin{cases} (\mathrm{Iref}-\mathrm{IL})\cdot\left(\mathrm{Kp}_{pv}+\frac{\mathrm{Ki}_{pv}}{s}\right) & soc \geq SOC_{max} \\ D_{mppt} & soc < SOC_{min} \end{cases} \quad (1)$$

where IL—converter output current, Iref—the reference value of output current, SOC—state of charge along with its minimum value SOC_{min} and maximum value SOC_{max}, and D_{mppt}—duty cycle in the case of the MPPT mode. Furthermore, the parameters of the current controller are signified by Kp_{pv} and Ki_{pv}, as revealed in Figure 3.

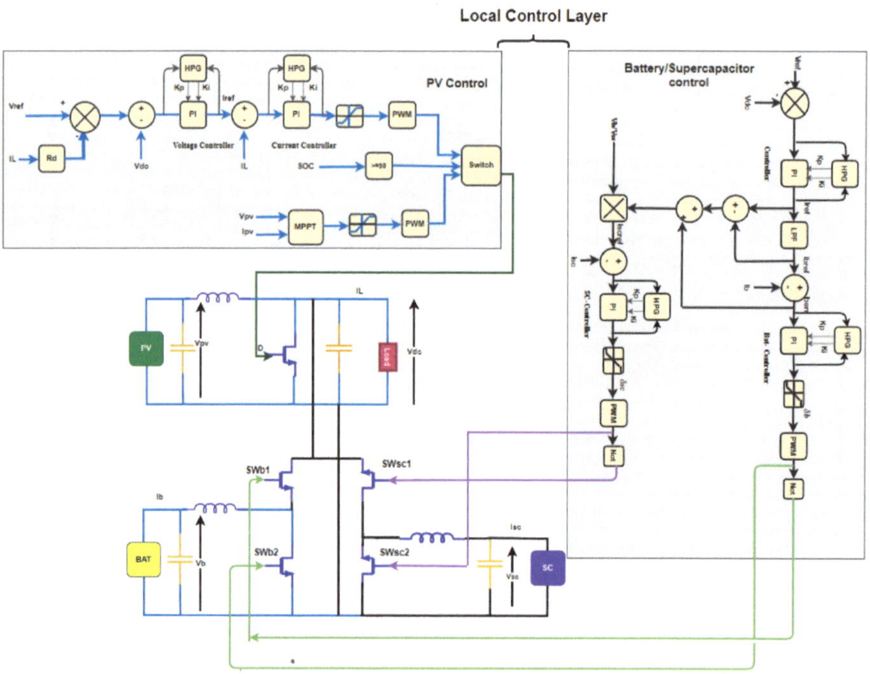

Figure 3. Proposed control method.

It can be noted from Equation (1) that voltage stabilization can be controlled effectively under different operation conditions, based on the threshold value of the SOC of the battery.

2.3. Hybrid Battery/SC Side Control Strategy

The intermittent nature of RESs such as PV and WTG makes it challenging to cater to load demand without ESSs. Power balance and voltage adjustment are essential functions of a battery storage system. The droop control approach is utilized in this paper to sustain the steadiness of the DC-bus voltage and the battery's automatic charge and discharge process [55]. Since the batteries feature a lower power density with a high energy density, which minimizes discharge/charge speed, SCs, which have a high power density, could be unified with batteries to realize the efficient operation of the ESSs in the system [38,39]. By adopting this hybridization, the batteries are exploited to generate the major energy supply, whereas the SCs are adopted to address the momentary fluctuations of the localized load in the MG and tackle power peaks during unpredicted circumstances [27]. In this article, hybrid ESS (HESS) is used to boost the efficacy of the MG under various loads and PV-generation scenarios. In order to reduce the stress on the dynamo battery and maintain a suitable DC bus voltage, the SC is offered to collect the high-frequency component. In addition, it prevents the batteries from overcharging and deep discharging, preserving their lifespan [52]. The hybrid battery/SC energy system is illustrated in Equations (2) and (3), as follows:

$$\delta_b = (I_{bref} - I_b)\cdot(Kp_{bat} + \frac{Ki_{bat}}{s}) \quad (2)$$

$$\delta_{sc} = (I_{scref} - I_{sc})\cdot(Kp_{sc} + \frac{Ki_{sc}}{s}) \quad (3)$$

where, I_{bref} and I_b refer to a reference value of battery current and real value of battery current, respectively, along with Kp_{bat} and Ki_{bat}, which indicate the parameters of the battery controller, as illustrated in Figure 3. While the SC current with its reference is denoted by I_{sc} and I_{scref}, respectively, in addition to Kp_{sc} and Ki_{sc}, which indicate the parameters of the supercapacitor controller, as illustrated in Figure 3. The difference between battery currents I_{bref} and I_b and supercapacitor currents I_{sc} and I_{scref} passes to the battery controller and the supercapacitor controller, as illustrated in Figure 3, to produce duty ratios δ_b and δ_{sc}, which are sent to pulse-width modulation (PWM) generators to generate switching pulses (SWb1 and SWb2) for the bidirectional converter of the battery and (SWsc1 and SWsc2) for the bidirectional converter of the supercapacitor.

3. Design Considerations of the DC–DC Power Converters

The DCMG, which is shown in Figure 2, has been simulated in the MATLAB environment. In this step, the specifications of PV arrays and hybrid batteries/SCs are necessary to be determined based on MGs requirements at the normal operating conditions. Moreover, directional DC–DC converters and bidirectional converters need to be designed based on the RESs, ESSs, and general system requirements. Therefore, in the following subsections, the design considerations of such converters are included to provide a clear idea about designing converters for other MGs.

3.1. Boost Converter

These converters are more prevalent in industrial applications because of their straightforward construction, ruggedness, ease of use, and relative inexpensiveness, so they are widely used in MGs [56]. These converters not only improve the output voltage of the PV system to the required limit but also accomplish MPPT control. In order to achieve an appropriate performance of this converter in the studied MG, its components, including the inductor L_{boost} and capacitance C_{boost}, need to be appropriately calculated based on the following, Equations (4) and (5) [56,57]:

$$L_{boost} = \frac{V_o D(1-D)}{2\Delta I_l \times F_s} \quad (4)$$

$$C_{boost} = \frac{V_i D(1-D)}{8 \times \Delta V_o \times {F_s}^2 \times L_{boost}} \quad (5)$$

$$V_o = V_i/(1-D) \quad (6)$$

$$I_o = \frac{P_L}{V_o} \quad (7)$$

where, I_o and ΔI_l refer to the output current and inductor current ripple, respectively. While the output power, switching frequency, input voltage, duty cycle, and output voltage with its ripple are denoted by P_L, F_s, V_i, D, V_o, and ΔV_o, respectively. It is important to mention that ΔV_o is selected to be 5% out of output voltage, which can be calculated based on Equation (6), and ΔI_l is required to be rated within 20% − 30% of the output current that is determined by using Equation (7). All these equations are essential to be considered in designing a boost converter, to obtain the proper DC-bus voltage of the system.

3.2. Buck-Boost Converter

A solar power system would be incomplete without some form of energy storage. A standard bidirectional DC/DC converter is used to coordinate the battery charging/discharging process in the MGs based on the system situation. In order to make this converter operate efficiently, the inductor L_{bat} and capacitor C_{bat} need to be calculated accurately based on the following Equations (8)–(10) [56,57].

$$L_{bat} = \frac{V_i \times D_{bt}}{\Delta I_l \times F_s} \quad (8)$$

$$C_{bat} = \frac{I_o \times D}{\Delta V_o \times F_s} \quad (9)$$

$$V_o = V_i\, D/(1-D) \quad (10)$$

where V_i and D_{bt} refer to the battery voltage and duty cycle at boost operation mode of the converter, respectively. Moreover, switching frequency and duty cycle are signified by F_s and D, which can be calculated by Equation (10), respectively. It should be noted that ΔV_o is chosen to be 5% of the output voltage (V_o) that can be determined by Equation (10), and ΔI_l has to be within 20%–30% of the output current (I_o). These formulas must be taken into account for designing a buck-boost converter.

4. Proposed Control Method

Metaheuristic optimization approaches such as GWO, explained in [49], and HPSO–GWO [46] are investigated in this study as an alternate strategy to realize optimal PI parameters, as illustrated in Figure 3. This figure includes boost converter with its control strategy, as explained in Section 2.2, and bi-directional converters with their control approaches, as illustrated in Section 2.3, to achieve reliable and stable operation of the PV system, battery, and supercapacitor, respectively. In this study, the accumulated errors between the nominal values of voltages/currents and their real values are used as a fitness function, with PI controllers' constraints to be applied in the hybrid PSO–GWO MATLAB code, which is explained in [46], to determine the best values of parameters of PI controllers employed in the local layer that can improve the performance of DC microgrids.

4.1. HPSO–GWO Algorithm

The HPSO–GWO is a novel type of swarm-based metaheuristic that has numerous benefits, including easy implementation and minimal memory usage [46]. The fundamental concept is to bring together PSO's exploitation ability with GWO's exploration ability to create both variations' strength and memory consumption. Therefore, it is coevolutionary since both variants do not utilize one after other. In other ways, they run in parallel. Instead of utilizing traditional mathematical formulae, the exploitation and exploration of the first three agents' sites in the search space are updated in HPSO–GWO. The mathematical expressions are shown in Equations (11)–(17).

$$\vec{D}_\alpha = \left|\vec{C}_1 \cdot \vec{X}_\alpha - w * \vec{X}\right| \quad (11)$$

$$\vec{D}_\beta = \left|\vec{C}_2 \cdot \vec{X}_\beta - w * \vec{X}\right| \quad (12)$$

$$\vec{D}_\delta = \left|\vec{C}_3 \cdot \vec{X}_\delta - w * \vec{X}\right| \quad (13)$$

$$\vec{X}_1 = \vec{X}_\alpha - \vec{A}_1 \cdot \vec{D}_\alpha \quad (14)$$

$$\vec{X}_2 = \vec{X}_\beta - \vec{A}_2 \cdot \vec{D}_\beta \quad (15)$$

$$\vec{X}_3 = \vec{X}_\delta - \vec{A}_3 \cdot \vec{D}_\delta \quad (16)$$

$$X \rightarrow (t+1) = \frac{\vec{X}_1 + \vec{X}_2 + \vec{X}_3}{3} \quad (17)$$

where it is important to mention that $\vec{C}_1, \vec{C}_1$, and $\vec{C}_1$ along with $\vec{A}_1$, $\vec{A}_2$, and $\vec{A}_3$ denote coefficient vectors of the best three wolves while $\vec{X}_\alpha$, $\vec{X}_\beta$, and $\vec{X}_\delta$ refer to the places of the best three wolves with respect to the respective prey in the search space, and the location of the current solution is represented by $\vec{X}$. Moreover, the inertia constant (w) is used to regulate the exploration and exploitation of the grey wolf in the search space. All the previous information can be used to calculate the exact distances between the current allocation of the best grey wolves and the respective prey in the space, which are represented as ($\vec{D}_\alpha$, $\vec{D}_\beta$, and $\vec{D}_\delta$), based on Equations (11)–(13). Furthermore, $\vec{X}_1$, $\vec{X}_2$, and $\vec{X}_3$ indicate that the final position of such wolves can be calculated by adopting Equations (14)–(16). It is worth mentioning that the calculated parameters in Equations (14)–(16) are applied in Equation (17) to determine the estimated position of the prey. The velocities and positions of the wolves, which are signified by $v_i{}^k$ and $x_i{}^k$, can be updated by using the PSO approach as follows (Equations (18) and (19)):

$$v_i{}^{k+1} = w*((v_i{}^k + r_1c_1(x_1 - x_i{}^k.) + r_2c_2(x_2 - x_i{}^k.) + r_3c_3(x_3 - x_i{}^k)) \quad (18)$$

$$x_i{}^{k+1} = x_i{}^k + v_i{}^{k+1} \quad (19)$$

where $v_i{}^{k+1}$ and $x_i{}^{k+1}$ refer to the updated values of velocity and position of the best three grey wolves, while w represents the inertia, constantly generated randomly in [0, 1]; besides, r_1, r_2, and r_3 are random values in [0, 1]. Moreover, x_1, x_2, and x_3 indicate the position of the best three wolves, which are obtained by using Equations (14)–(16). Furthermore, c_1, c_2, and c_3 refer to optimization parameters, which are selected to be 0.5, whereas the current position of the particle is signified by $x_i{}^k$.

4.2. Problem Formulation

This paper aims to optimize the parameters of PI controllers employed in the local control level of the studied MG. Consequently, proper voltage adjustment and precise power transfer can be realized, which may enhance the performance of the simulated MG. The proposed control approach optimizes PI controllers' parameters based on the objective function in Equation (20). In the literature, there are many types of objective functions that have been used to tune PI controllers, including the integral time absolute error (ITAE), the integral time square error (ITSE), the integral absolute error (IAE), and the integral square error (ISE). The objective function of ITAE has been considered in this article, as it gives shorter settling time, overshoots, and rising time than the other objective functions used in the literature [58], with the constraints stated in Equation (27).

$$\mathrm{ITAE} = \frac{\sum_{\mathrm{Er}}^{\mathrm{N}} \int_0^{\infty} \mathrm{t}|\mathrm{Er(t)}|\mathrm{dt}}{\mathrm{N}} \tag{20}$$

Subject to:

$$\mathrm{Er1} = (\mathrm{Vref} - \mathrm{Rd} * \mathrm{IL}) - \mathrm{Vdc} \tag{21}$$

$$\mathrm{Er2} = \mathrm{Iref} - \mathrm{IL} \tag{22}$$

$$\mathrm{Er3} = \mathrm{Vref} - \mathrm{Vdc} \tag{23}$$

$$\mathrm{Er4} = \mathrm{Ibref} - \mathrm{Ib} \tag{24}$$

$$\mathrm{Er5} = \mathrm{Isref} - \mathrm{Isc} \tag{25}$$

$$\mathrm{Er} = \mathrm{Er1} + \mathrm{Er2} + \mathrm{Er3} + \mathrm{Er4} + \mathrm{Er5} \tag{26}$$

where ITAE—integral of time-weighted absolute error; Er(t)—disparity between starting points and variables to be managed—the simulated time; Rd—virtual droop resistance, which can be determined by dividing the maximum voltage deviation (5% of Vdc) by the output current (IL); Vref—the reference voltage of the DC microgrid, which is set to be 48 V; and Vdc—the output voltage of the DC microgrid. N indicates the number of the obtained errors from PI controllers. It can be noticed that errors resulting from differences between the reference and actual values of voltages and currents in the microgrid, including the output voltage (Vdc), with its reference value (Vref) and virtual droop voltage (Rd ∗ IL); the output current (IL), with its reference value (Iref); battery current (Ib), with its reference value (Ibref); and the SC current (Isc), with its reference value (Iscref), as shown in Equations (21)–(26) and Figure 3, are substituted in Equation (20) to determine the optimal values of the PI controllers parameters that achieve minimum error in the system. It is essential to mention that the parameters of the local control layer have been set by using the hybrid PSO–GWO, as shown in Figure 3. The maximum and minimum values of PI parameters have been included in the HPSO–GWO Matlab code, to determine the best values based on the system requirements, as illustrated in Equation (27):

$$\begin{cases} \mathrm{Kp}_{min} \leq \mathrm{Kp} \leq \mathrm{Kp}_{max} \\ \mathrm{Ki}_{min} \leq \mathrm{Ki} \leq \mathrm{Ki}_{max} \end{cases} \tag{27}$$

where Kp_{min} and Kp_{max}, along with Ki_{min} and Ki_{max}, refer to the minimum and maximum values of PI controllers' parameters used in the local control of the DC microgrid, as illustrated in Figure 3. The main objective of limiting such values is to enable the hybrid PSO–GWO algorithm to search for proper values of Kp and Ki, within a certain range based on the system requirements.

To get the best values of Kp and Ki in the local control layer for this investigation, a number of steps have been used, including:

1. Initialize the grey wolf's populations, X_1, X_2, X_3, etc., which indicates that each wolf (X) represents Kp and Ki.
2. Initialize parameters $\vec{A}$, $\vec{C}$, and $\vec{a}$, as their capabilities for exploration and development may be leveraged to achieve a better balance in the GWO algorithm.
3. Compute the fitness value of each agent (grey wolf) to determine the best three wolves.
4. The placements of the best three wolves regarding targeted prey can be determined, based on Equations (11)–(17).
5. The locations and velocities of the best wolves are updated, based on Equations (18) and (19), respectively.
6. In case the current iteration is less than the maximum iterations limit, based on step 3, all other wolves (ω) will update the positions. Otherwise, the optimal values of X agents (Kp and Ki) will be obtained to be applied in the system.
7. Based on the first condition in step 4, $\vec{A}$, $\vec{C}$, and $\vec{a}$ will be updated accordingly. Then, the value of each search agent (wolf) is recalculated.

8. Based on the previous updates, the best position is updated. This process continues until the best values of Kp and Ki are obtained.
9. Figure 4 depicts all these steps.

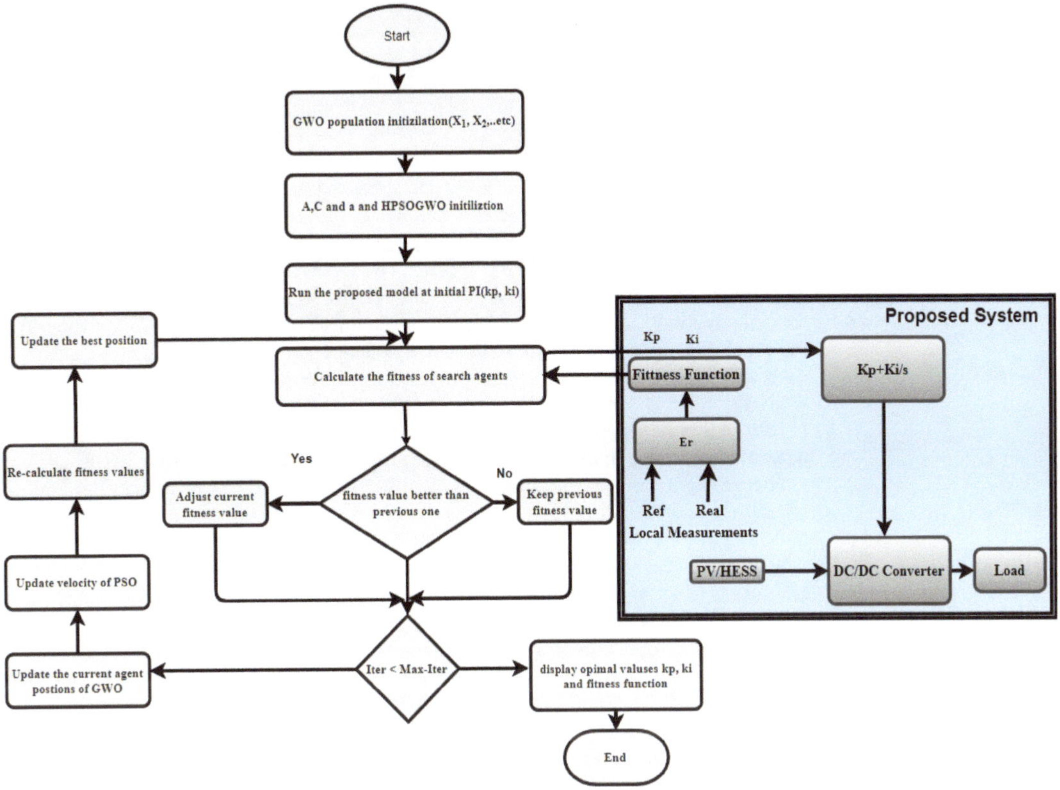

Figure 4. HPSO–GWO flowchart.

It is worth mentioning that the list of PI control parameters obtained by hybrid PSO–GWO is presented in Table 1.

Table 1. Optimal values of PI controllers.

Controller	Kp	Ki
PI controller 1	10	200
Battery controller	50	166.4
SC controller	80	350
Voltage controller (PV)	4.2956	0.6284
Current controller (PV)	100	250

Figure 4 explains the process of selecting PI controller parameters using the hybrid PSO–GWO algorithm. In this regard, a number of populations with the GWO and HPSO parameters are initialized to assign the initial Kp and Ki of the system. These initial values of PI controllers are applied in the local control layer to assess the microgrid's performance. In case its performance is not acceptable, which means less-accurate power sharing among sources in the microgrid and unregulated DC voltage, along with a high ripple of the battery current, are realized, the errors that result from the difference between the actual values of voltages and currents in the microgrid are substituted in the fitness function (Equation (20)), to be used in the hybrid PSO–GWO code to recalculate the new values of Kp and Ki for the local control layer. The main objective is to achieve minimum error, which may boost the performance of the local control level in the studied MG. It is worth mentioning that this process continues until the best performance of the microgrid is achieved.

5. Results and Discussion

In this study, the rating of the components of DC MG, which comprise of a PV array (6 panels, 120.7 W each) to provide 724.6 W, 14 Ah, and 24 V battery with 50% as a state of charge (SOC), 32 V, 29 F SC, and 48 V, 300 W load are taken from [39]. In order to validate the effectiveness of the proposed strategy in improving the performance of the local control layer, compared to that of the existing techniques in [38,39], different scenarios, including PV-generation variations and load fluctuations, are adopted in this study. It is important to mention that the durability of the suggested method has been evaluated using the rate of overshooting/undershooting in the voltage of MG, power sharing, battery current tracking, and system responsiveness. These scenarios have been tested, first with the conventional control approach, as depicted in Figure 5a–e. In this case, load changes by the rate of 50%, 38%, and 32%, and the fluctuations of solar irradiance simultaneously occur at 2 s, 3.5 s, and 4 s, respectively. The PV system can only generate 471 W at 0 s; therefore, most of this power is used to cater to demand, which is set to 300 W, while the remaining power is exploited to charge the battery to increase its SOC to an allowable threshold. This amount of PV-generated power remains 470 W until 2 s, so there is no extra power that can be employed to charge the battery at the time interval (1 s–2 s), due to the load demand going up to 500 W. Thus, the DC-link voltage drops sharply at 1 s, as shown in Figure 5a. It can also be noticed that PV generation increases to 580 W at 2 s, which leads to an increase in DC-link voltage up to 50.25 V. As previously mentioned, load demand increases by dissimilar rates at 1 s, 3 s, and 4 s, and this causes the bus voltage to diverge unacceptably from the standard limit of 5% at these time intervals. Based on the obtained results, this strategy is ineffective in facing critical operating conditions such as load-generation uncertainty. It is also worth noting that, even though the battery current is closely tracked in its reference value, some unwanted ripples are still involved. Furthermore, one of the most critical aspects to note is the system's poor responsiveness in rejecting imposed interruptions, implying that the system would be neither stable nor trustworthy.

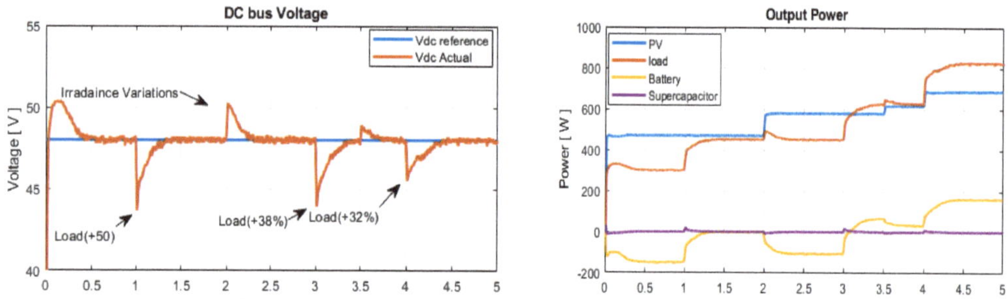

(a) (b)

Figure 5. *Cont.*

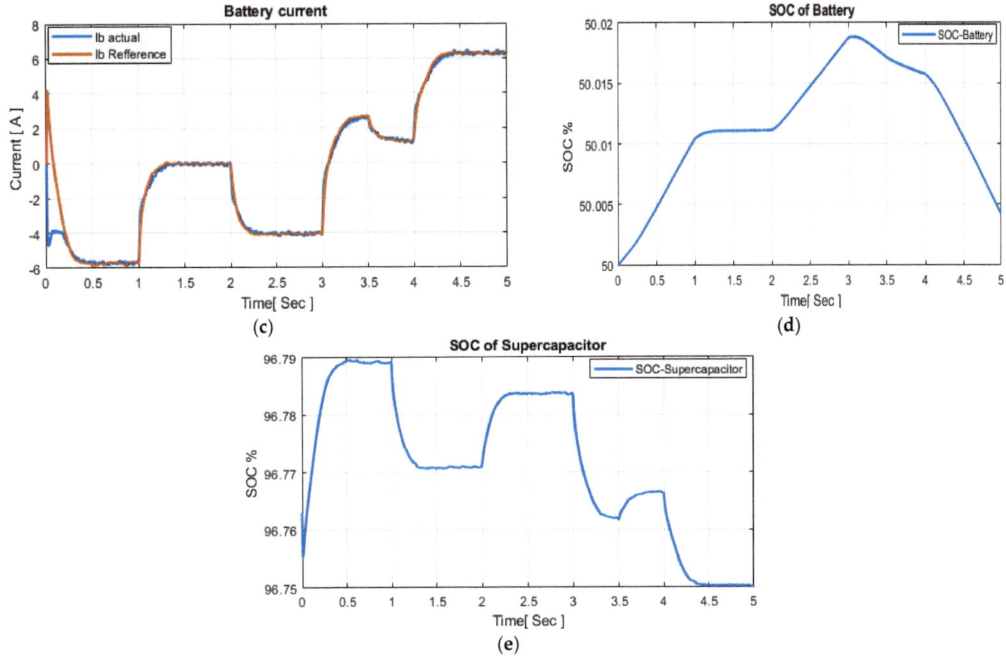

Figure 5. Results with the conventional PI method: (**a**) DC-link voltage; (**b**) power exchange; (**c**) battery current; (**d**) SOC of battery; (**e**) SOC of supercapacitor.

In order to tackle the main issues of the conventional control approach, GWO algorithms are adopted to enhance the microgrid's performance under such critical operating situations. The simulation results reveal that although the same critical operating conditions are applied with the proposed control scheme, the DC-bus voltage is preserved within the standard limit, and proper power-sharing of the studied DC MG is realized. To be more specific, by employing GWO, the voltage of the simulated MG rises from 43.66 V to 47.016 V, with a 50% increase in load demand at 1 s as well as a voltage-overshoot decrease from 50.25 V to 48.7 V at 2 s, as seen in Figure 6a.

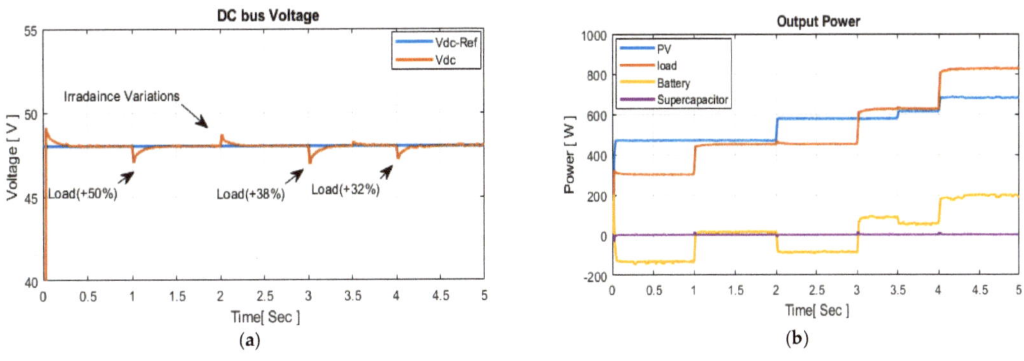

Figure 6. *Cont.*

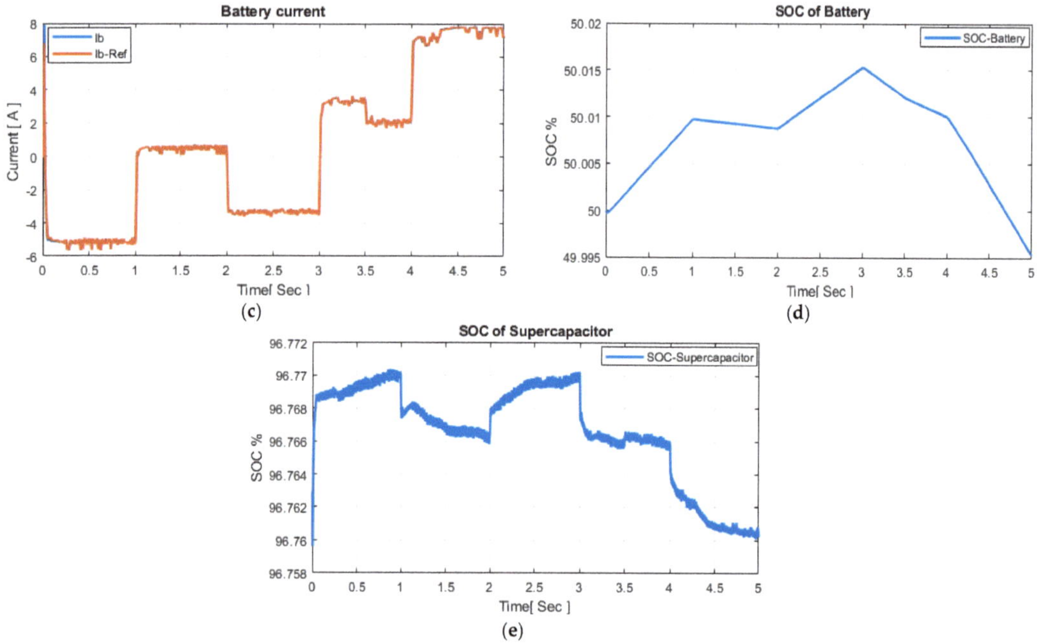

Figure 6. Results with the GWO: (**a**) DC-link voltage; (**b**) power exchange; (**c**) battery current; (**d**) SOC of battery; (**e**) SOC of supercapacitor.

From the previous discussion, it is found that the microgrid's performance needs to be further improved, especially the system's responsiveness in rejecting disturbances, power sharing, and battery current tracking, to be more proper for real-life implementation. Thus, the proposed hybrid PSO–GWO is utilized in this article to improve the traditional control method and the GWO employed in the simulated DC MG. The obtained results in Figure 7 demonstrate that although the studied MG has been subjected to the same critical operating conditions of both load and power-generation variations simultaneously, several advantages in the local control layer, including maintaining the bus voltage at the acceptable level (1.67%), optimal power sharing, and optimal battery current tracking with its reference value are achieved, as shown in Figure 7a–c. It is also noted that the response of the respective system in rebuffing the imposed disruptions is very quick, making the system more dependable and stable with no additional battery power delivered to support the system under such fluctuations, as illustrated in Figure 7a,b, respectively, which are better than [30,38,39]. This confirms that the proposed control method is more robust than the abovementioned studies under most operating conditions, implying that MG becomes more reliable and steadier in providing uninterrupted power to consumers under crucial operating conditions. It is also seen that the proposed control method achieves fewer overshoots/undershoots in the DC bus voltage, fewer current/voltage ripples, and less settling time in comparison with the conventional control approach used in the local control layer of the DC microgrid, as shown in Figure 7a, which enhances the reliability of the microgrid. In summary, it is clear that the proposed HPSO–GWO achieves a better voltage regulation, power-sharing, and settling time than the conventional control approach and GWO, as illustrated in Figures 7a–e and 8.

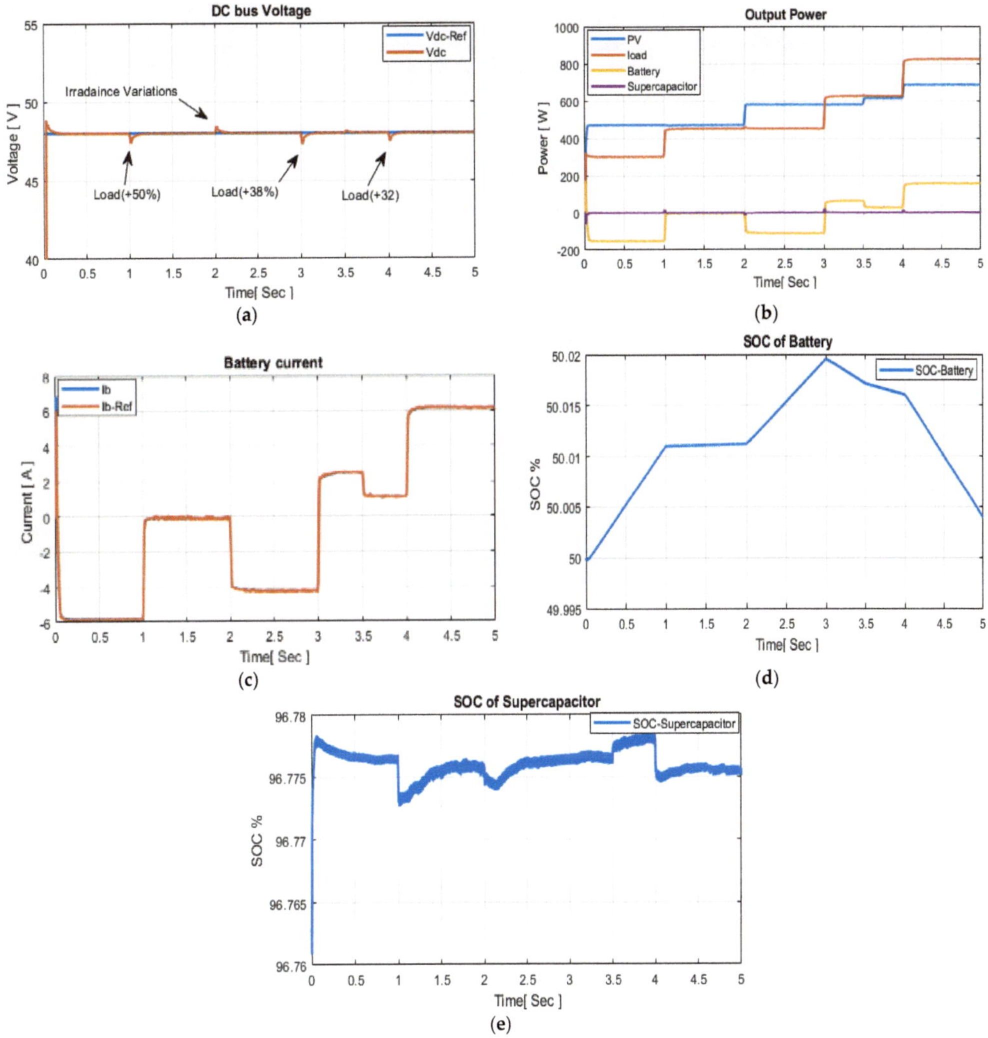

Figure 7. Results with the hybrid PSO–GWO: (**a**) DC-link voltage; (**b**) power exchange; (**c**) battery current; (**d**) SOC of battery; (**e**) SOC of supercapacitor.

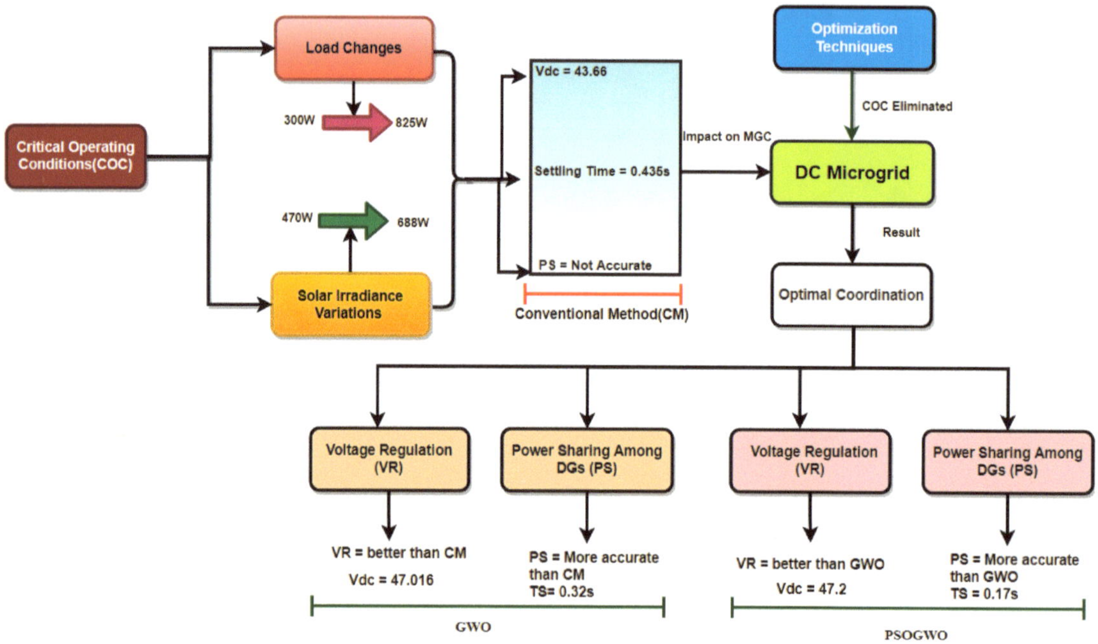

Figure 8. Summary of the main results of the proposed control technique.

Figure 8 illustrates that the large disturbances, including an increase in the load demand (300–825 W) along with the PV-generation fluctuations (470–680 W), are minimized by adopting the HPSO–GWO and GWO optimization approaches. Based on the obtained results, this control method may be suitable for being implemented in real-life scenarios, because the issues noted in previous articles, including transient stability issues, DC-bus voltage destabilization, power losses, and unregulated power sharing among participating sources in the microgrid, are solved in this article. Thus, this article may contribute to increasing the dependency on microgrids to generate reliable and uninterrupted power for consumers under crucial operating circumstances.

6. Conclusions

This study focuses on adopting HPSO–GWO to boost the performance of the local control layer. It is examined under different load and PV-generation scenarios. Although the load changes by 50%, 38%, and 32% at different time intervals, accompanied by PV-generation fluctuations, precise power transfer among DGs and voltage regulation are realized. Furthermore, the battery current can keep a consistent relationship with its reference value. Additionally, system responsiveness is effectively improved to resist any perturbation that may arise without causing any power dissipation. It is noticed that the proposed technique achieves fast voltage recovery with less settling time, overshoot/undershoot, and rising time, so the system operation becomes more reliable and stable under the abovementioned critical operating conditions, and this ensures the robustness of this control technique.

Author Contributions: Conceptualization, methodology, writing—original draft preparation, Z.H.A.A.-T.; writing—review and editing, supervision, investigation, visualization, T.T.L., G.F. and F.B. All authors have read and agreed to the published version of the manuscript.

Funding: This research received no external funding.

Institutional Review Board Statement: Not applicable.

Informed Consent Statement: Not applicable.

Data Availability Statement: Not applicable.

Acknowledgments: The authors appreciate Al-Furat Al-Awsat Technical University's financial support, and the research work is also part of the MBIE SSIF AETP, entitled Future Architecture of the Network.

Conflicts of Interest: The authors declare no conflict of interest.

References

1. Sanjeev, P.; Padhy, N.P.; Agarwal, P. Autonomous Power Control and Management between Standalone DC Microgrids. *IEEE Trans. Ind. Inform.* **2018**, *14*, 2941–2950. [CrossRef]
2. Abhishek, A.; Ranjan, A.; Devassy, S.; Verma, B.K.; Ram, S.K.; Dhakar, A.K. Review of hierarchical control strategies for DC microgrid. *IET Renew. Power Gener.* **2020**, *14*, 1631–1640. [CrossRef]
3. Gholami, M.; Pisano, A. Model Predictive Operation Control of Islanded Microgrids under Nonlinear Conversion Losses of Storage Units. *Electricity* **2022**, *3*, 33–50. [CrossRef]
4. Ali, S.; Shengxue, T.; Jianyu, Z.; Ali, A.; Nawaz, A. An Implementation of parallel buck converters for common load sharing in DC microgrid. *Information* **2019**, *10*, 91. [CrossRef]
5. Rezaei, O.; Mirzapour, O.; Panahazari, M.; Gholami, H. Hybrid AC/DC Provisional Microgrid Planning Model Considering Converter Aging. *Electricity* **2022**, *3*, 236–250. [CrossRef]
6. Al-Ismail, F.S. DC Microgrid Planning, Operation, and Control: A Comprehensive Review. *IEEE Access* **2021**, *9*, 36154–36172. [CrossRef]
7. Sahoo, S.K.; Sinha, A.K.; Kishore, N.K. Control Techniques in AC, DC, and Hybrid AC–DC Microgrid: A Review. *IEEE J. Emerg. Sel. Top. Power Electron.* **2018**, *6*, 738–759. [CrossRef]
8. Al-Tameemi, Z.H.A.; Lie, T.T.; Foo, G.; Blaabjerg, F. Control Strategies of DC Microgrids Cluster: A Comprehensive Review. *Energies* **2021**, *14*, 7569. [CrossRef]
9. Zhang, L.; Zhang, W.; Zeng, F.; Yang, X. A Review of Control Strategies in DC Microgrid. *J. Phys. Conf. Ser.* **2018**, *1087*, 042035. [CrossRef]
10. Dahale, S.; Das, A.; Pindoriya, N.M.; Rajendran, S. An overview of DC-DC converter topologies and controls in DC microgrid. In Proceedings of the 2017 7th International Conference on Power Systems (ICPS), Pune, India, 21–23 December 2017; pp. 410–415. [CrossRef]
11. Li, M.; Zhang, D.; Lu, S.; Tang, X.; Phung, T. Differential evolution-based overcurrent protection for DC microgrids. *Energies* **2021**, *14*, 5026. [CrossRef]
12. Augustine, S.; Quiroz, J.E.; Reno, M.J.; Brahma, S. *DC Microgrid Protection: Review and Challenges*; Sandia National Lab. (SNL-NM): Albuquerque, NM, USA, 2018.
13. Dragicevic, T.; Lu, X.; Vasquez, J.C.; Guerrero, J.M. DC Microgrids—Part II: A Review of Power Architectures, Applications, and Standardization Issues. *IEEE Trans. Power Electron.* **2016**, *31*, 3528–3549. [CrossRef]
14. Lu, X.; Guerrero, J.M.; Sun, K.; Vasquez, J.C.; Teodorescu, R.; Huang, L. Hierarchical control of parallel AC-DC converter interfaces for hybrid microgrids. *IEEE Trans. Smart Grid* **2014**, *5*, 683–692. [CrossRef]
15. Bharath, K.R.; Krishnan Mithun, M.; Kanakasabapathy, P. A review on DC microgrid control techniques, applications and trends. *Int. J. Renew. Energy Res.* **2019**, *9*, 1328–1338.
16. Elavarasan, R.M.; Ghosh, A.; Mallick, T.K.; Krishnamurthy, A.; Saravanan, M. Investigations on performance enhancement measures of the bidirectional converter in PV–wind interconnected microgrid system. *Energies* **2019**, *12*, 2672. [CrossRef]
17. Yaqub, R. Phasor Measurement Unit Assisted Inverter—A Novel Approach for DC Microgrids Performance Enhancement. *Electricity* **2021**, *2*, 330–341. [CrossRef]
18. Sahoo, S.; Mishra, S.; Fazeli, S.M.; Li, F.; Dragicevic, T. A Distributed fixed-Time secondary controller for DC microgrid clusters. *IEEE Trans. Energy Convers.* **2019**, *34*, 1997–2007. [CrossRef]
19. Sechilariu, M.; Wang, B.C.; Locment, F.; Jouglet, A. DC microgrid power flow optimization by multi-layer supervision control. Design and experimental validation. *Energy Convers. Manag.* **2014**, *82*, 1–10. [CrossRef]
20. Karimi, Y.; Oraee, H.; Golsorkhi, M.S.; Guerrero, J.M. Decentralized Method for Load Sharing and Power Management in a PV/Battery Hybrid Source Islanded Microgrid. *IEEE Trans. Power Electron.* **2017**, *32*, 3525–3535. [CrossRef]
21. Sun, K.; Zhang, L.; Xing, Y.; Guerrero, J.M. A distributed control strategy based on DC bus signaling for modular photovoltaic generation systems with battery energy storage. *IEEE Trans. Power Electron.* **2011**, *26*, 3032–3045. [CrossRef]
22. Mingsheng, Z.; Peilei, F.; Hesong, W.; Wenkui, W.; Pengcheng, C. Hierarchical Control Strategy for Microgrid. In Proceedings of the 2018 2nd IEEE Advanced Information Management, Communicates, Electronic and Automation Control Conference (IMCEC), Xi'an, China, 25–27 May 2018; pp. 1528–1532. [CrossRef]
23. Shuai, Z.; Fang, J.; Ning, F.; Shen, Z.J. Hierarchical structure and bus voltage control of DC microgrid. *Renew. Sustain. Energy Rev.* **2018**, *82*, 3670–3682. [CrossRef]

24. Meng, L.; Dragicevic, T.; Vasquez, J.C.; Guerrero, J.M.; Perez, J.R. Modeling and sensitivity analysis of consensus algorithm based distributed hierarchical control for DC microgrids. In Proceedings of the 2015 IEEE Applied Power Electronics Conference and Exposition (APEC), Charlotte, NC, USA, 15–19 March 2015; pp. 342–349. [CrossRef]
25. Shafiee, Q.; Dragicevic, T.; Vasquez, J.C.; Guerrero, J.M. Hierarchical control for multiple DC-microgrids clusters. In Proceedings of the 2014 IEEE 11th International Multi-Conference on Systems, Signals & Devices (SSD14), Barcelona, Spain, 11–14 February 2014; pp. 1–6. [CrossRef]
26. Yoshimoto, K.; Nanahara, T.; Koshimizu, G.; Uchida, T. New control method for regulating State-of-Charge of a battery in hybrid wind power/battery energy storage system. In Proceedings of the 2006 IEEE PES Power Systems Conference & Exposition, Atlanta, GA, USA, 29 October–1 November 2006; pp. 1244–1251.
27. Zhou, Y.; Obeid, H.; Laghrouche, S.; Hilairet, M.; Djerdir, A. Disturbance rejection control strategy of hybrid battery/super capacitors power system based on a single converter. In Proceedings of the 2019 8th International Conference on Renewable Energy Research and Applications (ICRERA), Brasov, Romania, 3–6 November 2019; pp. 534–539. [CrossRef]
28. Faria, J.; Pombo, J.; Calado, M.D.R.; Mariano, S. Power Management Control Strategy Based on Artificial Neural Networks for Standalone PV Applications with a Hybrid Energy Storage System. *Energies* **2019**, *12*, 902. [CrossRef]
29. Sathishkumar, R.; Kollimalla, S.K.; Mishra, M.K. Dynamic energy management of micro grids using battery super capacitor combined storage. In Proceedings of the 2012 Annual IEEE India Conference (INDICON), Kochi, India, 7–9 December 2012; pp. 1078–1083. [CrossRef]
30. Kollimalla, S.K.; Mishra, M.K.; Narasamma, N.L. Design and analysis of novel control strategy for battery and supercapacitor storage system. *IEEE Trans. Sustain. Energy* **2014**, *5*, 1137–1144. [CrossRef]
31. Cabrane, Z.; Kim, J.; Yoo, K.; Ouassaid, M. HESS-based photovoltaic/batteries/supercapacitors: Energy management strategy and DC bus voltage stabilization. *Sol. Energy* **2021**, *216*, 551–563. [CrossRef]
32. Dong, B.; Li, Y.; Zheng, Z.; Xu, L. Control strategies of microgrid with hybrid DC and AC buses. In Proceedings of the 2011 14th European Conference on Power Electronics and Applications, Birmingham, UK, 30 August–1 September 2011.
33. Gao, F.; Kang, R.; Cao, J.; Yang, T. Primary and secondary control in DC microgrids: A review. *J. Mod. Power Syst. Clean Energy* **2019**, *7*, 227–242. [CrossRef]
34. Keshta, H.E.; Saied, E.M.; Malik, O.P.; Bendary, F.M.; Ali, A.A. Fuzzy PI controller-based model reference adaptive control for voltage control of two connected microgrids. *IET Gener. Transm. Distrib.* **2021**, *15*, 602–618. [CrossRef]
35. Singh, P.; Lather, J.S. A PWM-based sliding mode voltage control of DC-DC boost converter for DC microgrid. In Proceedings of the 2018 IEEE 8th Power India International Conference (PIICON), Kurukshetra, India, 10–12 December 2018; pp. 1–5. [CrossRef]
36. Kumaravel, S.; Ashok, S. Adapted multilayer feedforward ANN based power management control of solar photovoltaic and wind integrated power system. In Proceedings of the IEEE PES Innovative Smart Grid Technologies—India (ISGT India), Kollam, India, 1–3 December 2011; pp. 223–228. [CrossRef]
37. Kumar, T.P.; Subrahmanyam, N.; Sydulu, M. CMBSNN for Power Flow Management of the Hybrid Renewable Energy—Storage System-Based Distribution Generation. *IETE Tech. Rev.* **2019**, *36*, 303–314. [CrossRef]
38. Singh, P.; Lather, J.S. Artificial neural network-based dynamic power management of a DC microgrid: A hardware-in-loop real-time verification. *Int. J. Ambient Energy* **2020**, *43*, 1730–1738. [CrossRef]
39. Singh, P.; Lather, J.S. Accurate power-sharing, voltage regulation, and SOC regulation for LVDC microgrid with hybrid energy storage system using artificial neural network. *Int. J. Green Energy* **2020**, *17*, 756–769. [CrossRef]
40. Kamboj, V.K. A novel hybrid PSO–GWO approach for unit commitment problem. *Neural Comput. Appl.* **2016**, *27*, 1643–1655. [CrossRef]
41. Gao, Z.-M.; Zhao, J. An improved Grey Wolf Optimization algorithm with variable weights. *Comput. Intell. Neurosci.* **2019**, *2019*, 2981282. [CrossRef]
42. Hu, P.; Pan, J.-S.; Chu, S.-C. Improved Binary Grey Wolf Optimizer and Its application for feature selection. *Knowl.-Based Syst.* **2020**, *195*, 105746. [CrossRef]
43. Liu, J.; Wei, X.; Huang, H. An improved Grey Wolf Optimization Algorithm and Its Application in Path Planning. *IEEE Access* **2021**, *9*, 121944–121956. [CrossRef]
44. Mirjalili, S.; Mirjalili, S.M.; Lewis, A. Grey Wolf Optimizer. *Adv. Eng. Softw.* **2014**, *69*, 46–61. [CrossRef]
45. Zhang, X.; Lin, Q.; Mao, W.; Liu, S.; Dou, Z.; Liu, G. Hybrid Particle Swarm and Grey Wolf Optimizer and its application to clustering optimization. *Appl. Soft Comput.* **2021**, *101*, 107061. [CrossRef]
46. Singh, N.; Singh, S.B. Hybrid Algorithm of Particle Swarm Optimization and Grey Wolf Optimizer for Improving Convergence Performance. *J. Appl. Math.* **2017**, *2017*, 2030489. [CrossRef]
47. Shaheen, M.A.; Hasanien, H.M.; Alkuhayli, A. A novel hybrid GWO-PSO optimization technique for optimal reactive power dispatch problem solution. *Ain Shams Eng. J.* **2021**, *12*, 621–630. [CrossRef]
48. Şenel, F.A.; Gökçe, F.; Yüksel, A.S.; Yiğit, T. A novel hybrid PSO–GWO algorithm for optimization problems. *Eng. Comput.* **2019**, *35*, 1359–1373. [CrossRef]
49. Al-Tameemi, Z.H.A.; Lie, T.T.; Foo, G.; Blaabjerg, F. Optimal Power Sharing in DC Microgrid under Load and Generation Uncertainties Based on GWO Algorithm. In Proceedings of the 2021 IEEE PES Innovative Smart Grid Technologies—Asia (ISGT Asia), Brisbane, Australia, 5–8 December 2021; pp. 1–5. [CrossRef]

50. Xu, Q.; Xu, Y.; Xu, Z.; Xie, L.; Blaabjerg, F. A Hierarchically Coordinated Operation and Control Scheme for DC Microgrid Clusters under Uncertainty. *IEEE Trans. Sustain. Energy* **2021**, *12*, 273–283. [CrossRef]
51. Shafiee, Q.; Dragičević, T.; Vasquez, J.C.; Guerrero, J.M. Hierarchical control for multiple DC-microgrids clusters. *IEEE Trans. Energy Convers.* **2014**, *29*, 922–933. [CrossRef]
52. Han, Y.; Chen, W.; Li, Q.; Yang, H.; Zare, F.; Zheng, Y. Two-level energy management strategy for PV-Fuel cell-battery-based DC microgrid. *Int. J. Hydrogen Energy* **2019**, *44*, 19395–19404. [CrossRef]
53. Cai, H.; Xiang, J.; Wei, W.; Chen, M.Z.Q. Droop Control for PV Sources in DC Microgrids. *IEEE Trans. Power Electron.* **2018**, *33*, 7708–7720. [CrossRef]
54. Jayachandran, M.; Ravi, G. Predictive power management strategy for PV/battery hybrid unit based islanded AC microgrid. *Int. J. Electr. Power Energy Syst.* **2019**, *110*, 487–496. [CrossRef]
55. Mendis, N.; Muttaqi, K.M.; Perera, S. Management of battery-supercapacitor hybrid energy storage and synchronous condenser for isolated operation of PMSG based variable-speed wind turbine generating systems. *IEEE Trans. Smart Grid* **2014**, *5*, 944–953. [CrossRef]
56. Behera, P.K.; Pattnaik, M. Design and Control of DC–DC Converters in a PV-Based LVDC Microgrid. In *DC—DC Converters for Future Renewable Energy Systems*; Springer: Singapore, 2022; pp. 1–29.
57. Zammit, D.; Staines, C.S.; Apap, M.; Micallef, A. Control of Buck and Boost Converters for Stand-Alone DC Microgrids. In Proceedings of the Eighth International Symposium on Energy, Aberdeen, Scotland, 24–27 June 2018.
58. Almabrok, A.; Psarakis, M.; Dounis, A. Fast tuning of the PID controller in an HVAC system using the Big Bang–Big Crunch Algorithm and FPGA technology. *Algorithms* **2018**, *11*, 146. [CrossRef]

electricity

MDPI

Article

Deduction of Strategic Planning Guidelines for Urban Medium Voltage Grids with Consideration of Electromobility and Heat Pumps

Shawki Ali [1,*], Patrick Wintzek [1], Markus Zdrallek [1], Julian Monscheidt [2], Ben Gemsjäger [2] and Adam Slupinski [2]

1 Institute of Power Systems Engineering, University of Wuppertal, 42119 Wuppertal, Germany
2 Siemens AG, 91058 Erlangen, Germany
* Correspondence: shawki.ali@uni-wuppertal.de

Abstract: With the evolution of electromobility and heat pumps in urban areas, distribution system operators find themselves facing new challenges in reinforcing their grids. With this evolution, the power demand is developing rapidly and grid reinforcement is urgently needed. The electromobility and heat pump loads are introduced by giving the assumed development scenarios in Germany and their corresponding nominal power assumptions. Furthermore, a method for load modeling in grid planning is explained. Subsequently, several grid planning approaches are presented while dividing them into conventional and innovative planning strategies. Among the investigated innovative planning strategies are three variants of load management that regulate different load types. By analyzing several urban medium voltage grids, this contribution deduces a solid basis for distribution system operators in the form of planning guidelines. The implemented grid planning method leading to the planning guidelines is presented in detail along the contribution.

Keywords: electromobility; heat pumps; medium voltage; planning guidelines; strategic grid planning

Citation: Ali, S.; Wintzek, P.; Zdrallek, M.; Monscheidt, J.; Gemsjäger, B.; Slupinski, A. Deduction of Strategic Planning Guidelines for Urban Medium Voltage Grids with Consideration of Electromobility and Heat Pumps. *Electricity* **2022**, *3*, 505–541. https://doi.org/10.3390/electricity3040026

Academic Editor: Pavlos S. Georgilakis

Received: 21 July 2022
Accepted: 12 September 2022
Published: 9 October 2022

Publisher's Note: MDPI stays neutral with regard to jurisdictional claims in published maps and institutional affiliations.

1. Introduction

Climate change is currently the main challenge facing governments, industries, and individuals alike. Even though there is already progress made to curb the expected climate changes, there is still a long way to go to transform the emission-intensive sectors such as transportation and heating into low-emission sectors [1]. A solution for that is to transform the energy source for transportation heating from the combustion of fossil fuels to electricity. On one hand, the electrification of these sectors can greatly reduce emissions, but on the other hand, it proposes an additional load on the power grids [2,3].

This contribution focuses on the impact of the electrification of the transportation sector—using charging infrastructure for electromobility—and the electrification of the household heating sector—using heat pumps (HPs)—on the electric medium voltage (MV) grids in terms of electric load development and the corresponding required reinforcement measures.

The electrification of the above-mentioned two sectors represents an unprecedented challenge for distribution system operators (DSOs) as they find themselves dealing with new load types, for which no large-scale measurement or historical data are available. Without historical data, the determination of the expected load development due to the charging infrastructure and HPs becomes complex. Furthermore, the penetration of these new loads into the grids in terms of location and number of units is unknown. Based on the government plans for a country-wide adoption of electromobility and HPs, the DSOs face the complexity of identifying the actual expected number of electric vehicles (EVs) and HPs in the specific grids.

The current grids are historically not designed to take on the load of the charging infrastructure and HPs. Therefore, the DSOs will have to reinforce their grids to ensure a reliable grid operation. Even though the so far applied conventional planning measures have proven their reliability and effectiveness, newly developed innovative technologies such as load management (LM) systems can offer an economically efficient alternative planning measure.

Even with the availability of historical data for the charging infrastructure and the HPs, their corresponding penetration levels and further evaluation of LM, performing grid planning for a complete urban grid area is an elaborate time-consuming task. Instead of that, DSOs tend to rely on planning guidelines (PGs) in reinforcing and expanding their grids.

Hence, this contribution aims to serve as a complete reference for grid planning under the consideration of these new loads. It displays scientifically backed load assumptions to overcome the challenge of the lack of historical data for the new loads. Furthermore, this contribution proposes a novel methodology to regionalize the country-wide development scenarios on specific grid areas in the MV level. Moreover, innovative planning technologies are considered such as LM, reactive power management (RPM) and energy storage (ES). The considered planning strategies then result in different planning variants, that are then assessed using an economic model as well as an alternative model. The newly proposed alternative model considers four extra assessment criteria in addition to the costs. Finally, the contribution derives seven PGs that can be directly applied by DSOs in their specific grid areas.

Although these new loads are mainly going to be connected to the low voltage (LV) level, the MV level must be investigated to successfully accommodate the accumulated load increase in the LV grids. On the MV level, two factors work in opposition to one another. The first factor is the fact that a MV grid supplies several LV grids and hence needs to adopt an accumulated large number of new loads. The second opposing factor is the decreasing demand factors with an increasing number of loads (explained in Section 2.4). Taking these two factors into consideration constitutes an added challenge to MV grid planning.

Since the density of the transportation and the heating sector in urban areas is much higher than in rural areas, this contribution focuses on the development of urban grids. In the context of this contribution, MV grids from six major German cities are utilized. The variance between the six cities helps to deduce PGs that are valid for any German urban MV grid or any urban MV grid having a grid topology similar to the investigated MV grids.

1.1. Related Work and the State of the Art

Strategic grid planning is as old as electric grids are. It constitutes an ongoing activity at DSOs since the electric grids are in a constant state of development. There are several publications discussing grid planning to a great extent [4–7]. Since this contribution focuses on the MV grid, references such as [8] are considered to determine the MV grid topology. Even though these references discuss strategic grid planning in detail, the impact of the charging infrastructure and HPs on the grids still needs to be investigated.

The evolution of electromobility has been addressed in terms of state of the art technologies either for EVs or charging methods in [9–11] including an overview of the wireless charging technologies in [12]. Especially relevant for the grids are the scenarios, load development, load demand, and load profiles that are discussed in several publications including [13,14]. Other publications, such as [15–17] discuss the impact of electromobility on the grid parameters concerning equipment loading and voltage profile for a specific grid as well as for a complete country. In addition, the effect of these new load types on the electric grid was investigated in emerging pilot projects such as [17]. Although these studies demonstrate the expected grid violations, they do not elaborate on the planning measures needed to remedy them. Besides the charging infrastructure, the impact of HPs on grids must also be analyzed.

The published work investigates the impact of HPs on the grid in different methods. Refs. [18,19] model the impact by generating load profiles of HPs in Germany and Great Britain and overlaps these profiles with load and generation profiles. Thus, the collective impact on the grids is analyzed for the total power demand in Great Britain in [18] and for an urban grid and a suburban grid in [19]. The collective impact of electromobility and HPs has also been investigated in [20–23]. Furthermore, several control strategies are proposed in [24–26] to reduce the grid violations caused by the integration of HPs into the grids. Although the above references have addressed the issue of HPs and electromobility, the impact of their integration on the MV level has not yet been studied.

Moreover, technological solutions for electromobility in terms of LM are introduced in different ways and methods in the literature [27]. Hence, the potential of LM to reduce grid reinforcements required due to electromobility has been investigated. For individual exemplary grids in [28,29]. Contributions such as [30] focus on managing the portfolio of the charging infrastructure to reduce their impact on the electric grids. In [31–33], different control strategies are proposed to curb the impact of the charging infrastructure on the grids. These publications, however, do not discuss the required grid reinforcement measures to absorb the increase in the load demand after the application of LM. Moreover, they do not consider the simultaneous regulation of charging infrastructure and HPs. In addition, the potential of regulating private charging points compared to regulating public charging points is still not investigated.

As for ES, Ref. [34] provides an overview of the ES system and its ability in reducing the grid violations. Its economic potential in comparison to LM or the conventional planning measure is yet not investigated.

Despite the extensive published and performed work, they lack concrete PGs for DSOs that recommend solid measures to be performed in the grid reinforcement. Publications such as [35–37] provide PGs with the consideration of distributed generation on an abstract level. These, however, do not take into account electromobility and HPs neither they are targeted for urban MV grids.

To sum up, although numerous publications have investigated the individual topics discussed in this contribution either solely or combined, no single publication can yet be found that considers these topics combined to the extent presented here and with a focus on urban MV grids.

Unlike the published work, this study investigates grids accumulated from several DSOs not only one. This guarantees that the final results can be considered generally valid for all urban MV grids as long as the grid has similar characteristics such as the voltage level, the topology, and the load density. Furthermore, a unified planning method is performed to reach generally valid results independent of the specific characteristics of the individual grids. In addition, more than one variant for the planning is investigated, such as a combination of several charging powers, the consideration of two development scenarios for each load type, three HP systems and three LM variants with six LM layouts (the variants stand for the regulated load and the layout expresses the configuration of the system) in addition to two ES capacities. Finally, seven tangible PGs for the MV level and three PGs for across voltage planning are deduced to simplify the grid planning process for the DSOs.

1.2. Objective and Structure of the Work

This contribution aims to provide a coherent instrument for strategic planning of urban MV grids while integrating electromobility and HPs. The goal here is to summarize all necessary knowledge to perform MV grid planning, including a set of PGs that guides DSOs while planning their grids while completing the given state of the literature. This contribution provides the following new findings:

1- Consideration of several charging powers for the charging infrastructure;
2- Differentiation in the charging infrastructure between private and public charging points;
3- Integration of charging hubs and charging points at customer substations;
4- Application of demand factors for different charging powers up to 500 charging points;
5- Analysis of three HP models different in the power rating;
6- Modeling and application of innovative technologies:
    i. LM system with three different regulation variants and six system layouts;
    ii. RPM systems;

  iii. ES with two independent storage capacities.

7- Consideration of two separate load development scenarios for each load type;
8- Application of grid planning to representative MV grid models from six major German cities;
9- Deduction of generally valid PGs:
  i. Introduction of concrete power values for the new load types for the two planning perspectives (feeder and substation);
  ii. Recommendation of standard equipment dimensions.

10- Establishment of a decision path for urban MV strategic grid planning;
11- Application of an alternative assessment model for the planning variants.

The contribution follows the steps performed during grid planning. It starts in Section 2 by explaining the technical preparation required in terms of the expected number of new loads, their localization in the grids and their assumed nominal power. Since there are no available large-scale measurement data for the new load types, the contribution proposes analytical methods to estimate the expected load development. It then continues in Section 3 with identifying the permissible limit to ensure a reliable operation of the grid. Section 3 also provides an overview of the so far applied planning measures labeled as the conventional planning strategy. As an extension to the conventional planning strategy, innovative planning technologies (LM, RPM and ES) are presented. After the technical preparation is complete, the contribution continues in Section 4 with a technical and economic assessment of the different planning variants. As a result of the assessment, seven PGs for the MV are derived and elaborately explained. Building on the MV PGs, a decision path for MV grid planning is presented. Furthermore, Section 4 introduces three across voltage PGs and ends with an alternative assessment analysis to concretize the recommended PGs for MV grids.

2. Integration of Electromobility and Heat Pumps

This section discusses the fundamental background required for analyzing the integration of electromobility and HPs. The main factors for precisely planning new loads are the determination of the number of loads, their localization in the grid and their expected power demand. In this context, the section starts with an overview of the assumed development scenarios for electromobility and HPs in Germany. Afterward, the methodology for the regionalization of country-wide scenarios down to street level is shortly explained to determine the number and the location of new loads in the MV grids. Furthermore, the load assumptions for electromobility and HPs are presented.

Electromobility represents an unpredicted new load type in the grids and is coupled with a big uncertainty regarding grid planning [9]. In the context of this contribution, electromobility will refer to the EVs used in private transportation apart from electrified public transport (such as electric buses). The aforementioned EVs are mainly charged at either private charging points (PrCPs) or public charging points (PuCPs). A charging point (CP) in the context of this contribution is defined as an outlet at which only one EV can be charged. PrCPs are meant to be privately owned and operated CPs that can only be accessed by a specific user (for example a CP in a private garage). Whereas, PuCPs are the unrestricted publicly accessed CPs typically found on the street side. The standards for charging infrastructure can be found in [10]. In addition, other charging concepts such as inductive charging [12] are not considered in this contribution because they do not change the power demand and therefore have no effect on grid planning.

As for HPs, the task of grid planning becomes much more complex. At a first glance, the development of HPs seems simple as they are solely integrated into households and—in a few cases—into commercial buildings (unlike CPs that can be found on the streetside). However, as the buildings differ in size, year of construction, thermal insulation, and correspondingly the heat demand, it becomes more and more difficult to precisely foresee the development of HPs. Therefore, this contribution focuses on individual HPs that are installed in standalone households, whereas high-power HPs used in industrial or commercial buildings are not investigated here, as they are considered to be individual planning cases.

2.1. Development Scenarios

The scenarios for electromobility and HPs differ greatly regarding the expected development over the upcoming years. Each published study [38–40] assumes different boundary conditions such as the progression of climate change regulations, consumer acceptance, technological progress, etc. Accordingly, the anticipated course of development varies from one study to another. Hence, two development scenarios are chosen per load type. A "progressive" scenario (referred to later on as prog.) is adopted that assumes an accelerated spread penetration of electromobility and HPs. In addition, a "conservative" scenario (referred to later on as cons.) is taken on which assumes a rather moderate limited penetration of new loads [41]. Figure 1 shows the scenarios considered in this contribution, which are chosen based on thorough literature research and are selected to be moderate scenarios among others.

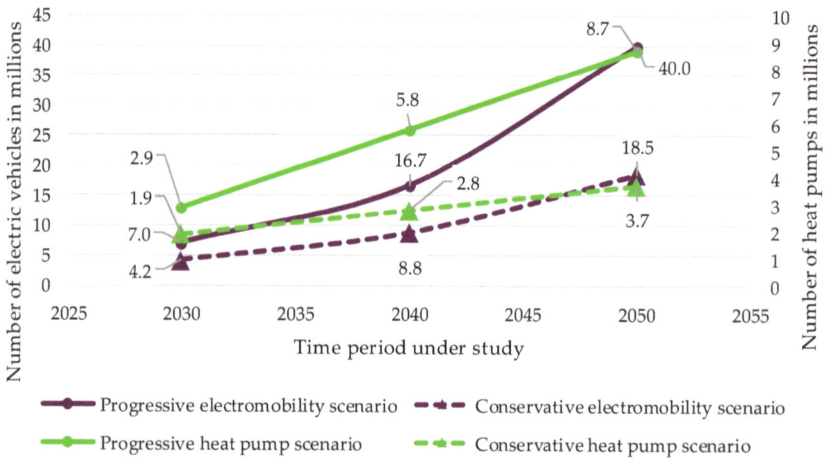

Figure 1. The applied progressive and conservative development scenarios for electromobility and heat pumps in Germany based on [41].

2.2. Regionalization

After determining the development scenarios, it becomes important to identify the number of new loads expected in an analyzed grid area. Therefore, a regionalization method is developed to calculate the number of new loads ending up in a grid area.

The regionalization methodology goes through five steps; starting at the country level, then down to the state level, after that to the city level, and afterward to the district level ending at the grid level. At each step, different regionalization factors are considered. For instance, for the step from the country level to the state level, the following regionalization factors are considered for electromobility for the respective state in comparison with other states in the country using a weighting term per regionalization factor: population, number of EVs, number of vehicles (combustion and electric), number of car owners (a single owner can own several cars), research budget for electromobility, and number of buildings. As for the same step for HPs, the number of existing HPs in the state is considered to be the regionalization factor [42].

By applying the regionalization methodology, the number of EVs and HPs are determined per MV grid. For validation, the regionalized numbers are cross-checked with the DSO-specific development plans. As further validation, these numbers are compared with the development of electromobility and HPs in specific grids. The expected number of EVs and HPs per grid are listed in Appendix A, Table A1.

2.3. Nominal Power Assumptions

As for the charging infrastructure, the available CPs on the market differ greatly in terms of their nominal charging power, which makes it difficult to assume a certain charging power for the complete charging infrastructure. By analyzing the public charging infrastructure register from the German federal network agency [43] the development of charging powers over the past years is deduced. It becomes clear that the charging powers for PrCPs are within the range of 3.7 kW to 11 kW, whereas the charging powers for PuCPs go up to 150 kW or even up to 350 kW. Since an exact spread of the charging powers for the next 30 years is feasibly not possible, as a first approach to the distribution of charging powers, the following ratios in Table 1 are assumed. The assumed distribution of private and public charging powers has been agreed upon with the six major DSOs in Germany. A perfect power factor of one is assumed for the charging infrastructure.

Table 1. The assumed distribution of charging powers for private and public charging infrastructure based on [41].

Charging Power	Private Charging Infrastructure			Public Charging Infrastructure		
	2030	2040	2050	2030	2040	2050
3.7 kW	10%	0%	0%	0%	0%	0%
11 kW	60%	65%	65%	5%	5%	5%
22 kW	30%	35%	35%	75%	20%	20%
50 kW	0%	0%	0%	15%	50%	50%
150 kW	0%	0%	0%	5%	25%	25%

As for the HPs, the approach to finding reasonable power assumptions is different, since, as mentioned previously, the heat demand varies greatly depending on each household and use case. The following steps are carried out to reach reasonable, generally valid power assumptions.

Firstly, the available HPs on the market are analyzed in terms of nominal power to find the widely spread power classes. Secondly, the average heat demand of standalone houses is analyzed to determine which power class can fulfill the required heat demand. This approach gives a first approximation of the required nominal power, which is 3 kW for the living space heating. Furthermore, an additional electric heating element for purposes such as warm water supply is analyzed, resulting in a big spectrum of nominal powers. Therefore, an average power assumption is adopted, giving an extra 6 kW for the electric heating element. This yields an HP with a total power of 9 kW (HP system + electric heating element) [44]. Since it is not clear whether all HPs are going to be installed with an extra heating element or not, the power assumptions are expanded to a hybrid system of 6.5 kW representing a distribution of the previous two HP systems. In summary, the three HP systems are as follows:

1- 3.0 kW (basic HP system);
2- 6.5 kW (hybrid distribution of HP systems 1 and 3);
3- 9.0 kW (basic HP system + electric heating element).

2.4. Load Modeling

The first step for representing the future grids is to convert the scenarios into power demand to be integrated into the grid model. The existing loads such as household, commercial or industrial loads are summarized here with the term "conventional loads" and are discussed at the beginning. Subsequently, the modeling for the new load types, namely the electromobility and HPs, follows.

To determine the development of the conventional household loads, a load demand model is developed [41]. The model performs a two steps process to determine the future load demand of household loads. Firstly, a statistical model is fed with the grid data (such as MV/LV distribution stations, energy consumption per building connection, standard load profiles, etc.) as well as socio-economic data [45] to estimate the future development of household loads. Subsequently, a deterministic model scales the corresponding standard load profile for the specified load type and year. As for the conventional industrial loads (customer stations in the MV level), the load development is not identified, due to the heterogeneity of the loads and their distinction from one MV grid to the other. Hence, the industrial loads are kept constant over the investigated period of time (the year 2022 until the year 2050).

In contrast to conventional loads, electromobility has little to no historical data that can be used for load modeling. Therefore, a statistical analytical method is implemented to model electromobility as a load in the grid.

For calculating the demand power of electromobility several approaches can be found in the literature. In [46], the coincidence factors for domestic CPs are calculated based on the driving and plug-in behaviors recorded. Depending on these recorded data sources, a model was developed to calculate the demand power for the domestic charging powers. Even though this model is well thought out, it does not tackle the public infrastructure that is not used by a single household. Therefore, in this contribution, the demand power is calculated depending on the tool developed in [47]. This tool simulates daily driving profiles by utilizing the collected driving data from [48]. These driving profiles are then converted to charging profiles for CPs. By simulating the seven days of the week, a weekly charging profile can be generated for a specific charging power. The weekly charging profile is then simulated for several weeks for different charging powers using an incrementally increasing number of CPs. Then, these profiles are accumulated to generate the peak power demand for a certain number of CPs for a certain charging power. Simultaneously, the demand factors for each number of CPs for the chosen dominant charging powers (3.7, 11, 22, 50 and 150 kW) are calculated. The demand factor is defined to be: "the ratio, expressed as a numerical value or as a percentage, of the maximum demand of an installation or a group of installations within a specified period, to the corresponding total installed load of the installation(s)" [49]. The integration of different shares of the charging powers can result in a mean effective charging power that varies from the simulated chosen charging powers. Therefore, a curve-fitting algorithm is applied to deduce the demand factor per kW increments between 3.7 kW and 150 kW [50]. Using these demand factors (shown in Figure 2) and the nominal power assumptions in Table 1, the demand power for electromobility is calculated according to the method in [42] and integrated into the grids. The demand factors per kW increments have been faded out to enhance the visibility of the figure.

As for HPs, grid planning is performed for each of the HP models resulting in an individual planning variant. Since a combination of the HP models in the same grid is not considered, the HP load modeling considers simply multiplies the nominal power with the demand factor for the corresponding number of HPs in the grid/feeder. The demand factor curve of the HPs is shown in Figure 3.

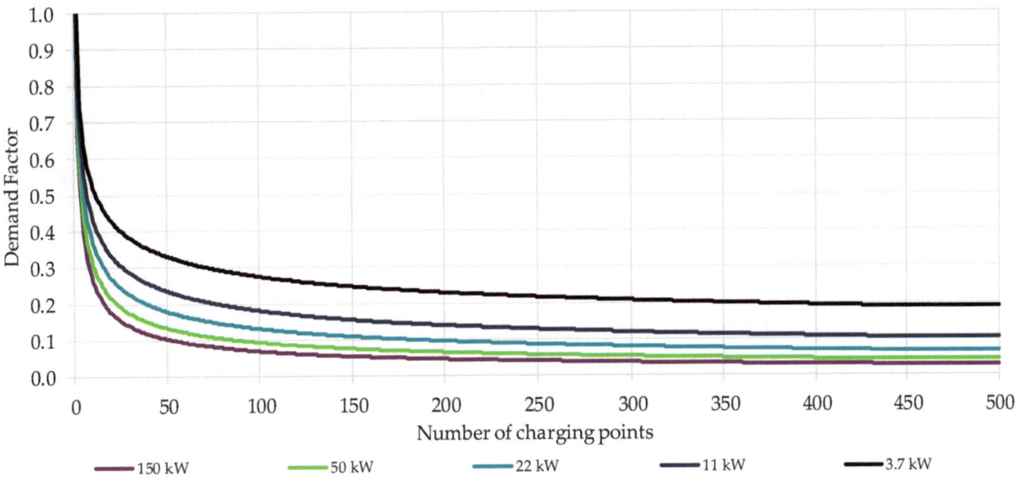

Figure 2. Demand factors for five dominant charging powers for up to 500 charging points based on [50].

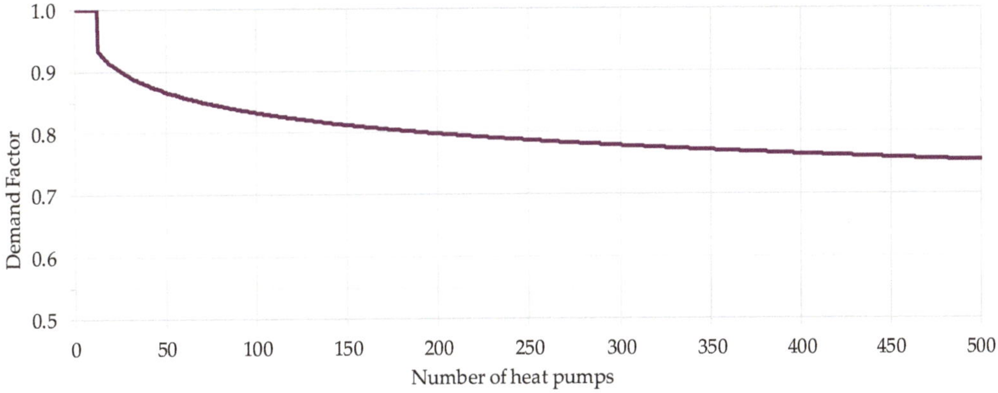

Figure 3. Demand factors for up to 500 heat pumps based on [51].

3. Methodology of Strategic Grid Planning

Grid planning depends on two main pillars. It starts with modeling the expected load and/or generation depending on the relevant operation point. Grid planning then continues with the implementation of strategic planning measure(s) in case limit violations are identified.

3.1. Identification of Grid Limits

In addition to the regionalization of new loads into the grids and their power assumptions, the identification of permissible grid limits constitutes a crucial step in strategic grid planning. Surpassing the grid limits leads to a hazardous operation of the equipment and may even lead to human losses. In the context of this contribution, the word equipment refers to current carrying operating equipment, i.e., lines—including overhead transmission lines and cables—and transformers. Apart from the DSO-specific grid limits, in the context of this contribution, the electric standards are taken as a reference to ensure universally applicable results.

3.1.1. Equipment Loading Limits

The general rule of thumb regarding the loading of equipment is not to exceed the thermal admissible current (I_{th}). This gives the maximum loading of $(I/I_{\text{th}}) \leq 100\%$ in normal operation (n-0) [52–54], where "n" corresponds to the total number of equipment in the grid. Since MV grids require an (n-1) level of reliability to ensure the supply of all connected loads in case equipment goes out of operation, the loading of equipment in (n-1) operation is the decisive factor for dimensioning MV equipment. The permissible loading limit used in the planning is $I/I_{\text{th}} \leq 120\%$ during (n-1) operation of the grid. Hence, it is assumed that for parallel transformers of equal capacity, a loading of $I/I_{\text{th}} \leq 60\%$ per transformer is allowed in (n-0) operation. Similarly, it is also assumed that for grids with an open ring structure, a maximum line loading of $I/I_{\text{th}} \leq 60\%$ is allowed per main feeder.

3.1.2. Permissible Voltage Range

Maintaining a constant node voltage V over a complete grid is extremely challenging. Therefore a voltage range is typically set per voltage level in order to ensure an admissible V at all grid nodes. According to [55], the admissible V ranges between $\pm$ 10% of the nominal voltage V_{n} over the entire grid. In this contribution, the voltage range is divided between the MV level and the downstream LV grids, resulting in a maximum admissible voltage drop ΔV = 4% V_{n}. in the MV level. The following Figure 4 shows the applied division of voltage range between MV and LV grids.

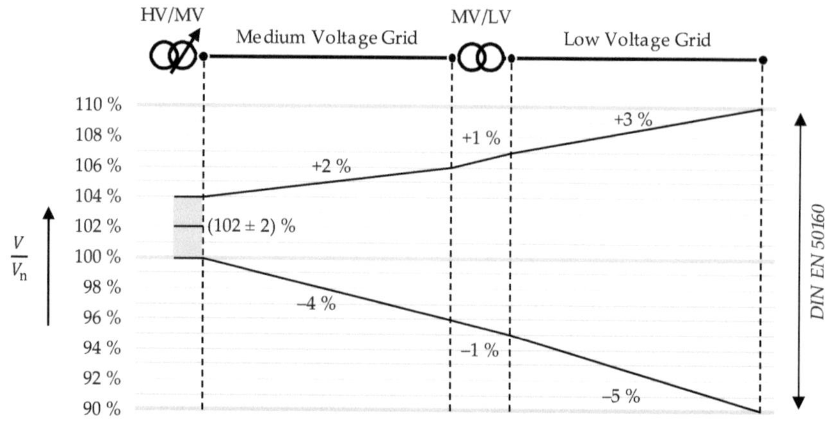

Figure 4. Assumed division of voltage range between medium voltage and low voltage grids.

3.2. Planning Strategies

After identifying the grid limit violations, the DSOs need to perform planning measures to remedy the violations and restore the grid within the recommended limits of operation. Traditionally, the DSOs can perform one of the approaches that have been established in the grid planning for years already such as cable reinforcement or transformer replacement. The approach of applying traditional measures is referred to as a "conventional planning strategy". However, with the emergence of new load types, innovative technologies have been developed to take advantage of the flexibility of these loads to reduce the required grid reinforcement(s). This approach is referred to later on as an "innovative planning strategy".

Depending on the identified grid violation, various planning measures can be implemented. The following Table 2 shows an overview of the use cases of the planning measures, which are explained later on. Starting with the two measures implemented within the conventional planning strategy, the cable and substation transformer measures remedy the grid violation locally without affecting other identified grid limit violations. For instance, when the substation transformer is overloaded in addition to several cable sections, the application of substation transformer measures remedies the local loading violation without having a significant impact on the overloaded cable sections. On the other side, the measures within the innovative planning strategy can have a grid-wide impact on limit violations. The measures are explained in detail in the upcoming section.

Table 2. Overview of the use cases for conventional and innovative planning strategies.

Measure	Application in Voltage Level	Use Case of the Measure		
		$I/I_{th} > 120\%$ *	$V/V_n < 96\%$	$V/V_n > 106\%$
Conventional Planning Strategy				
Cable	MV	✓	✓	✓
Substation transformer	MV	✓	-	-
Innovative Planning Strategy				
Load management	LV	✓	✓	-
Reactive power management	LV	✓	✓	✓
Energy storage	MV	✓	✓	✓

* loading in (n-1) operation. ✓: means the measure is having an influence on the grid violation.

3.2.1. Conventional Planning Strategy

This approach represents the traditional planning measures performed by DSOs and that have been established as standard planning options. These include the installation of new equipment such as cable or transformer or the application of a grid operational measure. These options are discussed in the following sections.

1. Cable measures

Cable measures are considered to be one of the main tools of grid planning. These cannot only remedy overloads in the grid but can also remedy voltage violations at certain nodes in the grid. Cable measures can generally be classified into two main categories; grid reinforcement and grid expansion.

As for grid reinforcement, the grid topology is maintained and the cable measures are performed at the already existing cable routes. In case the overloaded cable is identified to have a cable cross-section area smaller than the DSO's standard cable cross-section area or when the cable insulation type is considered to be old (e.g., NKBA), a cable replacement is performed. In this case, the new cable replaces the existing overloaded cable. However, if the overloaded cable has a cross-section area corresponding to the current standard cross-section area and has a relatively new insulation type (e.g., N2XS2Y), the cable is reinforced with a second new cable. In both cases, the newly constructed cable is laid in the same route as the overloaded cable.

As for grid expansion, the grid topology is changed by constructing a new cable route directly connecting the grid nodes. For instance, when the load significantly increases at a substation feeder, some of the load can be shifted to a newly constructed feeder. On one hand, this measure can drastically reduce the load on the existing feeder. However, on the other hand, the space for constructing new feeders may not be always available, especially at jammed substations in highly dense areas.

In the context of this contribution, only grid reinforcement measures are applied, since the investigated grids are collected from six different DSOs where the spatial boundaries of the individual substations differ remarkably. Furthermore, unified cable measures are performed for all grids, so that generally valid results can be deduced.

The applied cable type is NA2XS2Y with the cross-section areas (150 mm^2, 185 mm^2, 240 mm^2 and 300 mm^2). The wide range of applied cross-section areas helps, later on, to identify the most suitable cable cross-section for future planning measures. To ensure minimum costs, the planning tries to remedy the grid violation using the smallest cable cross-section area. When the violation persists, the next bigger cable cross-section is employed. When the violation persists after applying the biggest cable cross-section area (300 mm^2), two 150 mm^2 cables are laid. Furthermore, each cable measure is checked whether it will need to be re-planned in the following years up to the year 2050. If the same cable section becomes overloaded in one of the following supportive years, a bigger cable cross-section is laid in advance, so that the cable trench does not need to be dug frequently. Since the cables are not fully loaded and the modeled loading corresponds to the peak demand operating point, cable reduction factors are not considered.

2. Substation transformer measures

In contrast to the grid-wide cable measures, substation transformer measures are central. These measures are strictly constrained since a certain redundancy is required for substation transformers. The most common substation construction is a double power transformer of the same nominal power. The double transformer constellation ensures the supply of loads, in case one of the transformers breaks down. If the substation transformer is overloaded, the following three options can be performed.

- Transformer reinforcement

The main condition for this option is the space availability in the substation to construct a new transformer. The existing transformers can be reinforced with an extra transformer having the same nominal power as the existing transformers. As a result, several feeders can be shifted to the new transformer. Otherwise, the existing transformers can be loaded to higher levels while considering the newly added transformer as a reserve.

- Transformer replacement

For this option, the existing transformers are replaced with two (or more) identical transformers of a higher nominal power.

- Boosting the transformer loading

As the decisive criterion for loading the transformers is the temperature, their loading can be boosted by externally cooling them. The possible cooling concepts differ according to the power class of the transformer and its insulation type.

After the transformer measures are completed, the short-circuit currents in the substation must be calculated. Accordingly, the switchgear rating needs to be adjusted. Furthermore, all the transformers in the substation need to have the same nominal power and impedance to avoid rotating currents during parallel operation.

Since the existing transformer types and the space availability differ from one substation to another and from one DSO to another, "Transformer replacement" is the only considered transformer measure.

3.2.2. Innovative Planning Strategy

The innovative planning strategy aims to deploy new technologies that can reduce the required grid reinforcement measures or completely replace them. Since the new loads offer a degree of flexibility, the developed innovative technologies utilize this flexibility to either reduce or eliminate grid violations. In addition, the innovative technologies utilize the mentioned flexibility without compromising the comfort of the customers. Since this contribution focuses on increasing loads in the MV grids, several innovative technologies that are mentioned in the literature (e.g., active regulating transformer or feed-in management) are not mentioned here.

1. Decentralized automation systems

Decentralized automation systems (DASs) have been developed over the past few years to monitor and determine critical grid states as soon as they occur. In addition, a DAS can purposely regulate the grid state using regulators at the equipment to be controlled. According to the targeted application, DAS can serve as a prerequisite for the applied technology.

A DAS composes of a remote terminal unit, a communication link, sensors, and regulators. Regulators are assembled at the equipment to regulate their drawn active and reactive power. Sensors are installed at certain grid nodes to measure the current flowing at this node and the corresponding voltage. The communication link connects all DAS components to enable data transfer between them. The remote terminal unit process the measurements performed by the sensors and generates the command, by which the regulators control the equipment. The remote terminal unit generates the commands based on certain functionality. This functionality ranges from avoiding grid limit violations using load or generation management to optimizing the energy price by managing energy storage units [56].

In addition, this contribution assumes that the communication link occurs using the cellular phone network since the phone network is stable in urban grids that are solely analyzed here. The costs for SIM cards used for utilizing the cellular phone network are negligible and thus can be ignored. As for determining the gird state, sensors do not have to be mounted at each grid node. The determination of grid state using a few sensors mounted at specific grid nodes has already between discussed in the literature [57]. Using the developed techniques, the sensors are allocated in the grids. Finally, this contribution uses the presented DAS as a requirement for applying LM and reactive power management (RPM).

2. Load management

LM is growing in importance with the increased integration of flexible loads into the grids. This contribution considers LM, which regulates the power of CPs and HPs according to the grid state in regards to the voltage level and the loading. The regulation aims to reduce the grid limit violations or completely eliminate them. An energy market-oriented regulation is not considered here.

A crucial aspect of applying LM is a discrimination-free application. The regulation of the specified loads occurs independently of the load location and the load power. On one hand, the power of the CPs is regulated to a minimum of 3.7 kW, since a complete shut-off is not recommended. On the other hand, the HPs are shut off completely for certain time slots. In contrast to CPs, HPs can store heat that can be used during shut-off times. Table 3 shows the three LM variants considered in this contribution. The consideration of different LM variants helps to identify the advantage of each load regulation in terms of grid planning measures [58].

Table 3. Overview of the regulated load for each of the three load management variants ("✓": regulated, "-": not regulated).

Regulated Load	LM Variant 1	LM Variant 2	LM Variant 3
Heat Pumps	✓	-	-
Private Charging Points	✓	✓	-
Public Charging Points	-	-	✓

If a grid limit violation emerges at a single feeder, LM regulates the loads (according to the LM variant) connected to the feeder until the violation is remedied or the minimum load power is reached. However, if a substation transformer is overloaded, the LM regulates all corresponding loads in the grid to reduce the transformer loading. This regulation benefits the feeder loading but requires considerably more load regulation and the corresponding DAS components. Figure 5 shows a flowchart explaining the installation of DAS components according to the grid violation. It starts with checking whether a grid limit violation occurs. In this case, a second check for the extent of the grid violation is performed. The goal here is to install the DAS components solely at the required positions to reduce the installation costs. If the grid limit violation is restricted to particular feeders and does not cover the complete MV grid (including the transformer), the DAS components are to be installed at the violated feeders. Otherwise, a grid-wide installation of DAS components is necessary.

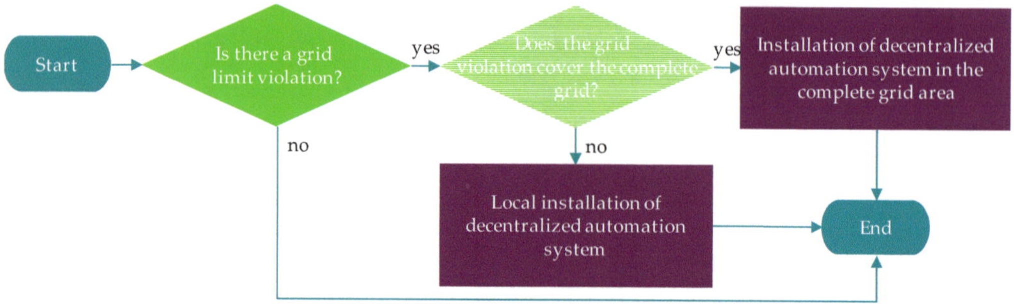

Figure 5. Flowchart for the installation of the components of the decentralized automation system for applying load management based on [59].

Since LM basically regulates loads connected to the downstream LV grids, the DAS needs to be installed at both voltage levels. The DAS in the MV level analyzes the grid state and sends signals to the DAS in the LV level to regulate their loads accordingly. With the ongoing roll-out of DASs, six LM layouts (shown in Table 4) are developed to analyze the effect of the layout costs on grid planning. The layouts differ in terms of the required measuring, information and communication technology (MICT) to operate LM. These layouts start with the assumption that neither the LV grids nor the MV grids are equipped with MICT. In this case, the MV level takes over the costs for complete MICT roll-out in both voltage levels. The layouts then end with the assumption that LM is already installed in the grid and can be used for no further costs. Figure 6 illustrates the differences between the six LM layouts for a single feeder. The shown number of sensors is then projected for a complete MV grid in case the transformer(s) and/or more than one feeder are overloaded.

Table 4. Specification of the six load management layouts based on [59].

Load Management Layout	Specifications
Total costs (MV + LV)	MICT is needed in the MV grid as well as all the LV grids
Half the costs (MV + 50% LV)	MICT is needed in the MV grid and half of the LV grids as the other half is already equipped
MV costs (MV)	MICT is needed in the MV grid, whereas the LV grids are already equipped with MICT
Reduced MV costs (Red. MV)	MICT is needed in a reduced coverage in the MV grid since it is already partially equipped with MICT
Base costs (B)	A remote terminal unit is needed to operate LM since MICT is fully constructed in the MV grid
No costs (0)	All the LM components are already constructed

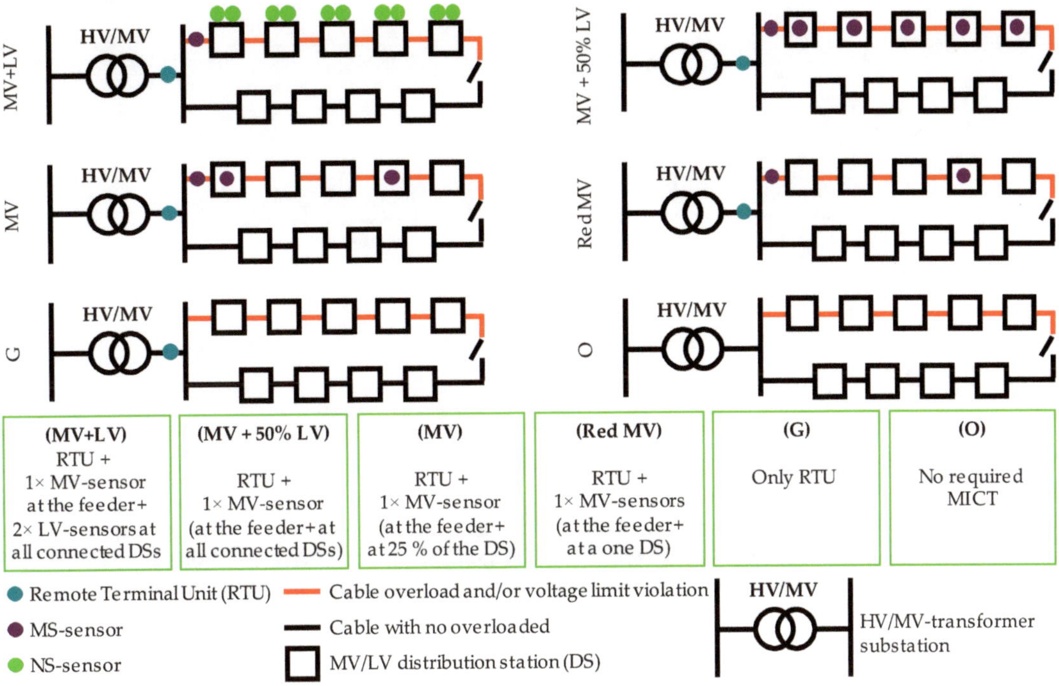

Figure 6. Illustration of the six load management layouts for one medium voltage ring based on [59].

Several challenges arise in implementing LM in the MV grids [60]. Since MV grids can span for several kilometers, a communication delay between the components may occur, thus leading to prolonged grid limit violations [61]. Several other challenges such as data privacy, measurement uncertainty, and the defense against cyber-attacks should also be considered before implementation.

1. Reactive power management

Basically, there are two application concepts of RPM. The first concept enforces a static RPM, which regulates the reactive power of loads on a predefined set point. The second concept applies a dynamic reactive power regulation according to the grid state. Obviously, the second principle requires a DAS to identify the grid state and establish the necessary RPM accordingly. The loads regulated by RPM are CPs since they offer the necessary flexibility for RPM.

By changing the angle between the voltage and the current drawn by the load, the reactive power drawn by the load can be adjusted. Changing the current to lagging or leading alters the direction of the current flow and can either increase or decrease the voltage at the node. This flexibility helps maintain the voltage level in its allowed range in case of an increase in load. The following Figure 7 displays the effect of lagging and leading currents on the node voltage.

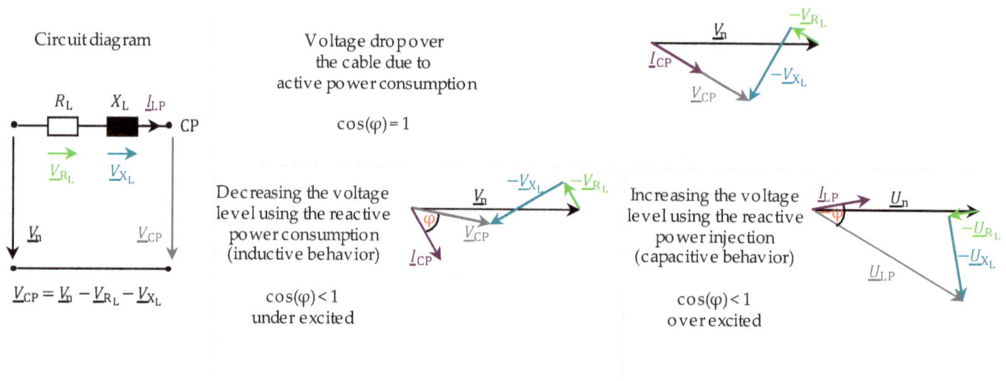

Figure 7. The principal operation of reactive power management based on [59].

In this contribution, RPM has proven to be ineffective in terms of grid planning due to two main reasons. The first reason is that with the integration of new loads, MV grids rarely suffer from voltage limit violations. In fact, it is recommended that the voltage limit distribution between LV and MV grids be reinvestigated for the grid planning (see PG 7 in Section 4.3). The second reason is that since RPM works on improving the voltage level by increasing the reactive power drawn (increasing the loading) and since the integration of new loads increases the equipment loading in the first place, applying the RPM merely increases the loading further. Thus, the RPM contributes to further equipment overloads rather than easing them and therefore, this technology is not investigated further.

2. Electrical Energy Storage

In this contribution, electrical ESs refer to an equipment that saves a certain amount of electrical energy to be used later on. An ES draws electric current from the grid, saves it using a chemical process and injects current afterward. Hence, the ES indicates in this context battery storage. Other ES devices such as flywheel energy storage or pumped hydro energy storage are not considered.

Technically, ESs can have two main purposes. On the one hand, it can be market-oriented. In this mode of operation, ES stores energy when the market signal indicates a low energy price. Afterward, the ES can inject current when the energy price increases. On the other hand, the ES can also be grid-oriented, in which it stores the current at off-peak times. Later on, the ES can inject the current at peak demand times to reduce the transformer and cable loading. In the context of this contribution, only the second purpose is considered.

Generally, ESs can be used to overcome overloads in the grid for various periods of time. Hence, the cost of ESs depends on the injected peak power in kilowatts and the period of injection in hours. The injected peak power is regulated by an inverter or a power conversion module, which is negligible in costs in comparison to the battery modules. Therefore, the cost of ESs significantly depends on the amount of energy saved or in other words the period of time in which an ES can inject its peak power. This contribution studies the costs for two ES; with one having a capacity to deliver peak power for 2 h and the other for 4 h.

4. Derivation of Strategic Planning Guidelines

After finishing the strategic planning for 11 MV grids, the results are consolidated in order to deduce generally valid PGs for all MV grids. A classification of the investigated MV grids is given in Appendix A, Table 2. The deduction of the PGs needs to find a balance between being specific enough to offer solid recommendations to the DSOs, but also broad enough to apply to the vast majority of MV grids. This chapter aims to find the exact balance between these two boundaries.

As a first step, the load development is calculated for the 11 analyzed MV grids in order to illustrate the expected load demand. Figure 8 shows the load development for 11 MV grids for the different load types. Base conventional loads (household and small commercial loads) are represented by distribution substations and the industrial loads are represented by customer stations. In contrast, the new loads are represented by charging infrastructure combining both private and public charging infrastructure and HPs with the different HP load assumptions. It can be seen that the conventional loads have a nearly constant load over the considered period of time. Since the penetration of charging infrastructure correlates strongly to the number of households in a grid (represented explicitly by the distribution substations), a high load development of charging infrastructure is noticed in the MV grids with a high load of distribution substations (e.g., G01 and G02). Furthermore, the same correlation is observed between the load development of HPs and the load of distribution substations (e.g., G08 and G11). On the contrary, a modest load development of charging infrastructure and HPs is expected in MV grids with a high industrial load taking G04 for instance. The corresponding grid parameters for the 11 MV grids are listed in the Appendix A, Table A1.

After modeling the load development for the analyzed MV grids, the grid value violations are identified. Accordingly, planning measures are performed and displayed in a consolidated form in Figure 9. The figure shows that the cable measures for the conventional planning are more than the cable measures for any of the LM variants. With a focus on the cable measures performed in the year 2030 (purple), it can be deduced that the cable measures remain constant irrespective of the applied planning strategy. This confirms that the short-term cable measures need to be performed regardless of the DSO's chosen strategy. These stand for the small dimensioned cable sections in the MV grids. As for LM-V1, there is no difference in the cable measures between the three HP models since the HPs are turned off in this LM variant. Figure 9 shows that the LM variants, LM-V1, LM-V2, and LM-V3, can reduce the cable measures and/or postpone them but cannot completely replace them. The same applies also to planning with the ES.

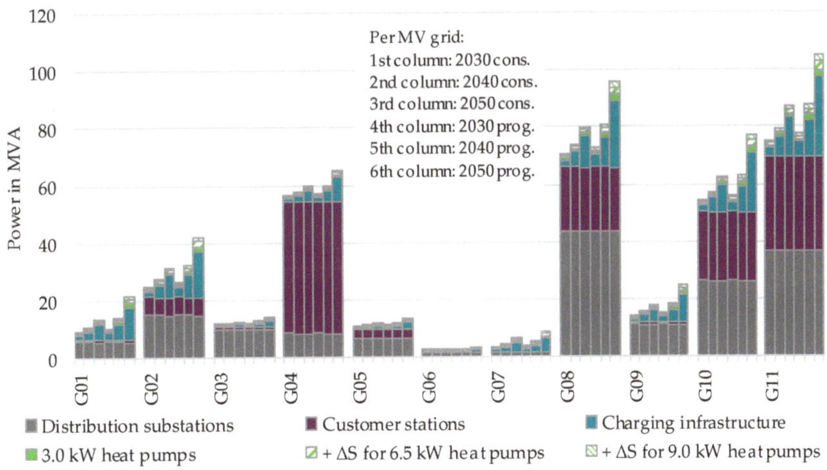

Figure 8. Load development for eleven medium voltage grids for the two development scenarios over the years 2030, 2040 and 2050 based on [59].

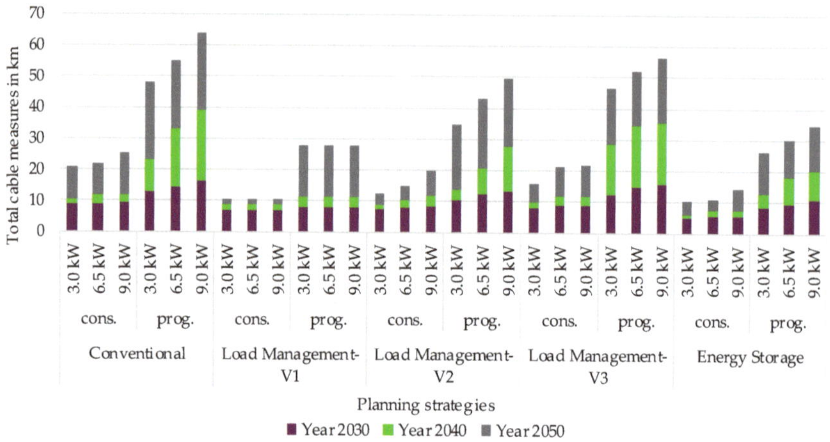

Figure 9. Consolidated cable measures for the medium voltage grids with different planning strategies over the investigated years for the two development scenarios and the three heat pump systems.

4.1. Technical and Economic Assessment

After planning, the costs for the planning measures are calculated to evaluate the different planning strategies. The net present value method is chosen for the cost calculation to contain all the investment and operating costs of the planning measures for the period under consideration. The advantage of this method is that it retraces the total spending back to the present, depending on the year in which the spending takes place. Hence, the comparability of the different planning strategies becomes easier. The presented costs are calculated according to the net present method using the assumed costs of the equipment presented in Appendix A, Table 3 [62].

The following Figure 10 shows the consolidated cable measures for the conventional planning, the three LM variants with the LM layout (MV) and the planning with two-hours and four-hour ESs. The costs constitute of capital expenditures (CapEx) and operational expenditures (OpEx). The residual value of the measures at the end of the considered timeframe is then deducted from the CapEx and the OpEx, thus giving the resulting costs of the planning. Figure 10 shows that the LM-V1 (MV) is cheaper than conventional planning. Whereas, LM-V2 (MV) costs nearly the same as conventional planning and the costs for LM-V3 (MV) slightly exceed the costs for conventional planning. The cost-efficiency of the different LM variants and layouts is discussed in detail in the fourth PG. The costs for the two ESs exceed the costs for conventional planning by far. Thus showing the previous unattractiveness of this technology.

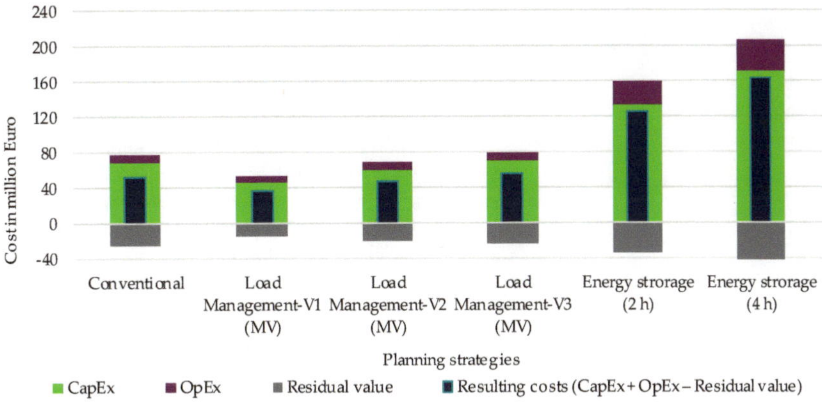

Figure 10. Consolidated costs in millions of Euros for all analyzed medium voltage grids for different planning strategies.

With an overview of the total costs, the savings potential of the planning strategies in comparison to the conventional planning is analyzed per planning variant and shown in Figure 11. It becomes clear that none of the planning variants with ESs is cheaper than conventional planning. In addition, a difference in the saving potentials of the three LM variants is clear. For instance, the planning variant "LM-V1 (MV)" is in 97% of all analyzed planning variants more economical than the conventional planning. Detailed usability of the different LM versions and layouts is investigated in the fourth PG.

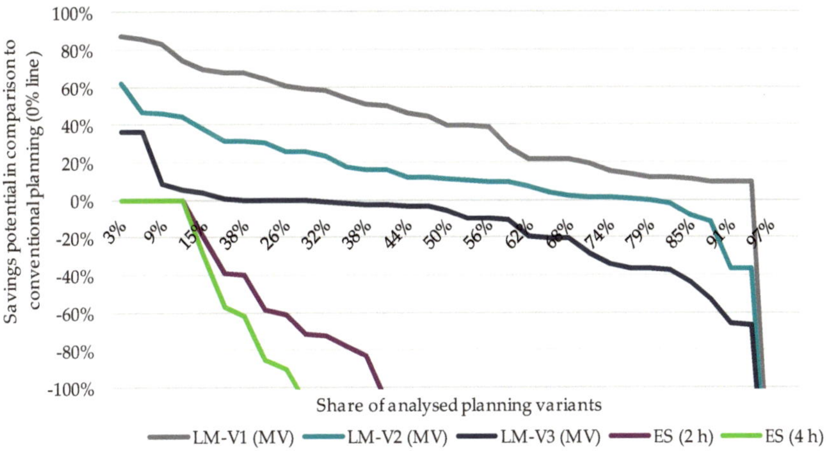

Figure 11. The savings potential for innovative planning strategies in comparison to conventional planning per planning variant.

4.2. Medium Voltage Strategic Planning Guidelines

Based on the results obtained from the MV grid planning, seven PGs are derived. The PGs serve as the main principles for the planning of MV grids. They, however, cannot replace the DSO-specific PGs.

The PGs are arranged in the same steps that a grid planner executes to complete grid planning. The PGs start with the power assumptions for the different load types mentioned previously in Section 2.4. They continue with the standard dimensioning of the equipment used in the planning. Furthermore, different digitization levels of the grid in the context of LM are investigated. The PGs proceed with characteristics of the different grid structures, voltage levels, and locations. They end with a prospect for a new voltage distribution between MV and LV grids.

4.2.1. First Medium Voltage Planning Guideline

A mean effective charging power of $P_{PrCP,SS}$ = [0.3; 2.4] kW for private charging points, of $P_{PuCP,SS}$ = [0.05; 0.8] kW for public charging points, an electric power of $P_{HP,SS}$ = [0.1; 0.5] kW for 3.0 kW heat pumps and of $P_{conv,SS}$ = 2 kW for conventional household loads are recommended per building connection for the dimensioning of substation transformer.

A mean effective charging power of $P_{PrCP,F}$ = [0.8; 2.7] kW for private charging points, of $P_{PuCP,F}$ = [0.1; 0.9] kW for public charging points, an electric power of $P_{HP,F}$ = [0.1; 0.5] kW for 3.0 kW heat pumps and of $P_{conv,F}$ = 2.4 kW for conventional household loads are recommended per building connection for the dimensioning of substation outgoing feeders.

The first step in performing grid planning is to model the grid with the proper load prognoses. In this regard, the power value assumptions play an essential role in the grid reinforcement measurements and the final planning. Since the demand factors differ between the two planning perspectives for dimensioning the substation transformers and the outgoing feeders, the power value assumption should be determined for the two perspectives separately.

The following Figure 12 shows the mean effective power value per building connection for the 3.0 kW HPs and the PrCPs and PuCPs for the two scenarios over the years up to the year 2050. The values are given for the two planning perspectives, namely substation transformer and outgoing feeder.

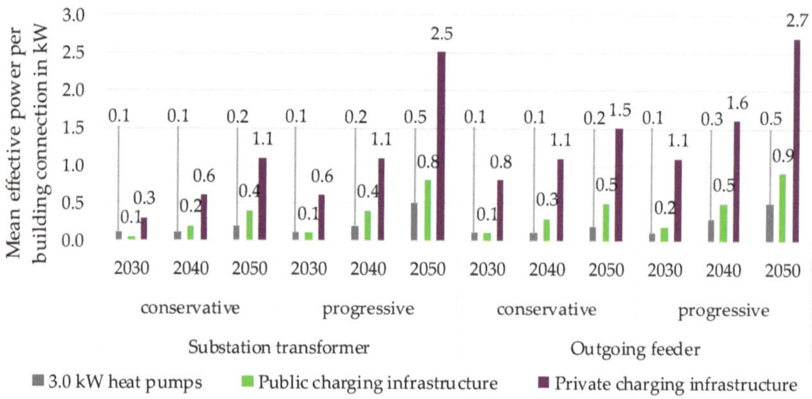

Figure 12. Mean effective power per building connection for dimensioning the substation transformer and outgoing feeders for private and public charging infrastructure and 3.0 kW heat pumps.

Since the number of CPs increase drastically up to the year 2050, the demand factor for CPs falls to a saturation value that nearly stays constant for an increasing number of CPs. This leads that the power assumptions for the progressive scenario in the year 2050 are nearly constant for the two planning perspectives (2.5 kW and 2.7 kW, respectively). Whereas, the power values for the conservative scenario in the year 2030 nearly triple depending on the planning perspective (0.3 kW and 0.8 kW, respectively).

In addition, Figure 12 shows that the mean effective power for PrCPs is always higher than the corresponding value for PuCPs, even though the charging power distribution in Table 1 assumes higher charging powers for the PuCPs than the PrCPs. The reason here lies in the higher number of PrCPs compared to PuCPs, since the number of PrCPs is expected to strongly exceed the number of PuCPs in the respective grids. Therefore, the effective power per building connection for PrCPs exceeds the corresponding value for PuCPs.

It is established that HPs exhibit a nearly constant demand factor. Even though the number of HPs out of the perspective of the substation transformer is greater than the number of HPs per outgoing feeder, the power values do not differ much between the two dimensioning perspectives for the respective year and scenario. A similar effect can be seen for PrCPs in the year 2050 for the progressive scenario. For the planning with the 6.5 kW or the 9.0 kW HPs, a factor of 2.2 or 3.0 is to be applied, respectively.

Apart from the power value assumptions for a complete MV grid, the power values per CP are needed when the number of CPs in a certain MV grid is available. The following Figure 13 shows the median values, as well as the arithmetic mean power values per CP for PrCPs and PuCPs for the two development scenarios until the year 2050. The figure shows that the charging power per CP decreases over the years for PrCPs. Since a nearly constant distribution of the charging powers is assumed over the years but with an increasing number of PrCPs, the demand factor decreases, and subsequently the charging power per PrCP. As for the PuCPs, their number also increases, but the distribution of charging powers suddenly jumps to higher charging power classes in the year 2030 to 2040 (see Table 1). Hence, a sudden increase in the charging power per CP is seen for the PuCPs from the year 2030 to 2040, which, in turn, then decreases—as expected—from the year 2040 to 2050.

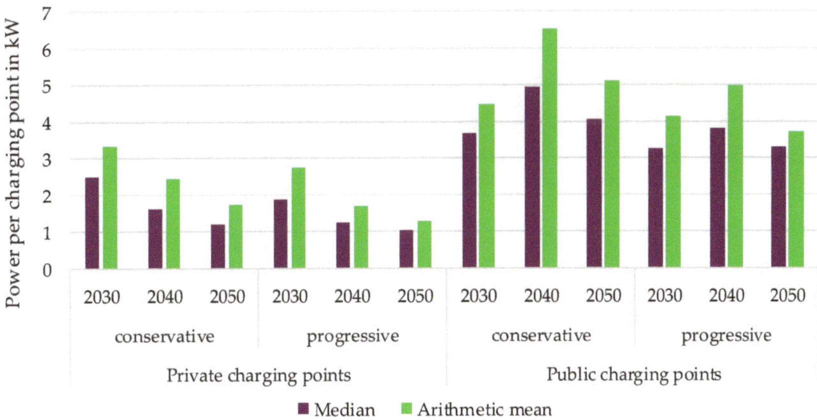

Figure 13. Effective charging power per charging point for private and public charging points for dimensioning the outgoing feeders for the different years and the two scenarios.

A summary of the power assumptions is summarized in Table 5 for the different load types. The given values represent the arithmetic mean for each given parameter over the considered time period and two scenarios. In addition to the aforementioned power values for PrCPs, PuCPs, and HPs, the power values for household loads are given. The power values for households do not differentiate between the different household buildings, such as one- or two-family houses and multi-family houses, since this level of granularity is no longer needed in the MV level. No power value assumptions for MV customer load stations are given since their power values differ greatly from one customer type to another. Hence, generally valid power assumptions cannot be deduced.

Table 5. Summary of the power value assumptions for the different load types and dimensioning views based on [59].

Load Type	Substation Transformer		Outgoing Feeder	
	kW per Building Connection	kW per Charging Point	kW per Building Connection	kW per Charging Point
Private charging points	[0.3 [1]; 2.4 [2]]	1.0	[0.8 [1]; 2.7 [2]]	[3.3 [1]; 1.3 [2]]
Public charging points	[0.05 [1]; 0.8 [2]]	0.3	[0.1 [1]; 0.9 [2]]	[4.5 [1]; 3.7 [2]]
Households	2.0	-	2.4	-
3.0 kW heat pumps	[0.1 [1]; 0.5 [2]]			
6.5 and 9.0 kW heat pumps	Factor: 2.2 and 3, respectively			

[1] Conservative scenario for the year 2030. [2] Progressive scenario for the year 2050.

4.2.2. Second Medium Voltage Planning Guideline

For 10 kV grids, the present standard cable cross-sections of c = 150 mm^2 or c = 185 mm^2 (Al) can be maintained. An upgrade with a further standard cable cross-section of c = 300 mm^2 is recommended.

After modeling the loads with the given power assumptions, the second step in the grid planning is performing reinforcement measures wherever necessary according to the available standard resources.

In the context of this contribution, two planning strategies are applied for planning overloaded cables. In addition, four cable cross-section areas starting from 150 mm^2 up to 300 mm^2 of the type NA2XS2Y are considered for the cable measures. The goal is to apply as many possible combinations of cable planning strategies and cable cross-sections as possible to deduce the most suitable standard cable cross-section.

The two planning strategies and the share of each cable cross-section to the total applied cable measures are shown in Figure 14. The first planning strategy (the outer ring in Figure 14) performs a cable replacement when the existing cable insulation material is considered old (such as NAKBA or NKBA) or the cable cross-section area is smaller than 120 mm^2. This is under the assumption that these aforementioned cable section(s) has a low economic residual value and can be replaced. Otherwise, a cable reinforcement is performed, in which a second cable is laid in parallel to the existing overloaded cable in the same cable trench.

As for the second planning strategy (inner ring in Figure 14), the overloaded cable is directly replaced, disregarding the existing cable insulation type or cross-section area. This strategy simulates the grid planning not only in terms of grid reinforcement but also in terms of grid expansion and strengthening.

Figure 14 shows that the most used cable cross-section for the first planning technique is 150 mm^2. However, the so far applied standard cable cross-section of 150 mm^2 (or 185 mm^2) is not sufficient for all the cable measures, and laying larger cable cross-sections is required. In addition, by applying the second cable technique, the share of the cable measures with the so far standard cable cross-sections (150 mm^2 and 185 mm^2) decrease, and larger cable cross-sections become necessary.

Therefore, with the planning strategy of cable reinforcement for the existing cable, the current standard cable cross-sections (be it 150 mm^2 or 185 mm^2) can still be applied. Furthermore, it is recommended to extend the standard cable cross-sections with a larger cross-section, namely 300 mm^2.

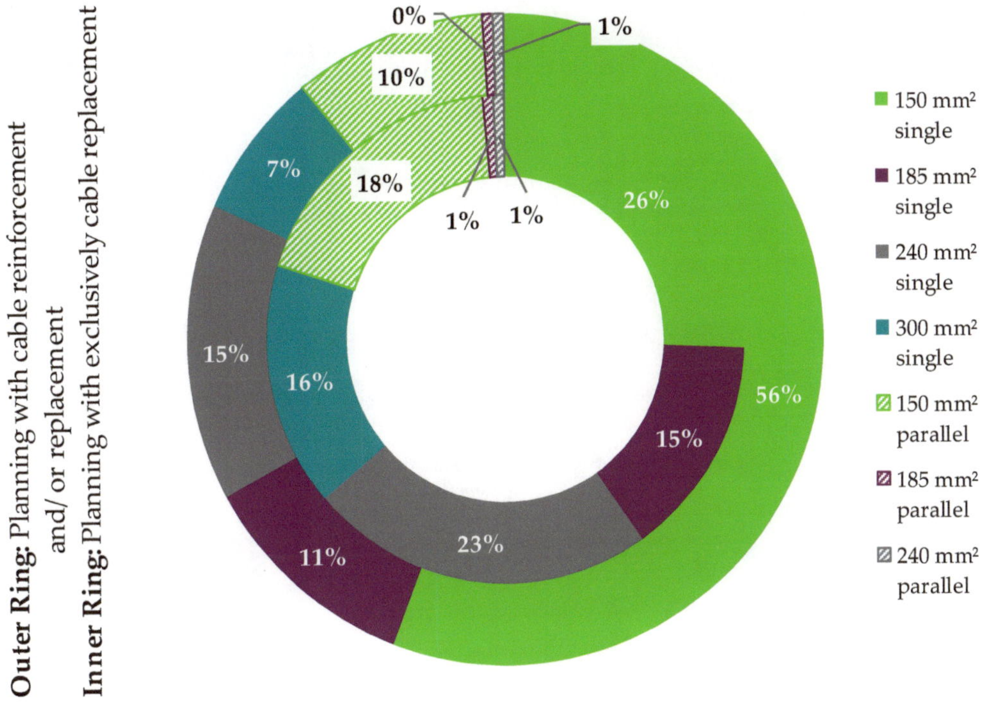

Figure 14. Share of cable measures for different cable cross-sections in relation to total cable measures for two planning techniques in the 10 kV grids.

4.2.3. Third Medium Voltage Planning Guideline

The dimensioning of substation transformers is to follow the load assumptions from the first planning guideline, as a standard transformer size cannot be deduced due to the heterogeneous load development per substation grid area.

The next step for grid planning is to reinforce the substation transformer(s) if necessary. Firstly, the total load in the MV level is to be determined. The following Figure 15 shows the load development for nine MV grids. It must be pointed out that the two MV grids, G06 and G07, are excluded from this analysis, as both of them represent a single MV ring and not a complete grid area. Figure 15 demonstrates that the load development varies significantly from one grid to another depending on the specific grid parameters. For instance, the load development for G01 exceeds 200% at the progressive scenario for the year 2050 whereas the load development for G05 does not exceed 50% for the same year. This reflects the heterogeneous nature of MV grids since they can vary greatly in terms of connected loads and grid parameters.

Furthermore, Figure 16 displays the loading of the substation transformers for the considered nine MV grids for the year 2050. Firstly, it can be seen that the installed transformers have different sizes depending on the grid area. The substation installed capacity ranges from 2 × 12.5 MVA to 3 × 40 MVA, thus making it unfeasible to deduce a current standard transformer size.

Therefore, it is recommended to analyze each MV grid individually. Using the power value assumptions supplied in the first PG, the expected load development can be calculated. Afterward, the required transformer size can be determined.

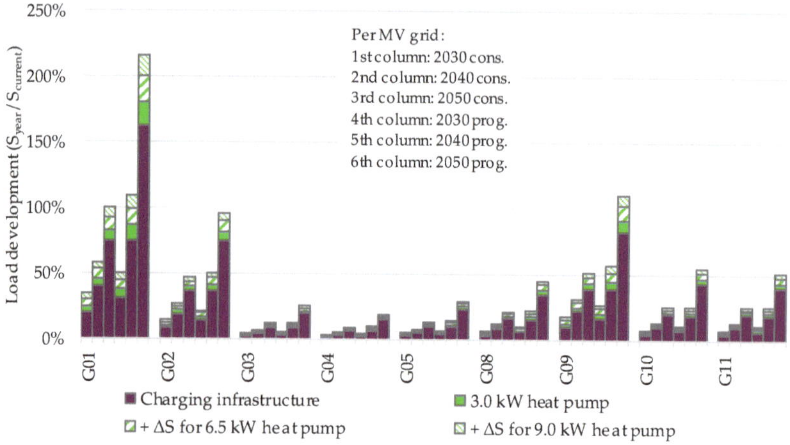

Figure 15. Load development for the two development scenarios and the years 2030, 2040, and 2050 in relation to the current load power based on [59].

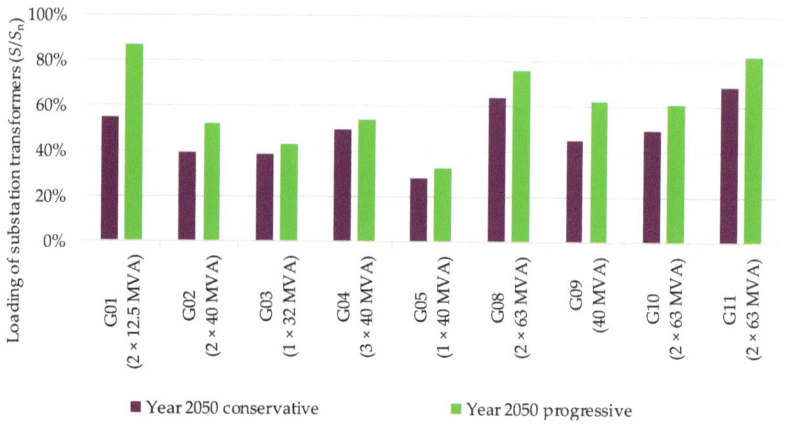

Figure 16. Loading of substation transformers for 9 medium voltage grids in the year 2050 for the two development scenarios.

4.2.4. Fourth Medium Voltage Planning Guideline

When the measurement, information, and communication technology in the medium voltage and the low voltage grids must be fully constructed, the conventional grid expansion is, in most cases, less expensive than a load management system and is recommended. If the measurement, information, and communication technology is available and can be used by the load management system, then the load management system becomes significantly cost-efficient and is recommended.

Since the electromobility and the HP loads are more flexible in terms of turn-on and shut-off times with a minimum impact on consumer comfort, LM has a big potential in the upcoming years to successfully integrate these new loads into the grids. As already shown in Figure 9, LM can either reduce or postpone the required cable measures over the years. However, contrary to the expectations, the usage of LM is not economical in all the investigated cases. As mentioned in Section 3.2.2, the application of LM requires a MICT to monitor the grid state and regulate the loads accordingly. Since MICT is still not integrated into most of the MV grids, the total cost of planning with MICT is strongly dependent on the costs required to implement MICT into the grids.

The costs for the six LM layouts are calculated for the three LM variants and demonstrated in Figure 17. The figure shows the costs for the conventional planning strategy along with the costs for the LM with the six layouts and the three LM variants. It can be seen that for LM-V1 (0) and LM-V2 (0), the total cost for planning is significantly lower than the costs for conventional planning. Whereas, when the costs for MICT need to be considered in the cases of LM-V1 (MV+LV), LM-V2 (MV+LV), and LM-V3 (MV+LV), the conventional planning becomes more economical than the LM.

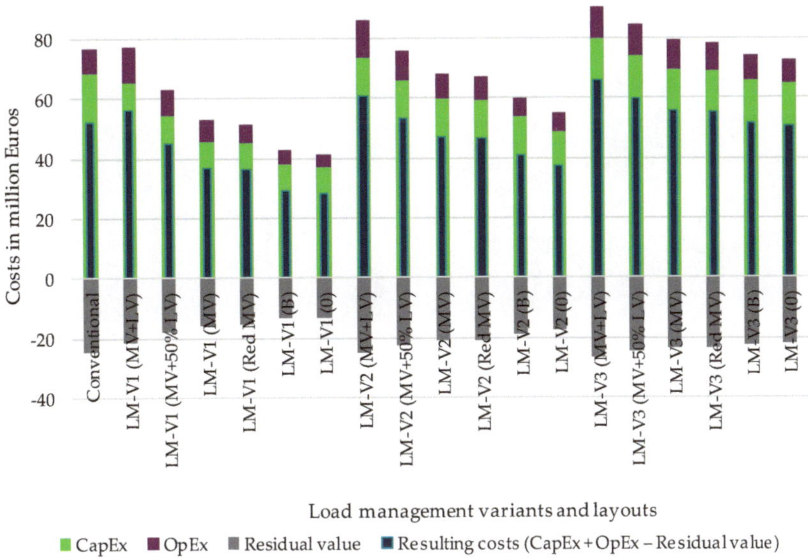

Figure 17. Consolidated costs over eleven medium voltage grids for conventional planning strategy and three load management variants with six layouts based on [59].

For analyzing the results per planning variant, Figure 18 shows the saving potential of the three LM variants with the six LM layouts in comparison to the conventional planning represented by the 0% line. Since each of the LM layouts assumes a certain degree of spread of MICT, the figure demonstrates the effect of these various degrees in terms of savings potential. It shows that when the MICT already exists in the grids and that the LM can use this existing infrastructure with no additional costs (represented by the layout "(0)"), then the planning strategy with LM becomes cost-efficient in comparison to conventional planning. However, for the MICT layout (MV+LV), the potential savings drops rapidly so that, in most cases, it becomes economical to apply a conventional planning strategy.

Moreover, Figure 18 demonstrates the effectiveness of the different LM regulation principles. It becomes clear that the regulation of PrCPs is more beneficial than the regulation of PuCPs. In addition, the HPs should be regulated if it is possible, as this can significantly reduce the necessary grid planning measures.

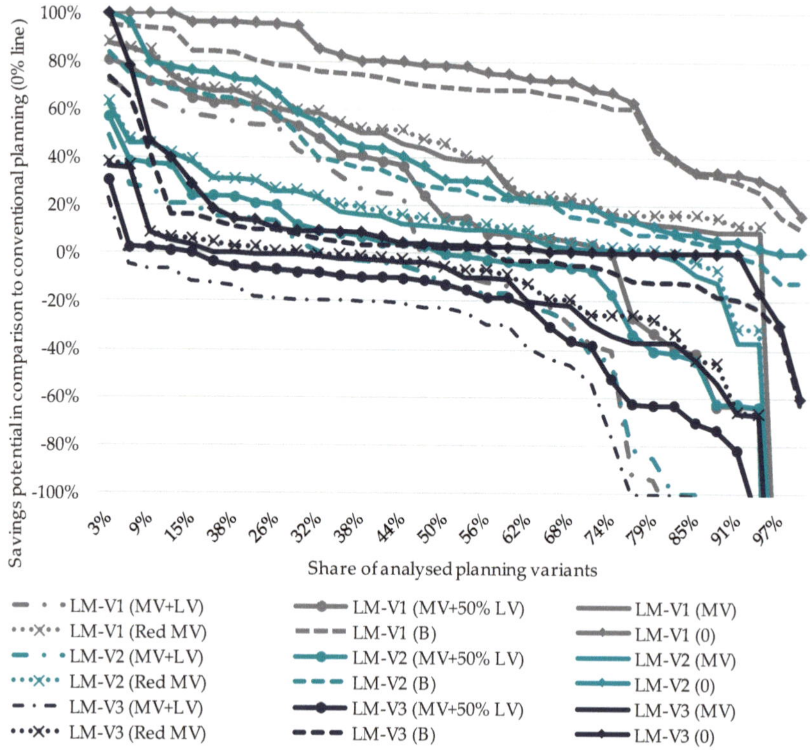

Figure 18. The savings potential for the three load management variants with the six layouts in comparison to conventional planning over all analyzed variants based on [59].

4.2.5. Fifth Medium Voltage Planning Guideline

Using conventional planning measures for 10 kV grids, a cable reinforcement of around 20% in suburban grids, 10% in semi-dense urban grids, and less than 10% in downtown grids is expected in relation to the total cable length of the grid.

With the rapid development of electromobility and HPs, the DSOs face a growing fear that the complete grid area will need to be reinforced. As the reinforcement of the complete grid area is very costly and exhausting, it is important to determine the required grid measures in advance so that the available capital is properly allocated.

The following Figure 19 shows the required share of cable measures in relation to the total grid cable length for the two development scenarios and the three HP models. The figure shows that the share of cable measures in suburban regions is around 20% which is represented by the grids G01 and G02. This share decreases for semi-dense urban grids down to around 10%, represented by the grids G09, G10, and G11. As for downtown grids, the expected share of cable measures drops to less than 10%, as shown in the grids G03, G05, and G08. The decline of expected cable measures from the outskirts going toward the city center relies on the fact that the integration of electromobility and HPs occurs mainly in suburban areas. Furthermore, it is important to consider the current cable loading, as the share of cable measures depends strongly on it. It is to be noted that the grids G04, G06, and G07 are excluded from this PG. The grid G04 is a 20 kV grid, for which the above-mentioned share of cable measures does not apply. The two grids, G06 and G07, are MV rings that do not represent a complete grid area and therefore are also excluded from this PG.

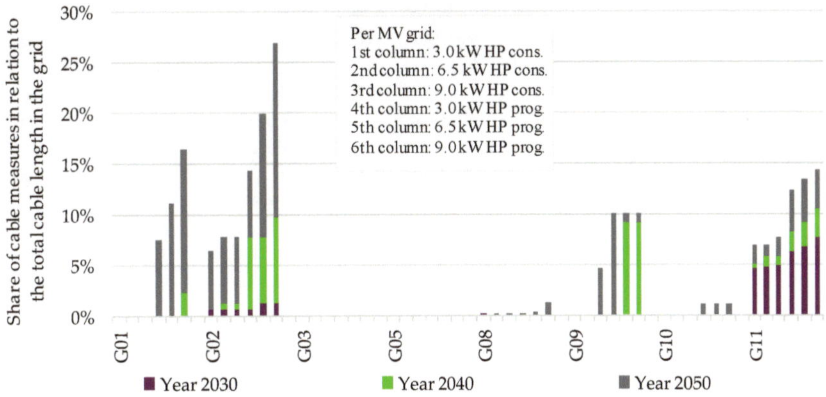

Figure 19. Share of the length of required cable measures in comparison to the total cable length in the grid for the two development scenarios over three years until 2050 based on [59].

4.2.6. Sixth Medium Voltage Planning Guideline

Grid reinforcements are hardly expected in 20 kV grids, as they are significantly steady for the integration of new loads in comparison to 10 kV grids.

As seen in the previous analysis and what is concretely shown in Figure 20, is that no cable overloads happen in the 20 kV grids, while there is a significant share of overloaded cables in the 10 kV grids.

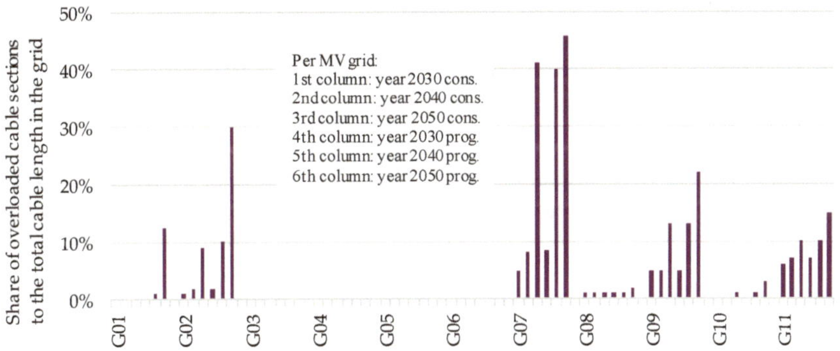

Figure 20. Share of the length of overloaded cable sections to the total cable length in the grid for the two development scenarios over three years until 2050 based on [59].

This becomes especially clear in the grids G08, G10, and G11, as these grids have both voltage levels (10 kV and 20 kV) in the same grid area. The three winding transformers of these three grids transform the voltage from 110 kV to 20 kV and from 110 kV to 10 kV. Thus, two separate grids with these two voltage levels (20 kV and 10 kV) supply the same city area. As the spread of electromobility and HPs is strongly dependent on the socio-economical structure and the building structure of the city area, it can be determined that the development of CPs and HPs in these three grids is identical. Furthermore, the current cable loading for the two voltage levels in these grids is nearly equal. By integrating the CPs and HPs in these grids, it is seen that overloads occur in the 10 kV in contrast to the 20 kV, where no overloads occur.

Obviously, while supplying the same load, the current decreases with increasing the voltage level. Thus, 20 kV grids can easily take up new loads into the grid without risking an overload.

4.2.7. Seventh Medium Voltage Planning Guideline

The allowed voltage range in the medium voltage grids is not fully utilized in most cases. Therefore, it is recommended to check the voltage range division between medium and low voltage grids in the grid planning and to modify it, if necessary.

Considering the voltage range division in Section 3.1.2 and the recommendations for cross-voltage level planning, the actual required voltage level in the MV level is investigated. Figure 21 displays the actual required voltage in the analyzed MV grids for the two scenarios, the three HP models, and the three planning years. The figure shows the maximum voltage drop calculated as the difference between V at the slack node and the minimum node voltage in the grid. On the other hand, the maximum voltage saving is shown, which represents the voltage range that can be shifted for the LV grid planning.

Figure 21. Maximum voltage drop and maximum voltage saving consolidated over eleven MV grids for the two development scenarios over three years until 2050 based on [59].

Figure 21 shows that the complete voltage range (4% V/V_n) is not utilized in any of the cases. In the year 2030, approximately half of the voltage range is needed. In the years afterward, the required voltage in the MV level increases gradually. This proves that the initially set voltage range between MV and LV grids is actually not needed at the MV level. Depending on the planning criteria in the LV level, a new voltage division between the two voltage levels is beneficial to avoid over-dimensioning the equipment in expectation of a voltage level violation that does not occur.

4.3. *Decision Path for Medium Voltage Grid Planning*

To summarize the aforementioned PGs, a flowchart for planning MV grids is shown in Figure 22. The process starts with checking the loading of the equipment before it checks the node voltages since voltage violations are not to be expected. If the overloaded element is simply a bottleneck meaning a short cable section and not grid-wide, conventional measures are recommended. However, if the overloaded cables are grid-wide or the substation transformer is overloaded, a further check for MICT should be performed. The fourth PG proves that in case the MICT is available, the LM is cheaper than conventional planning. However, if the MICT is not already available, conventional measures are directly recommended. When the overloads are remedied, the node voltages are checked. In case a voltage violation occurs, the nominal voltage value at the substation busbar is adjusted using voltage regulation at the substation transformer (abbreviated as VRS). If the voltage violation persists, conventional measures are recommended until the violation is remedied.

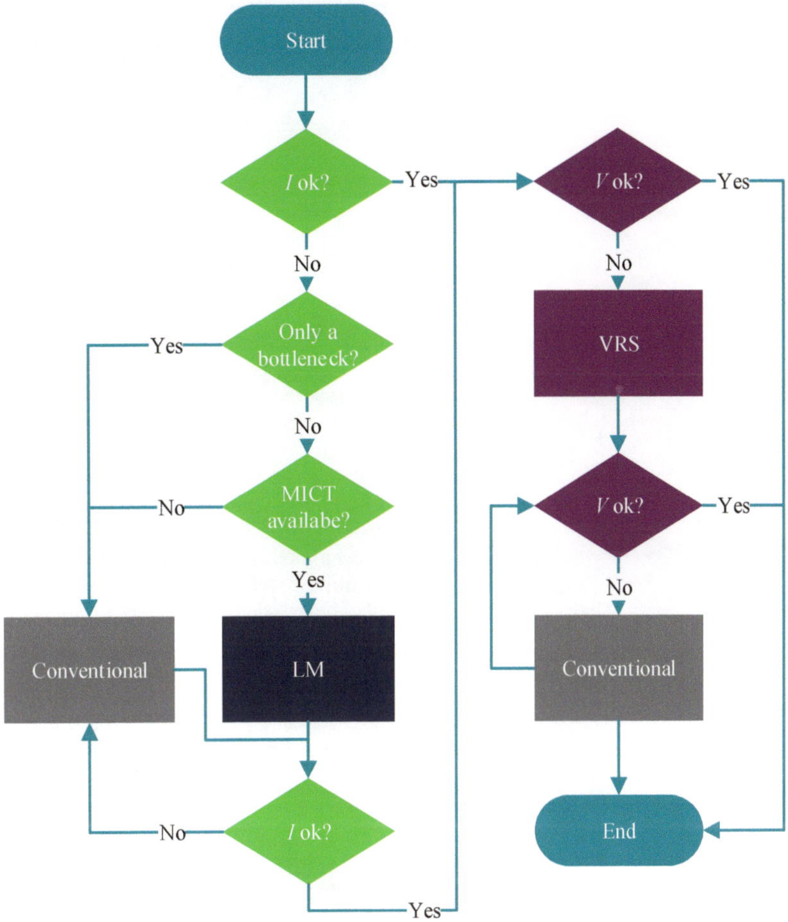

Figure 22. Flowchart for planning medium voltage grids.

4.4. *Across Voltage Level Planning Guidelines*

In extension to the aforementioned MV PGs, across voltage PGs are deduced with the help of results for the LV and high voltage levels. The across voltage PGs take into consideration that synergy effects between the voltage levels reduce the collectively required grid reinforcements irrespective of the corresponding voltage level [59].

4.4.1. First across Voltage Planning Guideline

Across voltage planning of the high, medium, and low voltage levels should be performed.

As shown in the seventh MV PG in Section 4.2.7, the originally set voltage range for the MV level is not fully utilized and a re-investigation of the voltage range is recommended. Furthermore, since the penetration of new loads primarily takes place at the LV level, the proposed LM and RPM for the MV level operate by accessing these loads at the LV level. By implementing such innovative technologies to benefit the MV level, a beneficial secondary effect can be aimed for, which is the reduction in the required reinforcement measures not only at the LV level but also at the high voltage level. Additional synergy effects can be aimed for while performing across voltage planning.

4.4.2. Second across Voltage Planning Guideline

Contrary to the voltage limit violations, the equipment overload is the main reason for grid reinforcement measures.

The results shown in this contribution demonstrate that the integration of the new loads leads to more severe equipment overloads rather than voltage limit violations. This is displayed for the MV level in Figure 21 and for the LV level in [63]. Thus, it is recommended that DSOs work on increasing the ampacity of their existing equipment. The newly installed equipment should have a higher nominal power than the so far used equipment to accommodate the increasing load demand. Moreover, the investments in voltage regulation measures (such as voltage controllers) can be reduced.

4.4.3. Third across Voltage Planning Guideline

Innovative planning strategies represent an economical solution in specific grids. For the remaining grids, conventional planning strategies are recommended.

Referring to the fourth MV PG in Section 4.2.4 and the previous across voltage level PG, it becomes clear that the challenges facing the MV grids are load-oriented. The three proposed LM variants have proven to be effective in remedying the expected equipment overloads, however, they cannot remedy all equipment overloads so cable measures are still needed. In addition, the economic efficiency of these LM variants depends strongly on the extent of the needed MICT infrastructure. These two factors combined conclude that innovative technologies such as LM are effective both technically and economically under specific conditions and for certain MV grids. For the remaining collectivity of MV grids, the conventional planning strategy has proven to still be an efficient grid planning strategy.

4.5. Results for the Application of an Alternative Assessment Model

In addition to the performed cost calculation to determine the attractiveness of the different planning variants, an alternative model is developed to assess the planning variants. The model considers not only the costs of the planning measures but also four further technical and non-technical criteria. The four further criteria and their method of calculation are as follows:

- Losses of the power grid: This criterion calculates the yearly energy losses in the power grid, hence representing energy costs that are paid by the DSO;
- Frequency of faults: This criterion calculates the frequency of equipment faults in the grid. Thus, it roughly depicts the amount of maintenance work that needs to be performed by the DSO;
- Stability of voltage: This criterion calculates the difference between the slack node voltage and the minimum node voltage in the grid. It shows how much voltage drop occurs in the grid and in turn how stable is the voltage level against unexpected voltage drops;
- Effort of construction: Since the performance of construction works constitutes an inconvenience for the residents and the users of the grid, this criterion is based on the required cable construction works per planning variant [62].

Figure 23 shows the consolidated ranking for the six planning variants over all investigated MV grids for each of the aforementioned assessment criteria. The LM results shown here are for the layout (MV). It becomes clear that by assessing different grid planning criteria, the planning variants can have different attractiveness. For instance, according to the "Cost of measures", the planning variant ES (4) has the worst ranking followed by ES (2 h). However, when the "Losses of the power grid" or "Frequency of faults" are considered, these planning variants increase in ranking.

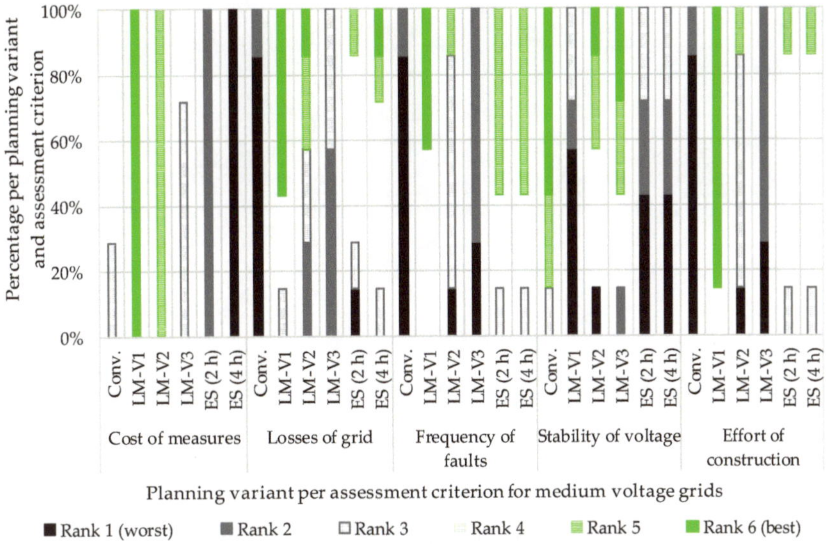

Figure 23. Ranking of the six planning variants for each assessment criterion.

Moreover, five weighting alternatives are developed to investigate the combination of the five aforementioned assessment criteria. The proposed weighting alternatives are: equal weighting, economically oriented, grid resilience, technically oriented and conservation of resources. These represent the different planning goals that can be aimed for by the DSO so as not to focus on only one of the assessment criteria but to consider the five assessment criteria collectively. According to the importance of each of the assessment criteria, the DSO can give a criterion a different weighting in the total score. The assumed individual specific weights for the weighting alternatives are listed in Table 6 [62].

Table 6. An overview of the weighting alternatives and their specific weights based on [59].

Assessment Criterion	Equal Weighting	Economically Oriented	Grid Resilience	Technically Oriented	Conservation of Resources
Cost of measures	20%	60%	10%	5%	10%
Losses of the power grid	20%	10%	10%	30%	35%
Frequency of breakdown	20%	10%	35%	30%	10%
Stability of voltage	20%	10%	35%	30%	10%
Effort of construction	20%	10%	10%	5%	35%

Figure 24 shows the results for the five weighting alternatives for the investigated MV grids. It is to be mentioned that in continuance with the previously shown results in Figure 23, the LM results here are for the layout (MV). The planning variants, here again, differ in attractiveness according to the applied weighting alternative. For instance, if a DSO is economically oriented in the context of the grid planning, then they would tend to the variant LM-V1. However, if the DSO considers the grid resilience as a decisive factor for grid planning, they may move toward applying LM-V2.

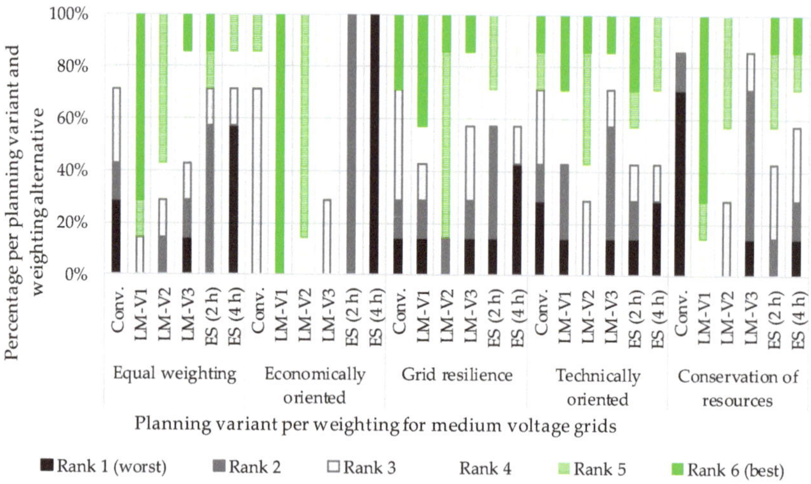

Figure 24. Ranking of the six planning variants according to the weighting alternatives.

5. Conclusions

This contribution provides a detailed explanation for the consideration of electromobility and HPs in the planning of urban MV grids. It started with an overview of their expected development scenarios and their regionalization into the specific grids. Subsequently, the contribution discussed their expected nominal power and their calculated actual power demand. In addition, the methodology of the performed grid planning was briefly stated. The methodology presented the identified grid limits and the applied planning strategies. Based on the planning, a technical and economic assessment of the resulting planning variants was performed. As a result, seven PGs were deduced, which recommend measures to be performed by the DSO. The complete planning process was finally summarized in a presented decision path.

The PGs represent a handy toolbox for the DSOs while planning their grids. They start with the power assumptions for the different loads, continue with standard dimensions for the equipment and show the feasibility of LM. The last three PGs give general expectations in regard to planning urban MV grids.

The results show that there is no way around conventional grid reinforcement such as cable reinforcement. The application of innovative technologies cannot solely resolve all the anticipated grid limit violations. A further main factor in successfully integrating the new loads into the grid is the construction of MICT. In most cases in which MICT is available in the grids, the deployment of LM in combination with cable measures is economical more than solely laying cables.

By proposing additional assessment criteria to the cost of equipment, the planning variants vary strongly in the final assessment. As the different assessment criteria focus on different aspects of grid planning, it becomes difficult to deduce a single optimum planning strategy. The consideration of five varying weighting alternatives widens the range of decisions even more. The combination of the five assessment criteria with the five proposed weighting alternatives offers the DSO a wide range of options for grid planning.

A study of the integration of photovoltaics in urban grids is recommended. As the grid reinforcement can be driven not only by the increasing load but also by the increasing decentralized generation, the opposing effect of increasing generation needs to be investigated in further work.

Author Contributions: Conceptualization, S.A., P.W., J.M. and B.G.; methodology, S.A. and P.W.; validation, M.Z. and A.S.; formal analysis, J.M.; resources, M.Z.; data curation, P.W. and S.A.; writing—original draft preparation, S.A.; writing—review and editing, P.W. and J.M.; supervision, M.Z.; project administration, M.Z.; funding acquisition, M.Z. All authors have read and agreed to the published version of the manuscript.

Funding: This research is a result of the project "PuBStadt" which was funded by the German Federal Ministry for Economic Affairs and Climate Action following the decision of the "German Bundestag, grant number 0350038A".

Supported by:

Federal Ministry for Economic Affairs and Climate Action

on the basis of a decision by the German Bundestag

Institutional Review Board Statement: Not applicable.

Informed Consent Statement: Not applicable.

Data Availability Statement: Not applicable.

Acknowledgments: The used data were provided by the six German DSOs: Stromnetz Berlin GmbH, Stuttgart Netze GmbH, Rheinische NETZGesellschaft mbH, SachsenNetze GmbH, Enercity Netz GmbH and Erlangener Stadtwerke AG.

Conflicts of Interest: The authors declare no conflict of interest.

Abbreviations

CapEx	**Capital Expenditures**
cons	Conservative
CP	Charging point
DAS	Decentralized automation system
DSO	Distribution System Operator
ES	Energy storage
HP	Heat pump
LM	Load management
LV	Low voltage
MICT	Measuring, information, and communication technology
MV	Medium voltage
OpEx	Operational Expenditures
PG	Planning guideline
PrCP	Private charging point
prog	Progressive
PuCP	Public charging point
RPM	Reactive power management

Appendix A

Table A1. Grid structure parameters of the investigated grids (values of the charging points and heat pumps for each grid: the first row is the conservative scenario, the second row is the progressive scenario, where the first column is the year 2030, the second column is the year 2040 and the third column is the year 2050).

Grid Voltage Level	Installed Transformer Power (MVA)	Total Cable Length (km)	No. of Distribution/Customer Stations	No. of Building Connections	No. of Metering Points	No. of Feeders	No. of Charging Points			No. of Heat Pumps		
G01 10 kV	2 × 12.5	40.9	44/11	3095	8797	6	1194	2175	3834	144	178	253
							1913	4072	8433	182	344	544
G02 10 kV	2 × 40	40.0	37/18	4041	15,982	14	1637	3343	6028	186	229	323
							2726	6319	12,014	235	444	708
G03 10 kV	32	16.6	41/3	484	4139	13	195	399	773	20	24	40
							324	758	1673	25	54	83
G04 20 kV	3 × 40	70.9	39/24	1815	6728	9	737	1512	2905	84	103	147
							1226	2849	6280	107	201	320
G05 10 kV	40	16.9	22/16	504	4221	16	203	420	803	24	29	41
							340	791	1745	29	55	87
G06 10 kV	-	4.6	7/-	182	2142	2	63	119	210	7	8	14
							98	231	483	14	21	35
G07 10 kV	-	17.0	9/-	2034	2601	2	657	1287	2385	90	117	162
							1125	2529	5409	117	225	360
G08 10/20 kV	2 × 63	183.4	170/62	5312	37,802	32	2156	4415	8517	240	298	425
							3623	8341	17,557	305	588	935
G09 10 kV	40	44.9	59/3	2663	14,400	15	1076	2070	3500	122	147	212
							1804	3707	7461	154	295	471
G10 10/20 kV	2 × 63	134.8	131/46	4619	48,424	26	1869	3854	7398	213	265	374
							3135	7229	15,961	275	517	816
G11 10/20 kV	2 × 63	174.9	171/66	6546	54,478	23	2686	5522	10,646	297	368	526
							4482	10,396	20,804	375	721	1151

Table 2. Classification of the investigated grids in terms of the building and urban structure (X = high share, O = low share, - = negligible or not present).

Structure	G01	G02	G03	G04	G05	G06	G07	G08	G09	G10	G11
One/Two-family houses	X	X	-	X	-	-	X	-	-	-	-
Multi-family houses	O	X	X	O	X	X	-	X	X	X	X
Industrial	-	O	O	X	O	-	-	X	O	O	O
Suburban	X	X		X			X				
Semi-dense urban								X	X	X	X
Downtown			X		X	X					

Table 3. Assumed costs for the applied medium voltage equipment.

Equipment	Parameter	Value	Unit
3-phase NA2XS2Y Cable 10/20 kV	Service life	45	[a]
	Operational costs	2.5	[% CapEx/a]
	Price increase	0.5	[%/a]
150 mm^2 single	Cable cost + installation	225,000	[Euro/km]
150 mm^2 parallel	Cable cost + installation	+50,000	[Euro/km]
185 mm^2 single	Cable cost + installation	237,500	[Euro/km]
185 mm^2 parallel	Cable cost + installation	+65,000	[Euro/km]
240 mm^2 single	Cable cost + installation	250,000	[Euro/km]
240 mm^2 parallel	Cable cost + installation	+80,000	[Euro/km]
300 mm^2 single	Cable cost + installation	275,000	[Euro/km]
300 mm^2 parallel	Cable cost + installation	+95,000	[Euro/km]
Energy storage	Service life	16	[a]
	Operational costs	2.5	[% CapEx/a]
	Basic cost	46,000	[Euro/unit]
	Power cost for 2 h capacity	550	[Euro/kW]
Decentralized automation	Service life	15	[a]
	Operational costs	2.5	[% CapEx/a]
	Basic cost	15,000	[Euro/unit]
	MV-sensor	8000	[Euro/unit]
	LV-sensor	3500	[Euro/unit]
HV/MV substation components	Service life	40	[a]
	Operational costs	2.5	[% CapEx/a]
	New construction	1,500,000	[Euro/unit]
	GIS switchgear	70,000	[Euro/unit]
	AIS switchgear	60,000	[Euro/unit]
	Disconnector	4500	[Euro/unit]
Transformer			
31.5 MVA	Transformer cost + installation	450,000	[Euro/unit]
40.0 MVA	Transformer cost + installation	500,000	[Euro/unit]
63.5 MVA	Transformer cost + installation	650,000	[Euro/unit]

References

1. Knobloch, F.; Hanssen, S.V.; Lam, A.; Pollitt, H.; Salas, P.; Chewpreecha, U.; Huijbregts, M.A.J.; Mercure, J.-F. Net emission reductions from electric cars and heat pumps in 59 world regions over time. *Nat. Sustain.* **2020**, *3*, 437–447. [CrossRef]
2. Ruhnau, O.; Bannik, S.; Otten, S.; Praktiknjo, A.; Robinius, M. Direct or indirect electrification? A review of heat generation and road transport decarbonisation scenarios for Germany 2050. *Energy* **2018**, *166*, 989–999. [CrossRef]
3. Nadel, S. Electrification in the Transportation, Buildings, and Industrial Sectors: A Review of Opportunities, Barriers, and Policies. *Curr. Sustain. Energy Rep.* **2019**, *6*, 158–168. [CrossRef]
4. Khator, S.; Leung, L. Power distribution planning: A review of models and issues. *IEEE Trans. Power Syst.* **1997**, *12*, 1151–1159. [CrossRef]
5. Seifi, H.; Sepasian, M.S. *Electric Power System Planning*; Springer: Berlin/Heidelberg, Germany, 2011. [CrossRef]
6. Hadjsaïd, N.; Sabonnadière, J.-C. (Eds.) *Electrical Distribution Networks*; John Wiley & Sons, Inc.: Hoboken, NJ, USA, 2013. [CrossRef]

7. Schlabbach, J.; Rofalski, K.-H. *Power System Engineering: Planning, Design and Operation of Power Systems and Equipment*, 2nd ed.; Wiley-VCH: Weinheim, Germany, 2014.
8. Thurner, L.; Kassel University Press GmbH. *Structural Optimizations in Strategic Medium Voltage Power System Planning*; University of Kassel: Kassel, Germany, 2018. [CrossRef]
9. Yong, J.Y.; Ramachandaramurthy, V.K.; Tan, K.M.; Mithulananthan, N. A review on the state-of-the-art technologies of electric vehicle, its impacts and prospects. *Renew. Sustain. Energy Rev.* **2015**, *49*, 365–385. [CrossRef]
10. Foley, A.M.; Winning, I.J.; Gallachóir, B.P.Ó. State-of-the-art in electric vehicle charging infrastructure. In Proceedings of the 2010 IEEE Vehicle Power and Propulsion Conference, Lille, France, 1–3 September 2010; IEEE: Piscataway, NJ, USA, 2014; pp. 1–6. [CrossRef]
11. Arif, S.; Lie, T.; Seet, B.; Ayyadi, S.; Jensen, K. Review of Electric Vehicle Technologies, Charging Methods, Standards and Optimization Techniques. *Electronics* **2021**, *10*, 1910. [CrossRef]
12. Ahmad, A.; Alam, M.S.; Chabaan, R. A Comprehensive Review of Wireless Charging Technologies for Electric Vehicles. *IEEE Trans. Transp. Electrification* **2017**, *4*, 38–63. [CrossRef]
13. Dulău, L.I.; Bică, D. Effects of Electric Vehicles on Power Networks. *Procedia Manuf.* **2020**, *46*, 370–377. [CrossRef]
14. Putrus, G.A.; Suwanapingkarl, P.; Johnston, D.; Bentley, E.C.; Narayana, M. Impact of electric vehicles on power distribution networks. In Proceedings of the 2009 IEEE Vehicle Power and Propulsion Conference, Dearborn, MI, USA, 7–10 September 2009; pp. 827–831. [CrossRef]
15. Konstantinidis, G.; Karapidakis, E.; Paspatis, A. Mitigating the Impact of an Official PEV Charger Deployment Plan on an Urban Grid. *Energies* **2022**, *15*, 1321. [CrossRef]
16. Garcia-Valle, R.; Peças Lopes, J.A. (Eds.) *Electric Vehicle Integration into Modern Power Networks*; Springer: New York, NY, USA, 2013. [CrossRef]
17. Kleiwegt, E.; Lukszo, Z. Grid Impact Analysis of Electric Mobility on a Local Electricity Grid. In Proceedings of the 2012 9th IEEE International Conference on Networking, Sensing and Control, Beijing, China, 11–14 April 2012; pp. 316–321. [CrossRef]
18. Love, J.; Smith, A.Z.; Watson, S.; Oikonomou, E.; Summerfield, A.; Gleeson, C.; Biddulph, P.; Chiu, L.F.; Wingfield, J.; Martin, C.; et al. The addition of heat pump electricity load profiles to GB electricity demand: Evidence from a heat pump field trial. *Appl. Energy* **2017**, *204*, 332–342. [CrossRef]
19. Arnold, M.P. *Integration of Residential Distributed Generators and Heat Pumps into the Low Voltage Grid from a Voltage Level Perspective*; Technischen Universitat Dortmund: Dortmund, Germany, 2016; Available online: https://core.ac.uk/download/pdf/46917314.pdf (accessed on 5 July 2022).
20. De Vos, L.; Leemput, N.; Refa, N.; Bernards, R.; Fidder, H.; de Rijke, F. Optimal Integration of Electric Vehicles, PV, Heat Pumps in Existing Distribution Grids in the Netherlands. In Proceedings of the CIRED 2019 Conference, Madrid, Spain, 3–6 June 2019.
21. Thormann, B.; Kienberger, T. Evaluation of Grid Capacities for Integrating Future E-Mobility and Heat Pumps into Low-Voltage Grids. *Energies* **2020**, *13*, 5083. [CrossRef]
22. Oliyide, R.O.; Cipcigan, L.M. The impacts of electric vehicles and heat pumps load profiles on low voltage distribution networks in Great Britain by 2050. *Int. Multidiscip. Res. J.* **2021**, *11*, 30–45. [CrossRef]
23. Blonsky, M.; Nagarajan, A.; Ghosh, S.; McKenna, K.; Veda, S.; Kroposki, B. Potential Impacts of Transportation and Building Electrification on the Grid: A Review of Electrification Projections and Their Effects on Grid Infrastructure, Operation, and Planning. *Curr. Sustain. Energy Rep.* **2019**, *6*, 169–176. [CrossRef]
24. Fischer, D.; Madani, H. On heat pumps in smart grids: A review. *Renew. Sustain. Energy Rev.* **2017**, *70*, 342–357. [CrossRef]
25. Nykamp, S.; Molderink, A.; Bakker, V.; Toersche, H.; Hurink, J.; Smit, G. Integration of heat pumps in distribution grids: Economic motivation for grid control. In Proceedings of the 2012 3rd IEEE PES Innovative Smart Grid Technologies Europe (ISGT Europe), Berlin, Germany, 14–17 October 2012; pp. 1–8. [CrossRef]
26. Kreuder, L.; Spataru, C. Assessing demand response with heat pumps for efficient grid operation in smart grids. *Sustain. Cities Soc.* **2015**, *19*, 136–143. [CrossRef]
27. Mahmood, A.; Javaid, N.; Khan, M.A.; Razzaq, S. An overview of load management techniques in smart grid. *Int. J. Energy Res.* **2015**, *39*, 1437–1450. [CrossRef]
28. Probst, A.; Siegel, M.; Braun, M.; Tenbohlen, S. Impacts of electric mobility on distribution grids and possible solution through load management. In Proceedings of the CIRED 2011—21st International Conference on Electricity Distribution, Frankfurt, Germany, 6–9 June 2011; p. 126.
29. Masoum, A.S.; Deilami, S.; Moses, P.S.; Masoum, M.A.S.; Abu-Siada, A. Smart load management of plug-in electric vehicles in distribution and residential networks with charging stations for peak shaving and loss minimisation considering voltage regulation. *IET Gen. Transmiss. Distrib.* **2011**, *5*, 877. [CrossRef]
30. Almutairi, A.; Alrumayh, O. Optimal Charging Infrastructure Portfolio for Minimizing Grid Impact of Plug-In Electric Vehicles. *IEEE Trans. Ind. Inf.* **2022**, *18*, 5712–5721. [CrossRef]
31. Khan, R.H.; Stüdli, S.; Khan, J.Y. A network controlled load management scheme for domestic charging of electric vehicles. In Proceedings of the 2013 Australasian Universities Power Engineering Conference (AUPEC), Hobart, TAS, Australia, 29 September–3 October 2013; pp. 1–6. [CrossRef]

32. Sawant, M.; Moyalan, J.; Bhagyashree, U.; Sheikh, A.; Wagh, S.; Singh, N. Power Grid Load Management using Electric Vehicles: A Distributed Resource Allocation Approach. In Proceedings of the 2019 27th Mediterranean Conference on Control and Automation (MED), Akko, Israel, 1–4 July 2019; pp. 428–433. [CrossRef]
33. Mahmud, K.; Hossain, M.J.; Ravishankar, J. Peak-Load Management in Commercial Systems with Electric Vehicles. *IEEE Syst. J.* **2018**, *13*, 1872–1882. [CrossRef]
34. Arteaga, J.; Zareipour, H.; Thangadurai, V. Overview of Lithium-Ion Grid-Scale Energy Storage Systems. *Curr. Sustain. Energy Rep.* **2017**, *4*, 197–208. [CrossRef]
35. Rackliffe, G. Guidelines for planning distributed generation systems. In Proceedings of the 2000 Power Engineering Society Summer Meeting (Cat. No.00CH37134), Seattle, WA, USA, 16–20 July 2000; Volume 3, pp. 1666–1667. [CrossRef]
36. Zhang, R.; Du, Y.; Yuhong, L. New challenges to power system planning and operation of smart grid devel-opment in China. In Proceedings of the 2010 International Conference on Power System Technology, Hangzhou, China, 24–28 October 2010; pp. 1–8. [CrossRef]
37. Harnisch, S.; Steffens, P.; Thies, H.H.; Cibis, K.; Zdrallek, M.; Lehde, B. New planning principles for low voltage networks with a high share of decentralized generation. In *CIRED Workshop 2016*; IET: Helsinki, Finland, 14–15 June 2016; pp. 1–4. [CrossRef]
38. Mosquet, X.; Zablit, H.; Dinger, A.; Xu, G.; Andersen, M.; Tominaga, K. *The Electric Car Tipping Point—The Future of Powertrains for Owned and Shared Mobility*; The Boston Consulting Group: Boston, MA, USA, 2018; Available online: https://web-assets.bcg.com/ef/8b/007df7ab420dab1164e89d0a6584/bcg-the-electric-car-tipping-point-jan-2018.pdf (accessed on 1 July 2021).
39. Kuhnert, F.; Stürmer, C. *Autofacts*; PricewaterhouseCoopers PwC: Toronto, ON, Canada, 2016.
40. RBC Electric Vehicle Forecast Through 2050 & Primer', RBC Capital Markets, May 2018. Available online: http://www.fullertreacymoney.com/system/data/files/PDFs/2018/May/14th/RBC%20Capital%20Markets_RBC%20Electric%20Vehicle%20Forecast%20Through%202050%20%20Primer_11May2018.pdf (accessed on 29 July 2021).
41. Monscheidt, J.; Gemsjäger, B.; Slupinski, A.; Yan, X.; Ali, S.A.; Wintzek, P.; Zdrallek, M. Energy demand model for electric load forecast in urban distribution networks. *CIRED Open Access Proc. J.* **2020**, *2020*, 31–33. [CrossRef]
42. Wintzek, P.; Ali, S.; Kotthaus, K.; Wruk, J.; Zdrallek, M.; Monscheidt, J.; Gemsjäger, B.; Slupinski, A. Application and Evaluation of Load Management Systems in Urban Low-Voltage Grid Planning. *World Electr. Veh. J.* **2021**, *12*, 91. [CrossRef]
43. Liste der Ladesäulen (Stand 1. Juli 2021)'. Bundesnetzagentur, 1 July 2021. Available online: https://www.bundesnetzagentur.de/SharedDocs/Downloads/DE/Sachgebiete/Energie/Unternehmen_Institutionen/E_Mobilitaet/Ladesaeulenkarte_Datenbankauszug.xlsx;jsessionid=EF0F9B1F6B36E1FA218A42606D92A6FE?__blob=publicationFile&v=8 (accessed on 1 July 2021).
44. Goebelsmann, F.; Wintzek, P.; Ali, S.; Zdrallek, M.; Monscheidt, J.; Gemsjaeger, B.; Slupinski, A. Potentials of a cost-optimized application of regulated distribution transformers in urban low-voltage grid planning with a high share of charging infrastructure, heat pumps and photovoltaic systems. In Proceedings of the PESS 2021—Power and Energy Student Summit, online, 25–26 November 2021; pp. 1–6.
45. *Population Structure Data'*; GfK SE Growth from Knowledge: Nuremberg, Germany, 2018.
46. Bollerslev, J.; Andersen, P.B.; Jensen, T.V.; Marinelli, M.; Thingvad, A.; Calearo, L.; Weckesser, T. Coincidence Factors for Domestic EV Charging from Driving and Plug-In Behavior. *IEEE Trans. Transp. Electrification* **2021**, *8*, 808–819. [CrossRef]
47. Uhlig, R.; Stotzel, M.; Zdrallek, M.; Neusel-Lange, N. Dynamic grid support with EV charging management considering user requirements. In Proceedings of the CIRED Workshop 2016, Helsinki, Finland, 14–15 June 2016. [CrossRef]
48. Follmer, R.; Gruschwitz, D.; Jesske, B.; Quandt, S.; Lenz, B.; Nobis, C.; Köhler, K.; Mehlin, M. *Mobilität in Deutschland 2008*; Ergebnisbericht: Struktur—Aufkommen—Emissionen—Trends; Federal Ministry for Digital and Transport: Bonn and Berlin, Germany, 2010; Available online: http://www.mobilitaet-in-deutschland.de/pdf/infas_MiD2008_Abschlussbericht_I.pdf (accessed on 24 February 2022).
49. IEC 60050: 691-10-05. Available online: https://www.electropedia.org/ (accessed on 24 February 2022).
50. Ali, S.; Wintzek, P.; Zdrallek, M.; Böse, C.; Monscheidt, J.; Gemsjäger, B.; Slupinski, A. Demand Factor Identification of Electric Vehicle Charging Points for Distribution System Planning. In Proceedings of the CIRED 2021—The 26th International Conference and Exhibition on Electricity Distribution, Online Conference, 20–23 September 2021; pp. 2574–2578. [CrossRef]
51. Rehtanz, C.; Greve, M.; Häger, U.; Hagemann, Z.; Kippelt, S.; Kittl, C.; Klouber, M.; Pohl, O.; Rewald, F.; Wagner, C. Verteilnetzstudie für das Land Baden-Württemberg. Technische Universität Dortmund—ie3—Institut für Energiesysteme, Energieeffizienz und Energiewirtschaft, Stuttgart, Studie für das Ministerium für Umwelt, Klima und Energiewirtschaft Baden-Württemberg, April 2017. Available online: https://um.baden-wuerttemberg.de/fileadmin/redaktion/mum/intern/Dateien/Dokumente/5_Energie/Versorgungssicherheit/170413_Verteilnetzstudie_BW.pdf (accessed on 29 July 2021).
52. DIN EN 60076-1:2012-03 VDE 0532-76-1:2012-03 Power Transformers—Part 1: General (IEC 60076-1:2011); German version EN 60076-1:2011'. VDE, March 2012. Available online: https://www.vde-verlag.de/standards/0532078/din-en-60076-1-vde-0532-76-1-2012-03.html (accessed on 20 July 2022).
53. DIN VDE 0276-1000 VDE 0276-1000:1995-06 Power Cables; Current-Carrying Capacity, General; Conversion Factors'. VDE, June 1995. Available online: https://www.vde-verlag.de/standards/0276001/din-vde-0276-1000-vde-0276-1000-1995-06.html (accessed on 20 July 2022).
54. *DIN EN 50588-1:2019-12—Medium Power Transformers 50 Hz, with Highest Voltage for Equipment Not Exceeding 36 kV—Part 1: General Requirements*, German version EN 50588-1:2017'; Beuth Verlag GmbH: Berlin, Germany, December 2019.
55. *DIN EN 50160:2020-11, Voltage Characteristics of Electricity Supplied by Public Electricity Networks; GERMAN Version EN 50160:2010 + Cor.:2010 + A1:2015 + A2:2019 + A3:2019*; Beuth Verlag GmbH: Berlin, Germany, 2019. [CrossRef]

56. Dorsemagen, F.; Korotkiewicz, K.; Oerter, C.; Zdrallek, M.; Birkner, P.; Friedrich, W.; Neusel-Lange, N.; Stiegler, M. Decentralized, Integrated Automation System for Medium- and Low Voltage Grids. In Proceedings of the International ETG Congress 2015, Bonn, Germany, 17 November 2015; pp. 20–26.
57. Dorsemagen, F.; Oerter, C.; Zdrallek, M.; Antoni, J.; Birkner, P.; Stiegler, M. State Identification Methods for MV-Grid Automation with Special Regard to LV-Interconnections. In Proceedings of the 23rd International Conference on Electricity Distribution (CIRED), Lyon, France, 15–18 June 2015; paper 1126.
58. Ali, S.A.; Wintzek, P.; Talmond, F.; Zdrallek, M.; Mayer, D.; Monscheidt, J.; Gemsjäger, B.; Slupinski, A. Methodology and application of a dynamic load management system on a medium-voltage distribution power system. In Proceedings of the ETG-Kongress, Online Conference, 18–19 May 2021; Volume 163, pp. 320–325.
59. Wintzek, P.; Ali, S.A.; Monscheidt, J.; Gemsjäger, B.; Slupinski, A.; Zdrallek, M. Planungs- und Betriebsgrundsätze für städtische Verteilnetze—Leitfaden zur Ausrichtung der Netze an ihren zukünftigen Anforderungen. *Wuppertal* **2021**, *35*. Available online: https://d-nb.info/1252809050/34 (accessed on 20 July 2022).
60. Koutsopoulos, I.; Tassiulas, L. Challenges in demand load control for the smart grid. *IEEE Netw.* **2011**, *25*, 16–21. [CrossRef]
61. Fan, Z.; Kulkarni, P.; Gormus, S.; Efthymiou, C.; Kalogridis, G.; Sooriyabandara, M.; Zhu, Z.; Lambotharan, S.; Chin, W.H. Smart Grid Communications: Overview of Research Challenges, Solutions, and Standardization Activities. *IEEE Commun. Surv. Tutorials* **2012**, *15*, 21–38. [CrossRef]
62. Ali, S.; Wintzek, P.; Zdrallek, M.; Gemsjäger, B.; Monscheidt, J.; Slupinski, A. An alternative assessment model for electric power system planning. In Proceedings of the CIRED Porto Workshop 2022: E-mobility and power distribution systems, Porto, Portugal, 2–3 June 2022; Volume 2022, pp. 67–71. [CrossRef]
63. Wintzek, P.; Ali, S.A.; Zdrallek, M.; Monscheidt, J.; Gemsjäger, B.; Slupinski, A. Development of Planning and Operation Guidelines for Strategic Grid Planning of Urban Low-Voltage Grids with a New Supply Task. *Electricity* **2021**, *2*, 614–652. [CrossRef]

MDPI AG
Grosspeteranlage 5
4052 Basel
Switzerland
Tel.: +41 61 683 77 34

Electricity Editorial Office
E-mail: electricity@mdpi.com
www.mdpi.com/journal/electricity

www.ingramcontent.com/pod-product-compliance
Lightning Source LLC
Chambersburg PA
CBHW041405210926
51850CB00013B/3185